RENEWABLE ENERGY TECHNOLOGIES FOR WATER DESALINATION

# Sustainable Water Developments
## Resources, Management, Treatment, Efficiency and Reuse

*Series Editor*

Jochen Bundschuh
*University of Southern Queensland (USQ), Toowoomba, Australia*
*Royal Institute of Technology (KTH), Stockholm, Sweden*

ISSN: 2373-7506

# Volume 4

# Renewable Energy Technologies for Water Desalination

*Editors*

Hacene Mahmoudi

*Faculty of Technology, Hassiba Benbouali University, Chlef, Algeria*

Noreddine Ghaffour

*Water Desalination and Reuse Center (WDRC), King Abdullah University of Science and Technology (KAUST), Thuwal, Saudi Arabia*

Mattheus Goosen

*Office of Research & Graduate Studies, Alfaisal University, Riyadh, Saudi Arabia*

Jochen Bundschuh

*Deputy Vice Chancellor's Office (Research and Innovation) & Faculty of Health, Engineering and Sciences, Toowoomba, Queensland, Australia & Royal Institute of Technology (KTH), Stockholm, Sweden*

CRC Press
Taylor & Francis Group
Boca Raton London New York

CRC Press is an imprint of the
Taylor & Francis Group, an **informa** business

A BALKEMA BOOK

CRC Press/Balkema

Schipholweg 107C, 2316 XC Leiden, The Netherlands

e-mail: Pub.NL@taylorandfrancis.com

www.crcpress.com – www.taylorandfrancis.com

First issued in paperback 2020

© 2017 Taylor & Francis Group, London, UK

*CRC Press/Balkema is an imprint of the Taylor & Francis Group, an informa business*

No claim to original U.S. Government works

ISBN 13: 978-0-367-57354-6 (pbk)
ISBN 13: 978-1-138-02917-0 (hbk)

**Visit the Taylor & Francis Web site at**
**http://www.taylorandfrancis.com**

**and the CRC Press Web site at**
**http://www.crcpress.com**

Typeset by MPS Limited, Chennai, India

*Library of Congress Cataloging-in-Publication Data*

Applied for

# About the book series

Augmentation of freshwater supply and better sanitation are two of the world's most pressing challenges. However, such improvements must be done economically in an environmental and societally sustainable way.

Increasingly, groundwater – the source that is much larger than surface water and which provides a stable supply through all the seasons – is used for freshwater supply, which is exploited from ever-deeper groundwater resources. However, the availability of groundwater in sufficient quantity and good quality is severely impacted by the increased water demand for industrial production, cooling in energy production, public water supply and in particular agricultural use, which at present consumes on a global scale about 70% of the exploited freshwater resources. In addition, climate change may have a positive or negative impact on freshwater availability, but which one is presently unknown. These developments result in a continuously increasing water stress, as has already been observed in several world regions and which has adverse implications for the security of food, water and energy supplies, the severity of which will further increase in future. This demands case-specific mitigation and adaptation pathways, which require a better assessment and understanding of surface water and groundwater systems and how they interact with a view to improve their protection and their effective and sustainable management.

With the current and anticipated increased future freshwater demand, it is increasingly difficult to sustain freshwater supply security without producing freshwater from contaminated, brackish or saline water and reusing agricultural, industrial, and municipal wastewater after adequate treatment, which extends the life cycle of water and is beneficial not only to the environment but also leads to cost reduction. Water treatment, particularly desalination, requires large amounts of energy, making energy-efficient options and use of renewable energies important. The technologies, which can either be sophisticated or simple, use physical, chemical and biological processes for water and wastewater treatment, to produce freshwater of a desired quality. Both industrial-scale approaches and smaller-scale applications are important but need a different technological approach. In particular, low-tech, cost-effective, but at the same time sustainable water and wastewater treatment systems, such as artificial wetlands or wastewater gardens, are options suitable for many small-scale applications. Technological improvements and finding new approaches to conventional technologies (e.g. those of seawater desalination), and development of innovative processes, approaches, and methods to improve water and wastewater treatment and sanitation are needed. Improving economic, environmental and societal sustainability needs research and development to improve process design, operation, performance, automation and management of water and wastewater systems considering aims, and local conditions.

In all freshwater consuming sectors, the increasing water scarcity and correspondingly increasing costs of freshwater, calls for a shift towards more water efficiency and water savings. In the industrial and agricultural sector, it also includes the development of technologies that reduce contamination of freshwater resources, e.g. through development of a chemical-free agriculture. In the domestic sector, there are plenty of options for freshwater saving and improving efficiency such as water-efficient toilets, water-free toilets, or on-site recycling for uses such as toilet flushing, which alone could provide an estimated 30% reduction in water use for the average household. As already mentioned, in all water-consuming sectors, the recycling and reuse of the respective wastewater can provide an important freshwater source. However, the rate at which these water efficient technologies and water-saving applications are developed and adopted depends on the behavior of individual consumers and requires favorable political, policy and financial conditions.

Due to the interdependency of water and energy (water-energy nexus); i.e. water production needs energy (e.g. for groundwater pumping) and energy generation needs water (e.g. for cooling), the management of both commodities should be more coordinated. This requires integrated energy and water planning, i.e. management of both commodities in a well-coordinated form rather than managing water and energy separately as is routine at present. Only such integrated management allows reducing trade-offs between water and energy use.

However, water is not just linked to energy, but must be considered within the whole of the water-energy-food-ecosystem-climate nexus. This requires consideration of what a planned water development requires from the other sectors or how it affects – positively or negatively – the other sectors. Such integrated management of water and the other interlinked resources can implement synergies, reduce trade-offs, optimize resources use and management efficiency, all in all improving security of water, energy, and food security and contributing to protection of ecosystems and climate. Corresponding actions, policies and regulations that support such integral approaches, as well as corresponding research, training and teaching are necessary for their implementation.

The fact that in many developing and transition countries women are disproportionately disadvantaged by water and sanitation limitation requires special attention to this aspect in these countries. Women (including schoolgirls) often spend several hours a day fetching water. This time could be much better used for attending school or working to improve knowledge and skills as well as to generate income and so to reduce gender inequality and poverty. Absence of in-door sanitary facilities exposes women to potential harassment. Moreover, missing single-sex sanitation facilities in schools and absence of clean water contributes to diseases. This is why women and girls are a critical factor in solving water and sanitation problems in these countries and necessitates that men and women work alongside to address the water and wastewater related operations for improvement of economic, social and sustainable freshwater provision and sanitation.

Individual volumes published in the series are spanning the wide spectrum between research, development and practice in the topic of freshwater and related areas such as gender and social aspects as well as policy, regulatory, legal and economic aspects of water. It covers all fields and facets in optimal approaches to the:

• Assessment, protection, development and sustainable management of groundwater and surface water resources thereby optimizing their use.
• Improvement of human access to water resources in adequate quantity and good quality.
• Meeting of the increasing demand for drinking water, and irrigation water needed for food and energy security, protect ecosystems and climate and to contribute to a social and economically sound human development.
• Treatment of water and wastewater also including its reuse.
• Implementation of water efficient technologies and water saving measures.

A key goal of the series is to include all countries of the globe in jointly addressing the challenges of water security and sanitation. Therefore, we aim to a balanced choice of authors and editors originating from developing and developed countries as well as gender equality. This will help society to provide access to freshwater resources in adequate quantity and good quality, meeting the increasing demand for drinking water, domestic water and irrigation water needed for food security while contributing to social and economically sound development.

This book series aims to become a state-of-the-art resource for a broad group of readers including professionals, academics and students dealing with ground- and surface water resources, their assessment, exploitation and management as well as the water and wastewater industry. This comprises especially hydrogeologists, hydrologists, water resources engineers, wastewater engineers, chemical engineers and environmental engineers and scientists.

The book series provides a source of valuable information on surface water but especially on aquifers and groundwater resources in all their facets. Thereby, it covers not only the scientific and technical aspects but also environmental, legal, policy, economic, social, and gender aspects of groundwater resources management. Without departing from the larger framework of integrated groundwater resources management, the topics are centered on water, solute and heat transport in aquifers, hydrogeochemical processes in aquifers, contamination, protection, resources assessment and use.

The book series constitutes an information source and facilitator for the transfer of knowledge, both for small communities with decentralized water supply and sanitation as well as large industries that employ hundreds or thousands of professionals in countries worldwide, working in the different fields of freshwater production, wastewater treatment and water reuse as well as those concerned with water efficient technologies and water saving measures. In contrast to many other industries, suffering from the global economic downturn, water and wastewater industries are rapidly growing sectors providing significant opportunities for investments. This applies especially to those using sustainable water and wastewater technologies, which are increasingly favored. The series is also aimed at communities, manufacturers and consultants as well as a diversity of stakeholders and professionals from governmental and non-governmental organizations, international funding agencies, public health, policy, regulators and other relevant institutions, and the broader public. It is designed to increase awareness of water resources protection and understanding of sustainable water and wastewater solutions including the promotion of water and wastewater reuse and water savings.

By consolidating international research and technical results, the objective of this book series is to focus on practical solutions in better understanding ground- and surface water systems, the implementation of sustainable water and wastewater treatment and water reuse and the implementation of water efficient technologies and water saving measures. Failing to improve and move forward would have serious social, environmental and economic impacts on a global scale.

The book series includes books authored and edited by world-renowned scientists and engineers and by leading authorities in economics and politics. Women are particularly encouraged to contribute, either as author or editor.

Jochen Bundschuh
(Series Editor)

# Editorial board

Blanca Jiménez Cisneros    Director of the Division of Water Sciences, Secretary of the International Hydrological Programme (IHP), UNESCO, Paris, France

Glen T. Daigger    Immediate Past President, International Water Association (IWA), IWA Board Member, Senior Vice President and Chief Technology Officer CH2M HIL, Englewood, CO, USA

Anthony Fane    Director, Singapore Membrane Technology Centre (SMTC), Nanyang Technological University (NTU), Singapore

Carlos Fernandez-Jauregui    Director, International Water Chair-EUPLA, Director, Water Assessment & Advisory Global Network (WASA-GN), Madrid, Spain

ADVISORY EDITORIAL BOARD

AFGHANISTAN

Naim Eqrar (water resources; water quality; hydrology; hydrogeology; environmental awareness), Full Member of the Secretariat of Supreme Council of Water and Land of Afghanistan (SCWLA); Full Member of the National Hydrological Committee of Afghanistan (NHCA) & Geoscience Faculty, Kabul University, Kabul

ALGERIA

Hacene Mahmoudi (renewable energy for desalination and water treatment; membrane technologies), Faculty of Technology, Hassiba Benbouali University of Chlef (UHBC), Chlef

ANGOLA

Helder de Sousa Andrade (geomorphology; impact of land use on water resources (water courses, rivers and lakes); climate variability impacts on agriculture), Department of Geology, Faculty of Sciences, Agostinho Neto University (UAN), Luanda

ARGENTINA

Alicia Fernández Cirelli (water quality; aquatic ecosystems; aquatic humic substances; heavy metal pollution; use of macrophytes as biosorbents for simultaneous removal of heavy metals; biotransfer of arsenic from water to the food chain), Director, Center for Transdisciplinary Studies on Water Resources, Faculty of Veterinary Sciences, University of Buenos Aires (UBA) & National Scientific and Technical Research Council (CONICET), Buenos Aires

Carlos Schulz (hydrogeology; groundwater management planning; water chemistry; groundwater-surface water interactions; trace element contaminants), President IAH Chapter Argentina; Faculty of Exact and Natural Sciences, National University of La Pampa (UNLPam), Santa Rosa, La Pampa

# Table of contents

# List of contributors

Faheem Ahmed — College of Science & General Studies, Alfaisal University, Riyadh, Saudi Arabia

Rehan Ahmed — School of Engineering and Physical Sciences, Heriot-Watt University, Edinburgh, UK

Diego-César Alarcón-Padilla — Solar Platform of Almería (PSA-CIEMAT), Almería, Spain

Mai Ali — Department of Electrical Engineering, Alfaisal University, Riyadh, Saudi Arabia

Edreese Alsharaeh — College of Science & General Studies, Alfaisal University, Riyadh, Saudi Arabia

Meshael Alturki — College of Science & General Studies, Alfaisal University, Riyadh, Saudi Arabia

Yousef M. Alyousef — Energy Research Institute, King Abdulaziz City for Science and Technology, Riyadh, Saudi Arabia & Office of Research and Graduate Studies, Alfaisal University, Riyadh, Saudi Arabia

Muhammad Anan — College of Engineering, Alfaisal University, Riyadh, Saudi Arabia

Hassan Arafat — Institute Center for Water and Environment (iWater), Department of Chemical and Environmental Engineering Masdar Institute of Science and Technology, Abu Dhabi, United Arab Emirates

Nishat Arshi — College of Science & General Studies, Alfaisal University, Riyadh, Saudi Arabia

Jochen Bundschuh — Deputy Vice-Chancellor's Office (Research and Innovation) & Faculty of Health, Engineering and Sciences, University of Southern Queensland, Toowoomba, Queensland, Australia & Royal Institute of Technology, Stockholm, Sweden

Pedro Cabrera — Group for the Research on Renewable Energy Systems, University of Las Palmas de Gran Canaria, Las Palmas de Gran Canaria, Spain

Youssef Elakwah — Office of Research and Graduate Studies, Alfaisal University, Riyadh, Saudi Arabia

Nadimul H. Faisal — School of Engineering, Robert Gordon University, Aberdeen, UK

Eanna Farrell — Institute Center for Water and Environment (iWater), Department of Chemical and Environmental Engineering Masdar Institute of Science and Technology, Abu Dhabi, United Arab Emirates

Shaik Feroz — Caledonian Centre for Creativity & Innovation (CCCI), Caledonian College of Engineering, Sultanate of Oman

Lourdes García-Rodríguez — Department of Energetic Engineering, University of Seville, Seville, Spain

Noreddine Ghaffour — Water Desalination & Reuse Centre, King Abdullah University of Science and Technology (KAUST), Thuwal, Saudi Arabia

Jaime González — Group for the Research on Renewable Energy Systems, University of Las Palmas de Gran Canaria, Las Palmas de Gran Canaria, Spain

Mattheus F.A. Goosen — Office of Research and Graduate Studies, Alfaisal University, Riyadh, Saudi Arabia

Mamdud Hossain — School of Engineering, Robert Gordon University, Aberdeen, UK

Sheikh Z. Islam — School of Engineering, Robert Gordon University, Aberdeen, UK

Sai P. Katikaneni — Research and Development Centre, Saudi Aramco, Dhahran, Saudi Arabia

Nadejda Komendantova — International Institute for Applied Systems Analysis, Austria & Department of Environmental Systems Science, Institute for Environmental Decisions (ETH), Zurich, Switzerland

Hacene Mahmoudi — Faculty of Technology, Hassiba Benbouali University, Chlef, Algeria

Thomas Missimer — U.A. Whitaker College of Engineering, Florida Gulf Coast University, Fort Myers, Florida, USA

Kim Choon Ng — Water Desalination & Reuse Centre, King Abdullah University of Science and Technology (KAUST), Thuwal, Saudi Arabia

Baltasar Peñate — Water Department, Canary Islands Institute of Technology (ITC), Las Palmas, Spain

Patricia Palenzuela — Solar Platform of Almería (PSA-CIEMAT), Almería, Spain

Chua Leok Poh — School of Mechanical and Aerospace Engineering, Nanyang Technological University, Singapore

Kyriakos Rossis — Wind Energy Department, Centre for Renewable Energy Sources & Saving, Pikermi, Greece

Zheng Shuai — School of Mechanical and Aerospace Engineering, Nanyang Technological University, Singapore

Vicente Subiela — Water Department, Canary Islands Institute of Technology (ITC), Las Palmas, Spain

Abd-Elhamid M. Taha — Department of Electrical Engineering, Alfaisal University, Riyadh, Saudi Arabia

Eftihia Tzen — Wind Energy Department, Centre for Renewable Energy Sources & Saving, Pikermi, Greece

Zhao Yong — School of Engineering, Nazarbayev University, Astana, Republic of Kazakhstan

Muhammad Wakil Shahzad — Water Desalination & Reuse Centre, King Abdullah University of Science and Technology (KAUST), Thuwal, Saudi Arabia

Guillermo Zaragoza — Solar Platform of Almería (PSA-CIEMAT), Almería, Spain

# Foreword by Nidal Hilal

Today, the supply of freshwater presents one of the most pressing challenges ever faced by the human race. It is a vital resource for human life. Yet, population growth and enhanced living standards, together with the expansion of industrial and agricultural activities, are creating unprecedented demands on clean water supplies all over the world. The Organization for Economic Co-operation and Development (OECD) and the United Nations (UN) have reported that 0.35 billion people in 25 different countries are currently suffering from water shortage, and this will grow to 4 billion people (two-thirds of the world population) in 52 countries by 2025. In addition, a staggering 2.6 billion people have no access to proper sanitation.

When you consider the facts from the World Health Organization (WHO) and other agencies, the root cause of this mounting crisis becomes clear. The world's population tripled in the 20th century, and is expected to increase by another 40–50 percent in the next 50 years. The unavoidable fact is that there is no more freshwater in the world today than there was 1 million years ago, and water as a resource cannot be replaced in the way that alternative fuel sources can replace petroleum.

Areas affected by acute water shortage are often in the poorest, most underdeveloped countries, which lack the necessary power and water-delivery infrastructures. The lack of clean water also creates considerable health, energy and economic challenges to the populations of these countries. Because of a growing population and climate change, water shortages will eventually limit economic growth and food supplies. The provision of clean water therefore represents one of the most pressing challenges of the 21st century.

What are the sustainable solutions to this challenge? What are the priorities? More and more reservoirs, wells, pipelines and river transfers are not the answer. Yet, in arid countries around the world, where conventional water resources are scarce, impaired-quality (non-conventional) water sources are still available for exploitation, such as seawater, brackish water, wastewater (and various effluents) and stormwater runoff. In particular, desalination of seawater has increasingly proved itself to be the most practical – and in many cases the only possible – solution for many countries around the globe.

Today, a new generation of lower energy and sustainable technologies are needed urgently for the desalination of sea and brackish ground water, waste-water treatment, and recycle. Indeed, for many water stressed regions around the globe, tapping into the seas may be the only option available to address the gap in freshwater supply for the foreseeable future and meet the increased water demand due to population growth, expanding urbanization and industry.

The rapid development of water desalination technologies and its market in the last decade clearly reflects their growing importance. However, in spite of the significant technological advances and successes in reducing energy requirements (mainly in membrane-based processes), desalination remains an energy-intensive process, contributing to greenhouse gas (GHG) emissions and thus having a contributory effect on climate change. In some GCC/MENA countries, the situation is especially challenging in terms of the cost of energy use, because thermal-based processes such as multi-stage flash (MSF) and multi-effect distillation (MED) dominate the market; these processes depend directly on the availability and cost of fossil fuel. Furthermore, they rely on the availability of good prior experience gained in these technologies, and must often deal with challenging seawater quality, especially in the Gulf region. To meet these and other related challenges, there is a tremendous interest in relying on renewable energy to address these global problems of water security and sustainability of water resources at a lower-cost, while also being more environmentally friendly.

Thus, this book is extremely timely and provides recent trends and challenges in applications of renewable energy technologies for water desalination with an emphasis on environmental concerns and sustainable development. While the emphasis is focused on applications of renewable energy for water desalination, specific additional sections will deal with economics & scale-up issues, government subsidies & regulations as well as environmental concerns. The aim is to give a holistic or complete approach to help users, decision makers in both governments and industry in selecting appropriate technology solutions for each specific case. Renewable energy technologies are rapidly emerging with the promise of economic and environmental viability for desalination, which necessitate parallel plans and actions by institutions dealing with water-energy nexus.

The chapters in this volume show that there is great potential for the use of renewable energy in water desalination applications in many parts of the world. Solar, wind, wave and geothermal sources or a combination of some or all of them could provide a viable source of energy to power seawater and brackish water desalination plants. Lastly, it is extremely important to note that part of the solution to the world's water scarcity is not only to produce more water, but also to be able to recognize that this must to be done in an environmentally and economically sustainable way.

Professor Nidal Hilal
Editor-in-Chief, Desalination
Chair in Water Process Engineering
Director of Centre for Water Advanced
Technologies & Environmental Research (CWATER)
Swansea University
United Kingdom
January 2017

# Editors' foreword

Renewable energy technologies are rapidly emerging with the promise of economic and environmental viability for desalination. This book describes recent trends and challenges in applications of renewable energy technologies for water desalination with an emphasis on environmental concerns and sustainable development. While the emphasis is on applications of renewable energy, specific additional sections will deal with economics & scale-up, government subsidies & regulations, and environmental concerns. There will also be chapter on selecting the best combination of renewable energy and desalination systems. The aim is to give a holistic or complete approach to help, for example, decision makers in government, industry and academia.

The book is divided into fifteen chapters. The first part provides an overview of renewable energy technologies for desalination. This is followed by a section on the use of solar energy for small-scale autonomous desalination system. There are two related chapters on application of solar nano-photocatalysis in reverse osmosis pretreatment processes, and metal oxide nano-photocatalysts for water purification. Fuel cells as an energy source are also presented as well as a critical review on fuel cell commercialization. There are chapters on wind turbine electricity generation for desalination with emphasis on design, application and commercialization, and wind technology design and reverse osmosis systems for off-grid and grid-connected applications. Geothermal energy applications for desalination are likewise outlined. Fuel cost distribution and desalination technology selection based on exergy analysis are debated as well as wireless networks employing renewable energy sources for industrial applications. In the case of the latter, innovations, trade-offs and operational considerations are analyzed. The need for achieving food security in the desalination age is assessed. Renewable energy policy and mitigating the risks for investment are discussed as well as integrating renewable energy sources into smart grids: opportunities and challenges. The final chapter covers current trends and future prospects of renewable energy-driven desalination.

This book will help to accelerate the development and scale-up of novel water production systems from renewable energies. These technologies will help to minimize environmental concerns. The chapters in this volume will show that there is great potential for the use of renewable energy in many parts of the world. Solar, wind, wave and geothermal sources could provide a viable source of energy to power both seawater and brackish water desalination plants. Finally, it must be mentioned that the production of potable water to meet global needs has to be done in an environmentally and economically sustainable way. This is the key challenge facing us.

Hacene Mahmoudi
Hassiba Benbouali University, Algeria

Noreddine Ghaffour
King Abdullah University of Science and Technology, Saudi Arabia

Mattheus A. Goosen
Alfaisal University, Saudi Arabia

Jochen Bundschuh
University of Southern Queensland, Australia

# About the editors

**Hacene Mahmoudi** (1975, Algeria) is a full Professor at the University Hassiba Benbouali of Chlef, Algeria, where he is acting as Vice President for external relations. The doctoral degree of Dr Mahmoudi is in mechanical engineering from the University of Sciences and Technology of Oran (2008) Algeria. He is the author of over 60 publications and chapters in textbooks. Hacene is a member of the Editorial Board of several water related journals and books. Pr. Hacene Mahmoudi's research interests are in the areas of renewable energy, desalination, sustainable development and membrane preparation and characterization.

**Dr.-Ing. Noreddine Ghaffour** is a Research Professor at the Water Desalination & Reuse Center (WDRC) at King Abdullah University of Science and Technology (KAUST). He is the WDRC's Research Theme Leader on Desalination. Before leaving Montpellier University, France, where he spent seven years, he spent another seven years as R&D and Capacity Building Manager at the Middle East Desalination Research Center (MEDRC). Prof. Noreddine has over 25 years of experience in the field of drinking water treatment technologies, and has specialized in the area of membrane and thermal based desalination processes and its related fields. Over the years, he has made major contributions becoming an internationally recognized expert in desalination technologies and its related fields, such as renewable energy-driven and innovative energy-efficient desalination technologies. He obtained his PhD degree from Montpellier University, France, in 1995. He is the author of over 200 publications, several patents and chapters in textbooks. Noreddine is an Associate Editor of the *Desalination* Journal and a member of the Editorial Board of other water related journals. He is a member of the main international desalination associations, and a frequent speaker, including keynotes, and sessions co-chair in international conferences, seminars and workshops. He also has experience in scaling-up innovative desalination processes and start-ups.

**Mattheus (Theo) F. A. Goosen** (1950, Netherlands), who is a Canadian citizen, has played key roles in the development of new start up academic institutions. For the past nine years he has held the position of founding Associate Vice President for Research & Graduate Studies at Alfaisal University a private start-up non-profit institution in Riyadh, Saudi Arabia (www.alfaisal.edu). The doctoral degree of Dr. Goosen is in chemical & biomedical engineering from the University of Toronto (1981) Canada. Theo has more than 180 publications to his credit including over 135 refereed journal papers, 45 conference papers, 11 edited books and 10 patents. His h index is over 48 and he has well over 8200 citations on Google Scholar. On Scopus he has 135 publications with over 4400 citations and an h-index of 35. Dr Goosen's research interests are in the areas of renewable energy, desalination, sustainable development, membrane separations, spray coating technology and biomaterials.

**Jochen Bundschuh** (1960, Germany), finished his PhD on numerical modeling of heat transport in aquifers in Tübingen in 1990. He is working in geothermics, subsurface and surface hydrology and integrated water resources management, and connected disciplines. From 1993 to 1999 he served as an expert for the German Agency of Technical Cooperation (GTZ – now GIZ) and as a long term professor for the DAAD (German Academic Exchange Service) in Argentine.

From 2001 to 2008 he worked within the framework of the German governmental cooperation (Integrated Expert Program of CIM; GTZ/BA) as adviser in mission to Costa Rica at the Instituto Costarricense de Electricidad (ICE). Here, he assisted the country in evaluation and development of its huge low-enthalpy geothermal resources for power generation. Since 2005, he has been an affiliate professor of the Royal Institute of Technology, Stockholm, Sweden. In 2006, he was elected Vice-President of the International Society of Groundwater for Sustainable Development ISGSD. From 2009–2011 he was visiting professor at the Department of Earth Sciences at the National Cheng Kung University, Tainan, Taiwan.

Since 2012, Dr. Bundschuh is a professor in hydrogeology at the University of Southern Queensland, Toowoomba, Australia where he leads the Platform for Water in the Nexus of Sustainable Development working in the wide field of water resources and low/middle enthalpy geothermal resources, water and wastewater treatment and sustainable and renewable energy resources. In November 2012, Prof. Bundschuh was appointed as president of the newly established Australian Chapter of the International Medical Geology Association (IMGA).

Dr. Bundschuh is author of the books "Low-Enthalpy Geothermal Resources for Power Generation" (CRC Press/Balkema, Taylor & Francis Group) and "Introduction to the Numerical Modeling of Groundwater and Geothermal Systems: Fundamentals of Mass, Energy and Solute Transport in Poroelastic Rocks". He is editor of 16 books and editor of the book series "Multiphysics Modeling", "Arsenic in the Environment", "Sustainable Energy Developments" and "Sustainable Water Developments" (all CRC Press/Balkema, Taylor & Francis Group). Since 2015, he is an editor in chief of the Elsevier journal "Groundwater for Sustainable Development".

# CHAPTER 1

# A critical overview of renewable energy technologies for desalination

Hacene Mahmoudi, Noreddine Ghaffour,
Mattheus F.A. Goosen & Jochen Bundschuh

## 1.1 INTRODUCTION

Rapid population growth and industrialization as well as climatic change have placed increasing strains on global potable water supplies (Caldera *et al.*, 2016; Goosen *et al.*, 2016; Sahin *et al.*, 2016). In particular, the demand for this limited renewable resource is anticipated to intensify due to the requirements of the agricultural, manufacturing and urban sectors. The United Nations World Water Assessment Programme estimates that by 2030 only 60% of the worldwide water needs can be met (Connor, 2015). Additionally, while economic development opens up and advances economies, and creates new wealth, it can be argued that millions of people do not benefit directly from this financial progress (Goosen, 2013; Gottinger and Goosen, 2012). This poses new challenges to the effective governance of potable water resource systems.

Sustainable growth using renewable energy sources is now considered by many as being the model to follow (Goosen *et al.*, 2009a, 2009b; Misra, 2000). Water desalination, for example, has become a technologically and economically viable solution to tackle the challenges associated with increasing water shortages (Droogers *et al.*, 2012; El Kharraz *et al.*, 2012). Nevertheless, desalination is an energy-intensive process normally requiring high-capacity plants that utilize expensive and non-renewable fossil fuels, which in turn contribute to global warming and air pollution (Ghaffour *et al.*, 2013). Even though renewable desalination systems cannot currently compete with conventional technologies in terms of the cost of water produced, they remain applicable for remote and arid areas and are likely to represent a feasible solution at the large scale in the near future.

There is a complex interaction between the economics of renewable energy and desalination, the water requirements of society, long-term environmental protection, and sustainable resource management. Sahin *et al.* (2016), for example, developed a system dynamics model to simulate changes to water governance through the integration of supply, demand and asset management processes (Fig. 1.1). In particular, they were able to explore the behavior of water resource systems over the next 100 years under changes brought about by climate variation and population growth. It was found that the current supply-side oriented approach to water governance was ill-equipped to cope with these changes, leading to economic hardship and chronic water shortages. Reorganization of the system through new water governance practices was proposed.

Renewable energy has great potential for solving future energy demands for desalination. Studies have been performed, for example, to determine if it is feasible to meet the 2030 global water demand with seawater reverse osmosis (SWRO) desalination driven exclusively by renewable energy (Caldera *et al.*, 2016). This was done by estimating the unit cost of water production or the levelized cost of water (LCOW) for renewable-energy-powered SWRO plants in 2030 and comparing it with the costs of existing fossil-powered SWRO desalination plants. The least-cost system was found to be a combination of fixed-tilt photovoltaic (PV), single-axis tracking PV, wind energy, batteries and gas power plants (Fig. 1.2). The authors predicted that by 2030 SWRO plants powered by renewable energy will produce water at similar prices to that of today's

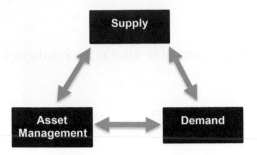

Figure 1.1.   Water governance through integration of supply, demand and asset management processes (Sahin *et al.*, 2016).

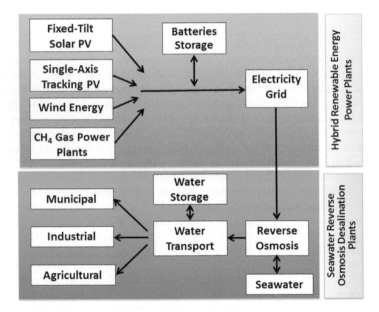

Figure 1.2.   A 2030 scenario where seawater reverse osmosis (SWRO) desalination plants are powered by hybrid renewable energy power that is cost optimized by storage (adapted from Caldera *et al.*, 2016). High voltage DC cables transport power to the desalination plants on the coast. The water produced is transported to meet the demands of the municipal, industrial and agricultural sectors. Water storage at the site of demand ensures constant water supply.

fossil-fuel-powered plants. Furthermore, the water transportation costs were found to contribute significantly to the LCOW. It can be argued that the results of Caldera *et al.* (2016) suggest that smaller, local desalination plants would be preferred economically due to their lower water transportation costs. Additionally, Figure 1.3a illustrates the water stress for the regions in a global map for the 2030 scenario (Caldera *et al.*, 2016). The resulting global desalination demand is presented in Figure 1.3b. The desalination demand for a region is equal to the installed SWRO capacity required for the region.

   This introductory chapter provides a critical overview of integrated approaches to using renewable energy such as solar and geothermal technologies for water desalination. Innovative and sustainable desalination processes that are suitable for integrated renewable energy systems are presented, along with the benefits of these technologies and their limitations. The market potential, environmental concerns, regulatory and socioeconomic factors are also evaluated, together with the need for accelerated development of renewable energy-driven desalination technologies.

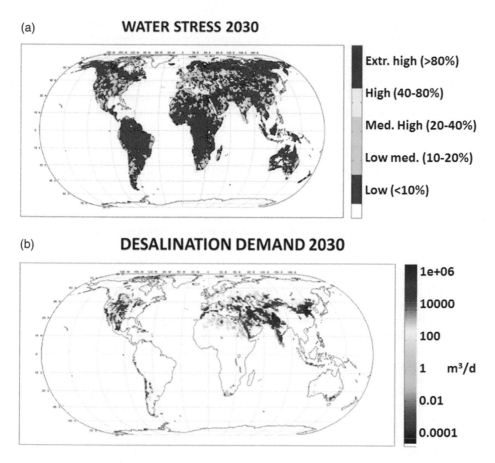

Figure 1.3. (a) Projected water stress for the 2030 scenario. Water stress is the ratio of total water demand in a region to the annual renewable water resources available in that region. (b) Desalination demand in the 2030 scenario. Adapted from Caldera *et al.* (2016).

## 1.2 WORLDWIDE DESALINATION CAPABILITY AND PROGRESS IN RENEWABLE ENERGY SYSTEMS

The total global desalination capacity stands at 81 million $m^3$ $day^{-1}$ and is expected to reach in excess of 100 million $m^3$ $day^{-1}$ by 2015 (Intelligence, 2013). On a global scale, 68% of desalinated water is manufactured by membrane, and 30% by thermal processes; the remaining 2% is produced by other technologies. The sizable decrease in desalination costs as a result of significant technological improvements has made desalinated water cost-competitive; resulting in an increase in desalination capacity (Reddy and Ghaffour, 2007).

There are numerous groupings of desalination and renewable energy technologies which are economically and technologically feasible. Selecting the most suitable renewable-driven desalination system depends on several factors, such as size of the plant, salinity of the feed water and required product, remoteness, existence of access to an electricity grid, technical infrastructure and the renewable source and its availability, potential and exploitation cost. The most commonly used arrangement is photovoltaic (PV) with reverse osmosis (RO). As practice has shown, there are no noteworthy technical obstacles in combining renewable energy and desalination technologies. Since heat losses are more significant in small thermal distillation units, large sizes are more attractive.

There is a strong relationship between water desalination and energy consumption (Abdel-Jawad, 2001; Alyousef and Abu-ebid, 2012). Furthermore, for arid countries such as Saudi Arabia there is an additional drain on energy resources since a large percentage of water consumed is desalinated water which also has to be conveyed over long distances to reach population centers (Alyousef and Abu-ebid, 2012). Alyousef and Abu-ebid (2012) claimed that in order to fulfill upcoming growth in water demand in Saudi Arabia, another 20–30 desalination plants with total capital outlay of (US) $50 billion will be needed by 2030. There is thus a strong need for effective policies for reducing energy consumption in the co-generation desalination sector.

## 1.3   ECONOMICS AND IMPROVEMENT OF THERMAL ENERGY STORAGE AND BACKUP SYSTEMS

To provide non-stop and year-round operation, concentrated solar power (CSP) processes can be combined with thermal energy storage and backup systems. CSP, which is an electricity generation technology that uses heat produced by solar radiation focused on a small zone, can be employed directly in desalination plants or it can be used indirectly to power a turbine and generate electricity. Widespread positioning of CSP plants has been hindered by cost and intermittency issues (Goosen et al., 2014). These problems may, though, be slightly eased with the addition of thermal energy storage employing, for example, molten salt or compressed air underground heat storage. The latter can be used in place of a natural gas boiler to provide backup energy for a concentrated solar thermal power plant during cloudy periods and night-time. Wagner and Rubin (2011) reported on the economic implications of thermal energy storage for concentrated solar thermal power. They noted that the additional equipment associated with a thermal energy storage system can add substantially to the already high capital cost of a plant.

The most established of all commercially operating CSP plants are based on parabolic trough collectors (PTCs) (Zhang et al., 2013). A PTC focuses sunlight onto an absorber tube that is mounted in the focal line of the parabola. The reflectors and the absorber tubes move in tandem with the sun. As in other thermal power generation plants, CSP requires water for cooling and condensing processes. Such water requirements are a significant complication for those locations with water scarcity. However, if the plant is located at the coast this problem does not occur as seawater can be used for cooling. As water cooling is more effective, operators of hybrid systems tend to use only dry cooling in the winter when cooling needs are lower, then switch to combined wet and dry cooling during the summer (Zhang et al., 2013). This is a factor that needs to be kept in mind when designing CSP plants for different locations.

Large scale deployment of solar power plants has been hindered by cost and intermittency issues (Goosen et al., 2014). Yet, these challenges may be lessened to some extent through the use of thermal energy storage by means of, for instance, molten salt. In thermal energy storage schemes, extra heat collected in the solar field is directed to a heat exchanger and warms a heat transfer fluid flowing from a cold tank to a hot tank (Zhang et al., 2013). As soon as required, the heat from the hot tank can be used for direct heat applications or be sent to make steam for electricity generation. Direct thermal storage systems, for example, can employ liquids such as mineral oil, synthetic oil and molten salts. Molten single salts, though, tend to be costly (Barlev et al., 2011). As an alternative, with indirect storage, a heat transfer fluid circulates heat, collected in the absorbers, which is then pumped to a thermal energy storage system, typically composed of a solid material. Additionally, to regulate production and to guarantee a nearly constant generation capacity, concentrated solar power (CSP) plants, with or without storage, are usually equipped with a fuel standby system (Zhang et al., 2013). Such plants fortified with backup systems are called hybrid plants, with burners providing energy to the heat transfer fluid, to the storage medium, or directly to the power block. One of the main advantages of the integration of a backup system with a CSP plant is the reduction in the required investments in the solar field and storage capacity.

Wagner and Rubin (2011) reasoned that if the aim is to inspire the extensive use of CSP plants, incentives such as investment tax credits are necessary to reduce the levelized cost of electricity

Table 1.1.   Estimates of levelized cost (US\$ MWh$^{-1}$) of electricity by source (adapted from DeCanio and Fremstad, 2011).

| Measure | Coal | Nuclear | Wind | Geothermal | Solar PV | Solar thermal |
|---|---|---|---|---|---|---|
| US\$ MWh$^{-1}$ (2008) | 100.4 | 119 | 149.3 | 115.7 | 396.1 | 265.6 |
| Levelized cost ranges in US\$ MWh$^{-1}$ | | | 90–130 | 100–160 | 250–350 | 240–290 |
| Levelized cost ranges in US\$ MWh$^{-1}$ | 78–144 | 107–138 | 84–140 | 85–120 | 212–296 | 199–325 |
| US\$ MWh$^{-1}$, annual real interest rate from 1–7% | | | | | 337–565 | |
| Typical energy cost, US\$ MWh$^{-1}$ | | | 50–90 (onshore) 100–140 (offshore) | | | |
| US\$ MWh$^{-1}$ (2007) US\$ MWh$^{-1}$ at sites with very good solar radiation | 43 | | 75 | | 280 | 200 150 |
| US\$ MWh$^{-1}$ | 28–75 (pulverized) | 33–74 | 50–156 (onshore) 71–134 (offshore) | | 226–2031 | 292 |
| US\$ MWh$^{-1}$, | 52–65 (pulverized) | 65–110 | 97–142 (onshore) 110–181 (offshore) | | 674–1140 | 220–324 |
| US\$ MWh$^{-1}$, | 64 (pulverized) | 73 | 91 | | | 175 |
| US\$ MWh$^{-1}$ | 54–120 ($r=5\%$) 67–142 ($r=10\%$) | 29–82 ($r=5\%$) 42–137 ($r=10\%$) | 48–163 (onsh. $r=5\%$) 101–188 (offsh. $r=10\%$) | | 215–333 (high load) 600 (low load) | 136–243 |
| Avg US\$ MWh$^{-1}$ | 79 | 84 | 112 | 99 | 491 | 225 |

(LCOE) and generate a positive annual profit. LCOE is frequently quoted as a suitable measure of the overall competiveness of diverse energy-producing technologies. It denotes the per-kilowatt-hour cost (in real dollars) of constructing and operating a generating plant over an expected financial life and duty cycle. Key inputs to calculating LCOE include capital costs, fuel costs, fixed and variable operations and maintenance costs, financing costs, and an assumed utilization rate for each plant type. Table 1.1, for example, gives a range of estimates of the LCOE from a variety of studies. The highest average value is from solar PV at 491 US\$ MWh$^{-1}$, with solar thermal next at 225 US\$ MWh$^{-1}$ (DeCanio and Fremstad, 2011) (see bottom of Table 1.1). It can be argued that the primary task at this time is to profoundly cut the levelized cost of solar thermal technologies to make them more viable.

The vast benefit of the geothermal option compared with other renewable energy sources is that it provides a constant stable energy supply (Ghaffour *et al.*, 2015). Solar and wind energy, in comparison, are intermittent and therefore require technically more complex capturing devices and costly energy storage devices that, furthermore, are limited in size so that scaling up to large units is hindered. High exploration costs, high investment risks and high installation costs are disadvantages of new geothermal reservoirs. Nevertheless, these large costs will be offset by freely available geothermal heat production during operation of the desalination unit. Moreover, there are several cases where the high exploration cost and investment risk can be eliminated. Heat provision from geothermal resources is generally lower cost than solar and can, therefore, be highly beneficial in areas where suitable geothermal resources are available. However, economic models must prove on a case-by-case basis which option is the more beneficial.

Ghaffour et al. (2015) contended that there are, so far, no large industrial-scale geothermal desalination plants owing to governmental, regulatory and policy reasons. They went on to conclude that future efforts must involve a more detailed assessment of the potential of low to moderate, but still economic, low-enthalpy geothermal fluids from onshore and offshore sources on a global scale. There is a need to perform detailed economic modeling for analysis of all the configurations possible for geothermal desalination plants. Provisions must be made for energy, water and environmental policies which favor the development of geothermal desalination technologies at small to large scales, facilitating market entry and wide market penetration, and a first commercial-scale desalination plant should be installed, which could showcase its economic viability and so pave the way for, and accelerate the development of, geothermal desalination as a competitive and sustainable option at the global level. These developments would contribute to freshwater and energy security and independency and, at the same time, to food security and environmental sustainability.

## 1.4    COMBINED RENEWABLE-ENERGY-DRIVEN LOW-ENERGY DESALINATION PROCESSES

Adsorption desalination (AD) and membrane distillation (MD) are two promising integrated renewable-energy-driven desalination processes which have great potential for scale-up, according to Ghaffour et al. (2015). It was reasoned that the main advantages of these more sustainable technologies, when compared with traditional technologies, are that they are simple, compact, scalable, operate at low temperatures and do not require continuous operation. At atmospheric pressure, AD and MD can function with intermittent energy supply (i.e. variable loads) without additional operating modifications and energy storage.

A fully automated solar-driven AD prototype system was commissioned in Saudi Arabia (Fig. 1.4) (Ng et al., 2009, 2013). It had a nominal capacity of $3\,m^3$ per ton of adsorbent (silica gel) per day, and was fully powered by solar energy using an array of $485\,m^2$ thermal collectors. In addition, the AD prototype could generate cooling suitable for air conditioning at a nominal capacity of 10 Rtons. Rtons, or refrigeration tons, is a technical term used by cooling engineers and 1 Rton is equivalent to 3.52 kW. The new technology produces distillate water and cold water simultaneously, and can be seasonally adjusted to favor either cooling or potable water production, which makes it particularly attractive in arid and semi-arid regions.

Membrane distillation (MD) is a thermally driven process that utilizes a hydrophobic, microporous membrane as a contactor to achieve separation by liquid-vapor equilibrium. Rather than a pressure, concentration or electrical potential gradient, the driving force for the MD process is the partial vapor pressure difference maintained across the two sides of a hydrophobic microporous membrane. As in the case for AD, one of the chief benefits of MD is that the process performance does not appear to be greatly influenced by high feed salinity (Alsaadi et al., 2014). Nevertheless, the major technology constraint is the need for better MD membranes with high flux and low wettability. In addition, the same authors reported that other challenges facing MD process scale-up include enhancing the water vapor flux, reducing conductive losses in order to maintain flux stability over time, enhancing thermal efficiency, minimizing temperature polarization, and avoiding pore wetting.

Another example of a combined renewable-energy-driven low-energy desalination technique is based on the humidification-dehumidification (HDH) process which exploits air as a carrier gas to evaporate water from a saline feed and produce freshwater by condensation. Advantages of this technique include flexibility in capacity, moderate installation and operating costs, simplicity, and the possibility of using new renewable energy sources, such as solar and geothermal as well

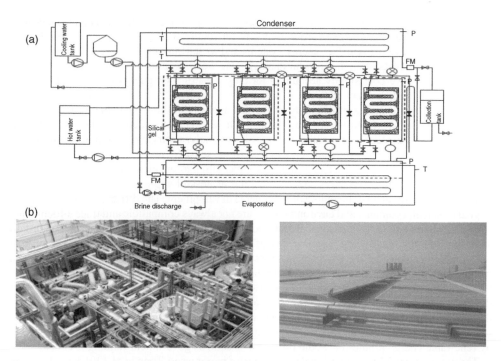

Figure 1.4.   (a) A schematic of the major components of an adsorption desalination cycle. (b) *LHS*: adsorption desalination with cooling prototype system at KAUST, Saudi Arabia; *RHS*: solar flat-plate collector system at KAUST, Saudi Arabia (Ng *et al.*, 2009, 2013).

as recovered energy or cogeneration (Davies and Paton, 2005; Mahmoudi *et al.*, 2010a). The HDH process functions at atmospheric pressure and low temperature, so the components are not submitted to mechanical pressures. The most important material characteristic required is resistance to corrosion. Water desalination by HDH has been the subject of many investigations (Davies and Paton, 2005; Mahmoudi *et al.*, 2010b). Different experimental data are available for using HDH at the pilot or industrial scale. One good example of the versatility of the technology is its application in a greenhouse which has been adapted for water desalination as well as crop production (Sablani *et al.*, 2003).

Geothermal energy also has unlimited potential in greenhouse desalination technology. Mahmoudi *et al.* (2010a) confirmed that such an energy source could be used for electricity production, including providing power for the greenhouse, as well as for commercial, industrial and residential direct-heating purposes. The condenser in the greenhouse desalination system constitutes the most critical component. For the greenhouse to be cost-effective the condenser has to be efficient, uncomplicated, inexpensive and low in maintenance. Mahmoudi *et al.* (2010b) proposed a new passive condenser in order to enhance its performance.

## 1.5   MARKET POTENTIAL, ENVIRONMENTAL CONCERNS, AND REGULATORY AND SOCIOECONOMIC ISSUES

Economic and political factors may influence large-scale deployment of renewable energy. In a critical review, Goosen *et al.* (2014) contended that in order to aid commercialization, different types of governmental policy instruments such as tax breaks and low-interest loans can be effective for diverse renewable energy sources. Broad-based policies, such as tradable energy certificates, are also likely to encourage innovation in technologies that are increasingly competitive with fossil fuels. Energy certificates issued under national legislation are normally used to provide evidence of compliance with an obligation on electricity producers, suppliers or consumers to use energy of a specific type or in order to qualify for financial support. There is also a need to eliminate subsidies for fossil fuel energy systems and to start taxing fossil fuel production and use to reflect the costs of environmental damage. One worrisome observation is that renewable energy sources have consistently accounted for only 13% of total energy use over the past 40 years (Moriarty and Honnery, 2009).

Reif (2008) performed a profitability in the risk analysis in the management of geothermal projects being implemented in Bavaria, Germany. It was recommended that the initiators of a project must run profitability simulations in order to analyze varying scenarios before implementing their plans. Reif's study also concluded that the sensitive response in the project's rate of return to changes in the parameters of their computer simulations made it clear that geothermal projects are financially risky. This suggests that the successful application of renewable energy technology requires an understanding of sustainable development. The difficulty is that three sets of goals – social, economic and environmental – are not always compatible and trade-offs are necessary. Grubert et al. (2014) reasoned that the connection between these goals is a complex affair, affecting both the quality and sustainability of the culture in which we live. As an example of this complexity, consider the government sponsorship of renewable energy projects in Germany, which is often mentioned as a model to be imitated, being based on environmental laws (e.g. the Renewable Energy Sources Act) that go back two decades. Frondel et al. (2010) maintained that the government's support mechanisms had essentially failed as a result of massive expenditures that showed little long-term promise for stimulating the economy, protecting the environment, or increasing energy security. They concluded that it is most likely that whatever jobs are generated by renewable energy promotion would disappear as soon as government funding is ended.

The non-sustainability of fossil fuel usage is also a major issue. Tonn et al. (2010) explored the question: is it possible for the United States to meet its energy needs sustainably without fossil fuels and corn ethanol? It can be reasoned that energy is one of the most pressing policy issues facing the United States today. The authors presented a scenario depicting life in the United States in the year 2050. The scenario was designed to represent energy sustainability: fossil fuels and corn ethanol were replaced by other sustainable and inexhaustible energy sources. The scenario involved the definition of a 2050 base case and described the disappearance of the suburbs, to be replaced by a mix of high-density urban centers and low-density eco-communities (Table 1.2). Tonn et al. (2010) concluded that the scenario appears plausible. The authors explained that supplementary examination is needed to more accurately establish the monetary costs to society and to build a well-organized and even-handed collection of policies to encourage this future vision.

In a related regulatory and policy study, Johnstone et al. (2010a, 2010b) examined the effect of environmental policies on technological innovation in the specific case of renewable energy. The analysis was conducted using patent data from a panel of 25 countries over the period 1978–2003. They found that public policy plays a significant role in determining patent applications. Different types of policy instruments were effective for different renewable energy sources. For example, broad-based policies, such as tradable energy certificates, were more likely to induce technological innovation.

Future food supply is intrinsically linked with energy, water and climate issues. For more than a century, fossil fuels have been the primary global energy source. However, the increasing scarcity of fossil fuels and the associated continuously high fuel prices and supply uncertainties, as well as the demand for significant reductions in greenhouse gas emissions, necessitate the search for new alternative and renewable energy sources. Implementation of renewable energy technologies, including those for powering thermal desalination technologies, is essential and governments and industries around the world are obliged to look into cutting-edge applications for renewable-energy-based water desalination, together with other applications of renewables.

In many countries, unsustainable fossil fuel use is subsidized by government. Such benefits must be removed. Economic incentives must be provided for investment in sustainable renewable desalination technologies that lead to significant market penetration. This is the only way to enable the private sector in the desalination industry to invest in large-scale deployment, which in turn

Table 1.2.    Scenario design and supporting technological and social change assumptions (Tonn *et al.*, 2010).

| Scenario design assumptions | Technology and social change assumptions |
| --- | --- |
| Liquid petroleum, coal and natural gas production/consumption in the USA are eliminated by 2050. | By 2050, electricity transmission losses will decrease by 25% due to advances in high-temperature superconducting lines and smart grid designs. |
| By 2050, approximately 50% of the population will live in super-urban highdensity areas and 50% will live in lowdensity, semi-self-sufficient areas. | By 2050, transportation energy efficiency will increase by 40% due to more efficient vehicles, and reductions in trip demand and trip length due to lifestyle and land use changes. |
| By 2050, transportation energy consumption will be 70% electricity and 30% biofuels. | By 2050, energy consumption in the residential, commercial and industrial sectors will decrease by 25% due to improvements in energy efficiency. |
| Energy consumed by the USA petroleum industry will fall to zero by 2050. | By 2050, energy consumption in the pulp and paper sector will decrease by another 20% due to decreased demand for paper and packaging. |
| Nuclear power production will be ∼26 quads by 2050. | By 2050, energy consumption in the food sector will decrease by another 20% due to more local production. |
| Wind power production will be ∼17 quads by 2050. | By 2050, energy consumption in the commercial sector will decrease by another 20% due to decreased need for commercial space. |
| Geothermal power production will be ∼5 quads by 2050. | By 2050, energy consumption in the residential sector will decrease by another 20% due to a decrease in the average size of homes and an increase in average household size (which will increase by 20%). |
| Solar power production will be ∼36 quads by 2050. | Requisite advances will be made in electric battery technologies, power storage technologies and smart grid technologies. |
| Unconventional hydro production will be ∼3 quads by 2050. | |
| Biofuels production, in multiple sectors, will increase to ∼17 quads by 2050. | |

will lead to cost reduction and wider market penetration. The end result will be that state regulation will no longer be necessary as renewable energy desalination will become cost-competitive with fossil fuels.

Finally, in recent years fuel cells have emerged as attractive alternative energy production systems because they are extremely efficient and do not produce pollutants when compared to fossil-fuel-driven technologies (Kim *et al.*, 2015). The largest markets for fuel cells for energy generation appear to be in stationary power, portable power, auxiliary power units and material handling equipment (DOE, 2013). Fuel cells convert chemical energy into electrical current and heat without combustion. According to Staffell (2015), fuel-cell-based combined heat and power (CHP) is the largest and most established market. In 2009, companies began selling thousands of units per year, marking the switch to mass manufacture. As reported by Staffell (2015), nearly 60,000 systems were sold in Japan during a four-year period. However, while hydrogen and fuel cells are not, in the strictest sense of the word, a renewable energy resource, they are very environmentally friendly when in operation. The question now is how CHP fuel cells can be employed to drive small-scale reverse osmosis systems. This is one area where further research and development is needed.

## 1.6  CONCLUDING REMARKS

Large-scale deployment of renewable energy is very dependent on economic and political influences as well as environmental and social factors. A crucial shortcoming of many renewable energy systems, such as wave, solar and wind, for example, is the variability in energy supply over time. Combining photovoltaic energy generation with other renewable resources such as wave energy in a hybrid power system can reduce supply interruption. An energy storage system such as a solar pond can be employed to run a small- to mid-size desalination plant continuously at constant load. On the other hand, when geothermal energy is used to power systems such as desalination plants the need for thermal storage is avoided.

Different types of governmental policy instruments, such as tax breaks and low-interest loans, can be effective for different renewable energy sources. This will aid in commercialization. There is also a need to eliminate subsidies for fossil fuel energy systems or to tax fossil fuel production and use to reflect the costs of environmental damage. One troublesome observation is that renewable energy sources have consistently accounted for only 13% of total energy use over the past 40 years. Another major concern is that, in some instances, government renewable energy policy may undermine and subvert market incentives, resulting in massive expenditures that show little long-term promise for stimulating the economy, protecting the environment, or increasing energy security. It is possible that whatever jobs are created by renewable energy promotion may vanish as soon as government support is terminated.

There is a need to find an equilibrium or compromise amongst three sets of aims: social, economic and environmental. The relationship between these areas is a complicated matter, affecting both the quality and sustainability of society. One way to help achieve a balance between them is to educate decision makers, students and industrial workforces, as this represents the primary vehicle available for catalyzing cultural change. This is critical for long-term economic and social sustainability.

To summarize, renewable energy technologies for desalination are quickly evolving to becoming economically and environmentally feasible. Even though there are concerns that government policy may undermine market incentives, there is great potential in the use of renewable energy in many parts of the world. Solar, wind, wave and geothermal sources, as well as fuel cell technology, could provide viable sources of energy to power both seawater and brackish water desalination plants. Finally, it should be noted that part of the answer to the global water scarcity faced by society is not only to produce more potable water, but to recognize that this needs to be accomplished in an ecologically sustainable manner.

## REFERENCES

Abdel-Jawad, M. (2001) Energy sources for coupling with desalination plants in the GCC countries. Economic and Social Commission for Western Asia (ESCWA), Beirut, Lebanon.

Alsaadi, A., Francis, L., Maab, H., Amy, G. & Ghaffour, H. (2014) Experimental & theoretical analyses of temperature polarization effect in vacuum membrane distillation. *Journal of Membrane Science*, 471, 138–148.

Alyousef, Y. & Abu-ebid, M. (2012) Energy efficiency initiatives for Saudi Arabia on supply and demand sides. In: Morvaj, Z. (ed) *Energy Efficiency – A Bridge to Low Carbon Economy*. InTech, Rijeka, Croatia. pp. 279–309.

Barlev, D., Vidu, R. & Stroeve, P. (2011) Innovation in concentrated solar power. *Solar Energy Materials and Solar Cells*, 95, 2703–2725.

Caldera, U., Bogdanov, D. & Breyer, C. (2016) Local cost of seawater RO desalination based on solar PV and wind energy: a global estimate. *Desalination*, 385, 207–216.

Connor, R. (2015) The United Nations world water development report: water for a sustainable world (Vol. 1). UNESCO, Paris.

Davies, P.A. & Paton, C. (2005) The seawater greenhouse in the United Arab Emirates: thermal modelling and evaluation of design options. *Desalination*, 173, 103–111.

DeCanio, S.J. & Fremstad, A. (2011) Economic feasibility of the path to zero net carbon emissions. *Energy Policy*, 39, 1144–1153.

DOE (2013) US DRIVE fuel cell technical team roadmap. Office of Energy Efficiency & Renewable Energy, U.S. Department of Energy, Washington, DC.

Droogers, P., Immerzeel, W.W., Terink, W., Hoogeveen, J., Bierkens, M.F.P., van Beek, L.P.H. & Negewo, B.D. (2012) Water resources trends in Middle East and North Africa towards 2050. *Hydrology and Earth System Sciences*, 16, 3101–3114.

El Kharraz, J., El-Sadek, A., Ghaffour, N. & Mino, E. (2012) Water scarcity and drought in WANA countries. *Procedia Engineering*, 33, 14–29.

Frondel, M., Ritter, N., Schmidt, C.M. & Vance, C. (2010) Economic impacts from the promotion of renewable energy technologies – the German experience. *Energy Policy*, 38, 4048–4056.

Ghaffour, N., Missimer, T.M. & Amy, G.L. (2013) Technical review and evaluation of the economics of water desalination: current and future challenges for better water supply sustainability. *Desalination*, 309, 197–207.

Ghaffour, N., Bundschuh, J., Mahmoudi, H. & Goosen, M.F. (2015) Renewable energy-driven desalination technologies: a comprehensive review on challenges and potential applications of integrated systems. *Desalination*, 356, 94–114.

Goosen, M.F.A. (2013) Institutional aspects of economic growth: assessing the significance of public debt, economic governance & industrial competition. *The Open Business Journal*, 6, 1–13.

Goosen, M.F.A., Laboy-Nieves, E.N., Schaffner, F. & Abdelhadi A. (2009a) The environment, sustainable development and human wellbeing: an overview. In: Laboy-Nieves, E.N., Schaffner, F., Abdelhadi, A. & Goosen, M.F.A. (eds) *Environmental Management, Sustainable Development and Human Health*. Taylor & Francis, London. pp. 3–12.

Goosen, M.F.A., Al-Obeidani, S.K.S., Al-Hinai, H., Sablani, S., Taniguchi, Y. & Okamura, H. (2009b) Membrane fouling and cleaning in treatment of contaminated water. In: Laboy-Nieves, E.N., Schaffner, F., Abdelhadi, A. & Goosen, M.F.A. (eds) *Environmental Management, Sustainable Development and Human Health*. Taylor & Francis, London. pp. 503–512.

Goosen, M., Mahmoudi, H. & Ghaffour, N. (2014) Today's and future challenges in applications of renewable energy technologies for desalination. *Critical Reviews in Environmental Science & Technology*, 44, 929–999.

Goosen, M.F.A., Mahmoudi, H., Ghaffour, N., Bundschuh, J. & Al Yousef, Y. (2016) A critical evaluation of renewable energy technologies for desalination. *Application of Materials Science and Environmental Materials (AMSEM2015), Proceedings of the 3rd International Conference, 1–3 October 2015, Phuket Island, Thailand*. pp. 233–258.

Gottinger, H. & Goosen, M.F.A. (eds) (2012) *Strategies of Economic Growth and Catch-up: Industrial Policies and Management*. Nova Science, New York.

Grubert, E.A., Stillwell, A.S. & Webber, M.E. (2014) Where does solar-aided seawater desalination make sense? A method for identifying sustainable sites. *Desalination*, 339, 10–17.

Intelligence, G.W. (2013) Market profile and desalination markets, 2009–2012 yearbooks and GWI website. Global Water Intelligence, Oxford.

Johnstone, N., Haščič, I. & Popp, D. (2010a) Renewable energy policies and technological innovation: evidence based on patent counts. *Environmental and Resource Economics*, 45, 133–155.

Johnstone, N., Haščič, I. & Kalamova, M. (2010b) Environmental policy design characteristics and technological innovation evidence from patent data. OECD Environment Working Papers, No. 16, OECD Publishing, Paris, France.

Kim, D.J., Jo, M.J. & Nam, S.Y. (2015) A review of polymer-nanocomposite electrolyte membranes for fuel cell application. *Journal of Industrial and Engineering Chemistry*, 21, 36–52.

Mahmoudi, H., Spahis, N., Goosen, M.F.A., Ghaffour, N., Drouiche, N. & Ouagued, A. (2010a) Application of geothermal energy for heating and freshwater production in a brackish water greenhouse desalination unit: a case study from Algeria. *Renewable and Sustainable Energy Reviews*, 14, 512–517.

Mahmoudi, H., Spahis, N., Abdul-Wahab, S.A., Sablani, S. & Goosen, M.F.A. (2010b) Improving the performance of a seawater greenhouse desalination system by assessment of simulation models for different condensers. *Renewable and Sustainable Energy Reviews*, 14, 2182–2188.

Misra, B. (2000) New economic policy and economic development. *IASSI Quarterly*, 18, 20–25.

Moriarty, P. & Honnery, D. (2009) Hydrogen's role in an uncertain energy future. *International Journal of Hydrogen Energy*, 34, 31–39.

Ng, K.C., Thu, K., Chakraborty, A., Saha, B.B. & Chun, W.G. (2009) Solar-assisted dual-effect adsorption cycle for the production of cooling effect and potable water. *International Journal of Low-Carbon Technologies*, 4, 61–67.

Ng, K.C., Thu, K., Kim, Y., Chakraborty, A. & Amy, G. (2013) Adsorption desalination: an emerging low-cost thermal desalination method. *Desalination*, 308, 161–179.

Reddy, K.V. & Ghaffour, N. (2007) Overview of the cost of desalinated water and costing methodologies. *Desalination*, 205, 340–353.

Reif, T. (2008) Profitability analysis and risk management of geothermal projects. *Geo-Heat Centre Bulletin*, 28, 1–4.

Sablani, S., Goosen, M.F.A., Paton, C., Shayya, W.H. & Al-Hinai, H. (2003) Simulation of freshwater production using a humidification-dehumidification seawater greenhouse. *Desalination*, 159, 283–288.

Sahin, O., Siems, R.S., Stewart, R.A. & Porter, M.G. (2016) Paradigm shift to enhanced water supply planning through augmented grids, scarcity pricing and adaptive factory water: a system dynamics approach. *Environmental Modelling and Software*, 75, 348–361.

Staffell, I. (2015) Zero carbon infinite COP heat from fuel cell CHP. *Applied Energy*, 147, 373–385.

Tonn, B., Frymier, P., Graves, J. & Meyers, J. (2010) A sustainable energy scenario for the United States: year 2050. *Sustainability*, 2, 3650–3680.

Wagner, S.J. & Rubin, E.S. (2011) Economic implications of thermal energy storage for concentrated solar thermal power. *World Renewable Energy Congress 2011, 8–13 May, Linkoping, Sweden.* pp. 81–95.

Zhang, H.L., Baeyens, J., Degreve, J. & Caceres, G. (2013) Concentrated solar power plants: review and design methodology. *Renewable and Sustainable Energy Reviews*, 22, 466–481.

# CHAPTER 2

## The use of solar energy for small-scale autonomous desalination

Patricia Palenzuela, Guillermo Zaragoza, Baltasar Peñate, Vicente Subiela,
Diego-César Alarcón-Padilla & Lourdes García-Rodríguez

### 2.1 INTRODUCTION

Solar desalination is proposed for sustainable desalination in many places of the world due to the usual coincidence between water shortage and high solar irradiation.

The use of solar energy for desalination follows two different approaches depending on the desalination technology chosen. There are two kinds of commercial desalination systems. The first ones were based on the use of thermal energy, with multi-stage flash (MSF) distillation giving way to the more efficient multi-effect distillation systems (MED). However, the most common desalination technology nowadays is a filtration system based on mechanical energy, that is, reverse osmosis (RO). In the first approach, the coupling with solar energy is performed by using solar thermal collectors, while in the second approach electricity is required and therefore photovoltaic (PV) panels are used.

There are no large-scale installations of solar desalination yet, but some pilot studies have been done for small-scale desalination (i.e. with production below $100 \, \text{m}^3 \, \text{day}^{-1}$). In this chapter we review the use of solar energy for desalination by considering a flagship case study for each technology, one for thermal desalination and another for reverse osmosis. In both cases, a fully autonomous system has been operated in real-world conditions over several years. The first case is a solar MED facility located at Plataforma Solar de Almería (PSA) in Tabernas (south-east Spain), and the second is a PV-RO system located in Ksar Ghilène (Tunisia). A thorough analysis of both plants and their performance is presented in the next two sections.

### 2.2 SOLAR MULTI-EFFECT DISTILLATION CASE STUDY: PLATAFORMA SOLAR DE ALMERÍA

#### 2.2.1 *Description of the system*

The MED-PSA plant is a forward-feed multi-effect distillation unit manufactured and delivered by Weir Entropie in 1987 within the Solar Thermal Desalination (STD) project (Zarza, 1991, 1994). The plant has a sequence of 14 evaporators, or "effects", in a vertical arrangement at decreasing pressures and temperatures from the first effect (at the top) to the fourteenth one. Since the MED plant is a forward-feed unit, it has 13 preheaters, each one located next to an evaporator, to increase the seawater temperature on its way to the first effect. Such increases take place when a small fraction of the vapor generated in each effect condenses on the outer surface of the preheater, transferring its phase change enthalpy to the seawater circulating inside.

The original configuration of the MED-PSA plant worked with low-pressure saturated steam generated from a low-pressure boiler connected to a parabolic trough solar collector field with a thermal storage system (Fig. 2.1). The temperature of the low-pressure steam was limited to 70°C (at 31 kPa) to avoid scaling problems. Heat was supplied to the desalination plant by a solar field consisting of east-west-aligned one-axis tracking collectors manufactured by Acurex. The field had a total aperture area of $2672 \, \text{m}^2$ and was not purposely dimensioned for the desalination plant,

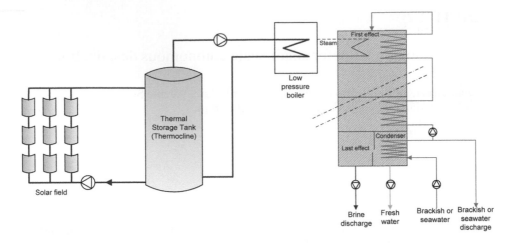

Figure 2.1. Schematic diagram of MED-PSA desalination system installed within the STD project.

Table 2.1. Design specifications of the MED-PSA plant.

| | |
|---|---|
| Number of effects | 14 |
| Feed seawater flow rate | $8\,m^3\,h^{-1}$ |
| Brine flow rate from the last effect | $5\,m^3\,h^{-1}$ |
| Total distillate output | $3\,m^3\,h^{-1}$ |
| Cooling seawater flow rate at 25°C | $20\,m^3\,h^{-1}$ |
| Cooling seawater flow rate at 10°C | $8\,m^3\,h^{-1}$ |
| Heat source energy consumption | 190 kW |
| Performance ratio | >9 |
| Vacuum system | Hydro-ejectors (seawater at 0.3 MPa) |
| Top brine temperature | 68.0°C |
| Last effect temperature | 35°C |

so about one fifth of its size would have been sufficient to supply the heat required. To increase the number of operating hours the solar field was connected to a single $115\,m^3$ thermocline thermal storage tank filled with the thermal oil circulating through the solar collectors. The thermal storage capacity was approximately $5\,MW_{th} \cdot h$ at charge/discharge temperatures of 300/225°C. Table 2.1 shows the technical specifications of the plant.

In 2005 the STD solar field was replaced by a new one within the framework of the AQUA-SOL project (Enhanced Zero Discharge Seawater Desalination using Hybrid Solar Technology, 2002–2006, European Commission program FP5-EESD). A new solar field composed of static compound parabolic concentrators (CPC) and a new water-based thermal storage system were installed at PSA for the supply of the thermal power required by the MED plant. The original horizontal tube bundle of the first effect was replaced by a new one able to work directly with the hot water coming from the thermal storage tanks (Alarcón-Padilla and García-Rodríguez, 2007). Also within the framework of AQUASOL, a prototype of a double-effect LiBr-$H_2$O absorption heat pump (DEAHP) was manufactured by Entropie and coupled to the MED-PSA unit (Alarcón-Padilla *et al.*, 2007, 2008, 2010). The DEAHP was integrated into the system to increase the overall plant efficiency by recovering the phase change enthalpy of the saturated steam produced in the last effect of the distillation unit. Figure 2.2 shows pictures of all the components of the solar MED facility at the PSA, and Figure 2.3 depicts the general layout of their integration. The solar collector field is composed of 252 stationary solar collectors (Ao Sol CPC, 1.12x) with an overall aperture area of roughly $500\,m^2$, arranged in four rows of 63 collectors each with

Figure 2.2. (a) Multi-effect distillation plant; (b) compound parabolic collector solar field; (c) thermal storage system; (d) double-effect LiBr-H$_2$O absorption heat pump.

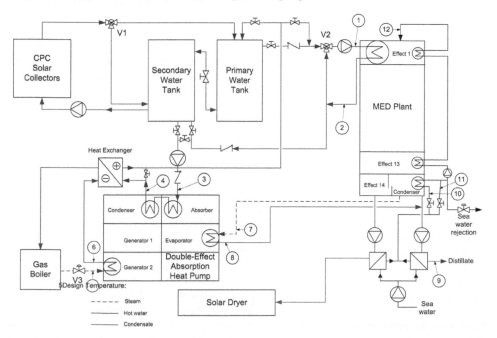

Figure 2.3. Layout of AQUASOL system.

Table 2.2.    Nominal conditions of the new MED-PSA plant's first cell.

|  | Desalination driven by solar collectors | Desalination driven by absorption heat pump |
|---|---|---|
| Heat source energy consumption | 200 kW | 150 kW |
| Inlet/outlet hot water temperature | 74.0/71.0°C | 66.5/63.5°C |
| Brine temperature (in the first cell) | 68.0°C | 62.0°C |
| Hot water mass flow rate | 12.0 kg s$^{-1}$ | 12.0 kg s$^{-1}$ |
| Pressure drop | 40 kPa | 40 kPa |
| Nominal plant production | 3 m$^3$ h$^{-1}$ | 2.7 m$^3$ h$^{-1}$ |

35° tilt angle and south-facing orientation. The solar field's hydraulic configuration corresponds to a reverse-feeding hydraulic layout which allows equally distributed flow rates in the four rows without further regulation. The thermal storage system consists of two interconnected 12 m$^3$ capacity water tanks. The use of two tanks enables the solar contribution to be increased over the year as well as obtaining the temperature stratification necessary for a constant hot water inlet temperature. A continuous control valve (V2) allows provision of a stable inlet temperature to the MED plant by mixing water from the primary tank with the return coming from the first effect heat exchanger tube bundle. The rest goes to the secondary tank, closing the loop.

As shown in Figure 2.3, the heat required for the first effect can be provided either with the CPC solar field (solar mode) or with the LiBr-H$_2$O DEAHP (fossil mode). The nominal conditions of the MED plant in the two modes are shown in Table 2.2.

The specific characteristics of the MED-PSA pilot plant are described below (see the layout of the plant in Fig. 2.4). A seawater feed of 8 m$^3$ h$^{-1}$ flow is pumped through the 13 preheaters, increasing its temperature before being sprayed over the horizontal tube bundle of the first effect. A perforated tray guarantees that seawater is distributed as a falling film, coating the surface of the tubes entirely. The first effect is the only one that receives an external supply of thermal power through the decrease in the enthalpy of the hot water flow circulating inside the tube bundle (in the AQUASOL layout) or the phase change enthalpy variation of the saturated steam (in the STD layout). The heat released in both configurations promotes the partial evaporation of the seawater sprayed over the tubes. The vapor generated goes first through a wire mesh demister in order to remove brine droplets and a small fraction of it condenses over the external surface of the preheater associated with the first effect. The condensate and the rest of the vapor flow naturally inside the tube bundle to the next effect. Here, the vapor transfers its latent heat to the seawater that was not evaporated in the previous effect (called brine), which is falling by gravity over its tube bundle. Therefore, on the one hand the vapor condenses, mixing with the distillate produced in the previous preheater, and on the other hand, part of the brine evaporates, increasing its salt concentration.

In the rest of the effects, as in the second one, an evaporation/condensation process takes place, and the heat source that drives this process is always the vapor produced in the previous effect. The vapor generated in the last stage is condensed through the so-called final condenser which is cooled by seawater. Due to the fact that, at nominal seawater temperature conditions, the cooling flow is greater than the required water feed flow, some of this seawater is rejected at a temperature approximately equal to the last effect temperature.

Due to the special geographical situation of PSA (30 km from the coast), the experimental facility is equipped with two water pools (PL-1 and PL-2) with a capacity of 5 m$^3$ and 20 m$^3$, respectively. Brackish water wells supply source water to PL-2 (Fig. 2.5a), where additional salts can be added until the salinity and composition required by a particular experiment are reached. The saline water flow required by the final condenser is taken from PL-2 and the rejected cooling water, plus the brine and the distillate output from the plant, are streamed to PL-1 (Fig. 2.5b). Due to the fact that the rejected seawater coolant, brine and distillate are at a higher temperature,

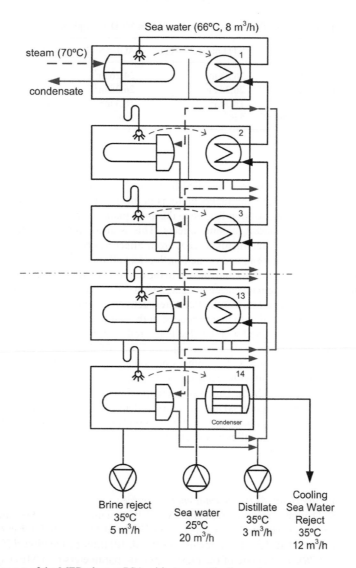

Figure 2.4.   Layout of the MED plant at PSA with steam as the thermal energy source.

Figure 2.5.   (a) PL-2 pool for the feed water to the plant; (b) PL-1 pool for the rejected streams from the plant.

Table 2.3.    Surface areas of tube bundles in MED-PSA plant.

|              | STD project layout | AQUASOL project layout |
|--------------|--------------------|------------------------|
| First effect | $26.28\,m^2$       | $24.26\,m^2$           |
| Effects 2–14 | $26.28\,m^2$       | $26.28\,m^2$           |
| Preheaters   | $18.3\,m^2$        | $18.3\,m^2$            |
| End condenser| $5\,m^2$           | $5\,m^2$               |

a water-cooled refrigeration tower is installed between PL-1 and PL-2 in order to reduce the temperature of the former to the nominal conditions of the experiment set in the latter.

One important energy optimization feature in this pilot plant sis the efficient use of the energy content of the distillate produced in some effects. This is accomplished as follows: from the first effect to the fourth one, the distillate goes from effect to effect enabling its sensible heat to promote additional evaporation, and in the fourth effect all the distillate is extracted with one portion directed to the seventh effect and the rest to the tenth. The process is similar in the seventh and tenth effects, splitting the condensate extracted between the tenth and the thirteenth effects, and between the thirteenth and fourteenth effects, respectively. Finally, in the thirteenth effect, all the distillate produced is mixed with that produced in the fourteenth, and this mixture enters the last condenser in which it is mixed with the distillate produced there.

The vacuum system consists of two hydro-ejectors, one of them connected simultaneously to the second and seventh effects and the other one connected to the final condenser. They are connected within a closed circuit with a tank and an electric pump that circulates seawater through the ejectors at a pressure of 0.3 MPa. This system creates the initial vacuum in the plant as well as removing the air (lack of air tightness) and the non-condensable gases during normal operation.

The tube bundles of the evaporators, preheater and condenser are made of 90/10 Cu-Ni tubes. The surface areas of each of the different tube bundles are indicated in Table 2.3.

### 2.2.2    Desalination system operation

#### 2.2.2.1    Low-pressure steam as the thermal energy source

In Phase I of the STD project, an evaluation of the MED-PSA plant was carried out during different experimental campaigns performed between October 1989 and May 1990. Each campaign aimed to study the influence of different parameters on the distillation production and on the plant performance ratio (*PR*). This is one of the most important parameters of a MED system, which is defined as the ratio between the mass of distillate (in kg) and the thermal energy supplied to the process, normalized to 2326 kJ (1000 Btu), which is the latent heat of vaporization of water at 73°C.

##### 2.2.2.1.1    *Influence of first effect's temperature and temperature gradient on the distillate production*
The test conditions were as follows:

- Single days were evaluated in order to exclude effects of differences in fouling rates, air leaks, salinities, etc.
- The condenser temperature was maintained at a constant value.
- Steady-state conditions were required.

The following test days satisfied the required conditions: 8, 9 and 11 January 1990; 20 April 1990. The end condenser temperatures varied slightly due to the rise in seawater pool temperature, despite the cooling tower being operational throughout the entire plant operation. Nevertheless, the deviations in the condenser temperature were very small and did not invalidate the tests.

Table 2.4.   Experimental test results with the average values of every test at steady-state periods.

| Date | $M_d$ [m³ h⁻¹] | $PR$ | $M_{oil}$ [kg s⁻¹] | $Q_{boiler}$ [kW] | $M_{fw}$ [m³ h⁻¹] | $T_{v1}$ [°C] | $T_{v,cond}$ [°C] |
|------|------|------|------|------|------|------|------|
| 20/04/90 | 1.9 | 9.5 | 1.3 | 126.5 | 8.6 | 56.4 | 29.9 |
| 20/04/90 | 2.0 | 9.5 | 1.4 | 135.5 | 8.5 | 58.0 | 30.2 |
| 20/04/90 | 2.3 | 9.6 | 1.6 | 150.0 | 8.6 | 60.6 | 30.2 |
| 11/01/90 | 1.5 | 9.3 | 0.6 | 99.3 | 10.3 | 44.5 | 19.5 |
| 11/01/90 | 2.1 | 10.1 | 0.9 | 126.4 | 10.1 | 50.5 | 19.1 |
| 11/01/90 | 3.0 | 10.2 | 1.5 | 186.4 | 10.0 | 63.6 | 21.4 |
| 09/01/90 | 1.8 | 9.3 | 0.7 | 120.0 | 10.4 | 52.5 | 25.0 |
| 09/01/90 | 2.8 | 10.4 | 1.2 | 171.9 | 9.0 | 66.7 | 27.3 |
| 08/01/90 | 2.2 | 9.8 | 0.9 | 139.7 | 10.3 | 52.9 | 19.7 |
| 08/01/90 | 3.1 | 9.9 | 1.5 | 202.0 | 10.3 | 64.6 | 20.1 |

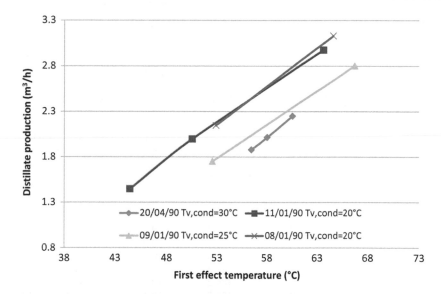

Figure 2.6.   Distillate production as a function of the first effect temperature for various test days.

Figure 2.6 shows the distillate production as a function of the first effect vapor temperature ($T_{v1}$) for different vapor condenser temperatures ($T_{v,cond}$). As can be seen, the distillate product increased with the first effect vapor temperature, for all condenser vapor temperatures.

Figure 2.7 shows the influence of the temperature gradient in the plant (temperature difference between the first effect and the condenser vapor temperatures) on the distillate production. It can be seen that, operating with a given temperature difference, the distillate production increased with increasing condenser temperature.

### 2.2.2.1.2   *Influence of the temperature gradient on the performance ratio*

The same test conditions as before were established to run the experiments. The resulting experimental data are shown in Table 2.5.

The relationship of *PR* to temperature difference is depicted in Figure 2.8. As can be seen, the data did not reveal any *PR* dependence on the temperature difference. The accuracy of the plant performance ratio calculations is strongly dependent on the accuracy of temperature and flow measurements, so a small error in these measurements yields a much higher error in the performance ratio calculation. This could be the reason why no dependence between the performance ratio and the temperature difference was observed. In conclusion, the MED plant showed a *PR* within the range of 9.4–10.4 when the plant operated with low-pressure steam.

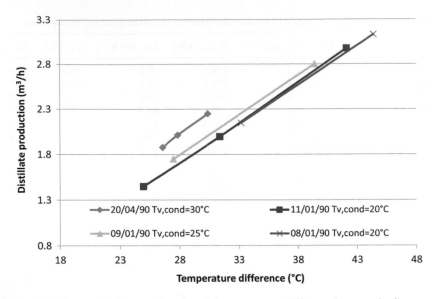

Figure 2.7. Distillate production as a function of the temperature difference between the first evaporator and the condenser for various test days.

Table 2.5. Experimental test results with the average values of every test at steady-state periods.

| Date | $M_d$ [m³ h⁻¹] | $PR$ | $M_{oil}$ [kg s⁻¹] | $Q_{boiler}$ [kW] | $M_{fw}$ [m³ h⁻¹] | $T_{v1}$ [°C] | $T_{v,cond}$ [°C] |
|------|------|------|------|------|------|------|------|
| 01/03/90 | 2.8 | 9.8 | 2.0 | 179.9 | 11.7 | 65.0 | 30.4 |
| 13/03/90 | 2.7 | 9.8 | 1.9 | 176.8 | 9.8 | 65.0 | 30.8 |
| 14/03/90 | 2.9 | 10.7 | 2.1 | 170.5 | 10.3 | 65.4 | 30.2 |
| 15/03/90 | 2.7 | 10.7 | 1.9 | 164.2 | 9.2 | 65.3 | 30.6 |
| 16/03/90 | 2.7 | 10.2 | 2.0 | 169.5 | 9.9 | 64.9 | 30.7 |
| 05/01/90 | 3.2 | 10.2 | 1.5 | 197.6 | 10.4 | 64.4 | 20.4 |
| 05/01/90 | 3.0 | 10.3 | 1.4 | 188.6 | 10.4 | 64.2 | 21.2 |
| 11/01/90 | 3.0 | 10.2 | 1.5 | 186.2 | 10.0 | 63.5 | 21.5 |
| 11/01/90 | 2.0 | 10.2 | 0.9 | 126.7 | 10.1 | 50.4 | 19.1 |
| 11/01/90 | 1.4 | 9.3 | 0.6 | 99.3 | 10.3 | 44.5 | 19.5 |

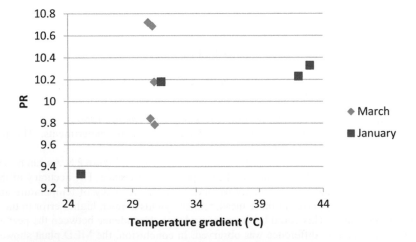

Figure 2.8. Performance ratio versus total temperature difference.

Table 2.6.   Experimental test results with the average values of every test at steady-state periods.

| Date | $M_d$ [m³ h⁻¹] | $PR$ | $M_{oil}$ [kg s⁻¹] | $Q_{boiler}$ [kW] | $M_{fw}$ [m³ h⁻¹] | $T_{v1}$ [°C] | $T_{v,cond}$ [°C] |
|---|---|---|---|---|---|---|---|
| 13/12/89 | 3.3 | 10.5 | 1.3 | 202.4 | 11.0 | 65.9 | 27.3 |
| 13/12/89 | 3.1 | 10.2 | 1.1 | 192.0 | 11.2 | 66.8 | 32.0 |
| 13/12/89 | 2.6 | 10.0 | 1.0 | 164.5 | 9.7 | 67.2 | 35.4 |
| 21/02/90 | 2.8 | 9.8 | 1.2 | 185.7 | 11.3 | 65.2 | 30.6 |
| 21/02/90 | 2.6 | 9.9 | 1.1 | 170.2 | 12.3 | 65.4 | 34.1 |
| 26/02/90 | 2.7 | 9.9 | 1.9 | 177.5 | 12.1 | 64.5 | 29.7 |
| 26/02/90 | 2.4 | 9.7 | 1.7 | 159.9 | 12.1 | 64.8 | 34.7 |
| 18/04/90 | 2.4 | 9.3 | 1.8 | 166.8 | 8.8 | 64.7 | 34.2 |
| 18/04/90 | 2.2 | 9.3 | 1.7 | 152.6 | 9.5 | 65.7 | 37.8 |

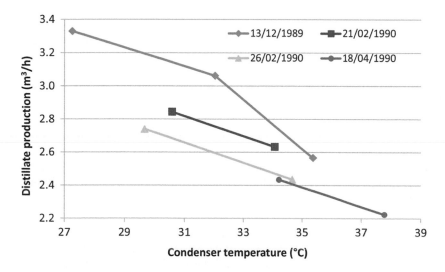

Figure 2.9.   Distillate production as a function of the condenser temperature.

### 2.2.2.1.3  *Influence of end condenser temperature on distillate production*
The following test conditions were established:

- The tests were performed on a single day in order to avoid effects from differences in fouling, different salinities or air leaks.
- The temperature of the first evaporator was kept constant.
- The condenser temperature was changed stepwise and kept constant until a steady state was reached.
- Steady-state conditions were required.

The following test days satisfied the required conditions: 13 December 1989; 21 and 26 February 1990; 18 April 1990. In contrast to the previous tests, it was difficult to identify tests with defined steps in condenser temperature. As previously mentioned, this is due to the fact that the cooling tower was not sufficient to provide continuous control of the seawater temperature at the inlet of the condenser, thus varying the end condenser temperature within a narrow range around the predefined set-point.

Measurement data for steady-state conditions are shown in Table 2.6, which summarizes the average values of each variable.

Figure 2.9 depicts the product flow as a function of the condenser temperature for the different test days. As expected, the distillate production decreased with increasing condenser temperature.

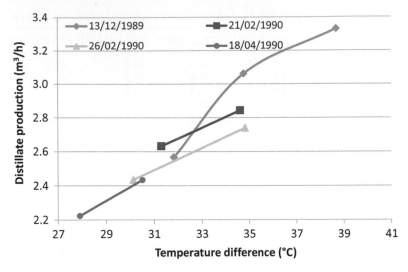

Figure 2.10.   Distillate production as a function of the temperature difference between the first evaporator and the condenser.

It can also be observed that the plant capacity decreased over time as the effects of fouling and scaling increased, lowering heat exchanger efficiency.

In order to eliminate the effect of fluctuations in the first effect temperature, Figure 2.10 shows the distillate production as a function of the temperature difference between the first evaporator and the condenser.

### 2.2.2.2   *Hot water as the thermal energy source*

During the AQUASOL project, the behavior of the MED plant outside nominal conditions was investigated in two experimental campaigns performed in 2007 (Blanco *et al.*, 2011) and 2010 (Fernández-Izquierdo *et al.*, 2012). The purpose of these studies was to find the best working conditions that maximize the *PR* and distillate production of the MED plant when the thermal energy is only provided by a static solar collector field. Both campaigns evaluated the *PR* and the distillate production at different hot water inlet temperatures and feed water flow rates.

### 2.2.2.2.1   *2007 test campaign*

The following experimental procedure was carried out in this campaign:

- The hot water inlet temperature from the primary water tank was varied over a wide range: from 63 to 75°C for each feed water flow rate tested (from 5 to 9 $m^3 h^{-1}$). The hot water inlet temperature was controlled by the three-way valve V2 (Fig. 2.3).
- The vapor temperature in the last effect was maintained at a constant of approximately 31°C. This was accomplished by varying the seawater flow rate at the inlet of the condenser using a manual valve. The cooling tower was switched on when necessary in order to keep the seawater temperature at the inlet of the condenser constant. The hot water flow rate was kept constant at the design value: 12 $L s^{-1}$.

Table 2.7 shows the operating conditions tested during this campaign, and the results obtained are shown in Figure 2.11, where it can be observed that the distillate production increased both with the hot water inlet temperature and as the feed water flow rate increases from 5 to 8 $m^3 h^{-1}$. Feed water flow rates greater than this did not have a positive effect on the distillate production. The maximum distillate production, which varied from 2.3 to 3 $m^3 h^{-1}$ depending on the hot water inlet temperature, was achieved at the nominal flow rate of 8 $m^3 h^{-1}$.

Table 2.7. Operating conditions of the 2007 test campaign.

| Test | Feed water flow rate [m³ h⁻¹] | Hot water inlet temperatures [°C] | | | | | | | | |
|------|------|------|------|------|------|------|------|------|------|------|
| 1 | 5 | 60 | 62 | 64 | 66 | 68 | 70 | 72 | 74 | 76 |
| 2 | 6.2 | 60 | 62 | 64 | 66 | 68 | 70 | 72 | 74 | |
| 3 | 7 | 60 | 62 | 64 | 66 | 68 | 70 | 72 | 74 | |
| 4 | 8 | 60 | 62 | 64 | 66 | 68 | 70 | 72 | 74 | |
| 5 | 9 | 60 | 62 | 64 | 66 | 68 | 70 | 72 | 74 | 75 |

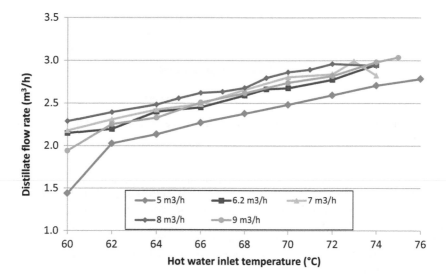

Figure 2.11. Distillate production at different feed water flow rates and hot water inlet temperatures.

Table 2.8. Experimental results at different inlet hot water temperatures from the primary water tank.

| Thermal energy input | $T_{h,in}$ [°C] | 75.0 | 73.0 | 71.0 | 69.0 | 67.0 | 65.0 | 63.0 |
|------|------|------|------|------|------|------|------|------|
| | $T_{h,out}$ [°C] | 70.5 | 68.8 | 67.1 | 65.3 | 63.6 | 61.9 | 60.1 |
| Vapor temperature and distillate | $T_{v1}$ [°C] | 68.4 | 66.7 | 65.0 | 63.3 | 61.6 | 59.9 | 58.2 |
| | $T_{v14}$ [°C] | 31.4 | 31.4 | 31.3 | 31.2 | 31.2 | 31.1 | 31.1 |
| | $M_d$ [m³ h⁻¹] | 3.1 | 3.0 | 2.9 | 2.8 | 2.7 | 2.6 | 2.5 |
| MED plant energy data | $Q_h$ [kW] | 221.2 | 206.6 | 191.9 | 182.2 | 167.6 | 152.9 | 143.1 |
| | $STC$ [kWh$_{th}$ m⁻³] | 70.4 | 68.2 | 65.5 | 64.6 | 61.8 | 58.8 | 57.5 |
| | $SEC$ [kWh$_e$ m⁻³] | 3.9 | 4.0 | 4.2 | 4.3 | 4.5 | 4.7 | 4.9 |
| | $PR$ | 9.2 | 9.5 | 9.9 | 10.0 | 10.4 | 11.0 | 11.2 |

Based on the results obtained in this campaign, all of the main MED plant parameters were analyzed against different hot water inlet temperatures from the primary water tank, using 8 m³ h⁻¹ as the feed water flow rate. Table 2.8 shows the average values obtained from the experimental results (Alarcón-Padilla *et al.*, 2010). In addition, the results in terms of distillate production and performance ratio are depicted in Figure 2.12.

The experimental results showed that the distillate production increased with the hot water inlet temperature as well as the specific thermal energy consumption, which caused the performance ratio to decrease with the rise in temperature of the energy source. According to the

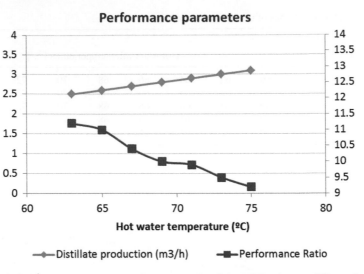

Figure 2.12.  Distillate production and performance ratio of the MED plant at different hot water inlet temperatures.

efficiency results, the best behavior was obtained for an input temperature between 65 and 67°C, as temperatures below 65°C led to a significant reduction in the distillate production.

It is important to highlight that the specific electricity consumption shown in these results cannot be considered as representative due to the pilot scale of the MED-PSA plant. The main electricity consumption of this plant was a result of creating the vacuum required at the beginning of each test day, which is not needed in a normal continuously operating plant that only needs energy to maintain the vacuum level.

#### 2.2.2.2.2  2010 test campaign
In this experimental campaign, the hot water and feed water flow rates were maintained at constant levels and the hot water and vapor temperature in the last effect were changed as follows:

• The hot water inlet temperature from the primary water tank was varied from 57 to 74°C.
• The seawater flow rate through the condenser was varied in order to maintain a temperature difference of 35°C between the first and last effects (resulting in a temperature gradient between effects similar to that of nominal conditions). As before, the use of the cooling tower was required.

Table 2.9 summarizes the main parameters of the different working conditions in the test campaign conducted in 2010.

Figure 2.13 shows a test of the MED plant working at nominal conditions (Fernández-Izquierdo *et al.*, 2012). The test was performed with a hot water inlet temperature of 74°C, which resulted in a level of distillate production very close to the nominal one (see Table 2.2). This could indicate a lack of fouling or scaling in the plant at that juncture.

Figure 2.14 and Figure 2.15 show the results in terms of distillate, *PR* and thermal consumption (determined by the hot water inlet mass flow rate and the difference between the inlet and outlet temperatures of hot water in the first effect) obtained in the tests performed at the different working conditions shown in Table 2.9. Table 2.10 shows the average values of these experimental test results (Fernández-Izquierdo *et al.*, 2012).

As can be seen from the results, the maximum *PR* obtained was 10, when the plant operated with a hot water inlet temperature of 68°C. Plant operation at a higher water temperature did not result in higher efficiency but did result in greater distillate production. The maximum distillate

Table 2.9.   Operating conditions of the 2010
test campaign.

| Test | Hot water inlet temperature [°C] |
|------|----------------------------------|
| 1 | 57 |
| 2 | 60 |
| 3 | 63 |
| 4 | 65 |
| 5 | 68 |
| 6 | 70 |
| 7 | 72 |
| 8 | 74 |

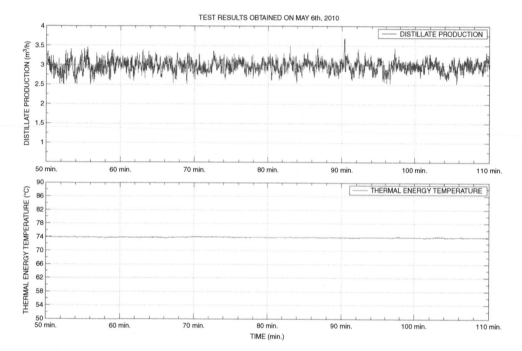

Figure 2.13.   Distillate production of the MED-PSA plant working at nominal conditions (adapted from Fernández-Izquierdo *et al.*, 2012).

production was the result of operation of the MED plant at nominal conditions (74°C). It was also observed that the impacts of partial load operation were greater on distillate production than on the *PR*.

### 2.2.3   *Operational conclusions*

The most significant results to be drawn from the operation of the MED-PSA plant were:

- The plant has a low thermal inertia, taking about 35 minutes to reach steady-state distillate production following plant startup.
- The specific electricity consumption varies from 3.3 to 5 kWh$_e$ m$^{-3}$ of distillate, depending on the load.

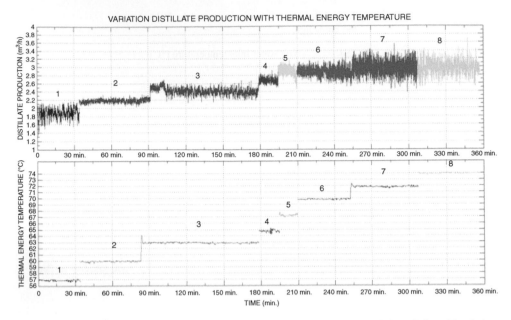

Figure 2.14.   Distillate production at different hot water inlet temperatures (adapted from Fernández-Izquierdo *et al.*, 2012).

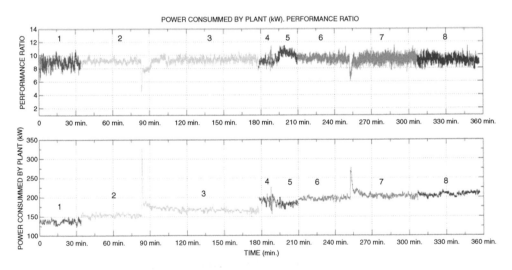

Figure 2.15.   Performance ratio and thermal power consumption at different hot water inlet temperatures (adapted from Fernández-Izquierdo *et al.*, 2012).

- The MED plant produced a *PR* within a range of 9.4–10.4 when operating with low-pressure steam.
- The specific thermal energy consumption of the plant when operating with hot water is in the range of 57.5 $KWh_{th}$ m$^{-3}$ (hot water inlet temperature 63°C) to 70.4 kWh$_{th}$ m$^{-3}$ (hot water inlet temperature 75°C).
- The distillate production of the plant when operating with hot water varies from 1.9 m$^3$ h$^{-1}$ to 3.1 m$^3$ h$^{-1}$, depending on the thermal energy source temperature.

Table 2.10.  Average values of the experimental results from 2010 test campaign.

| Test | Inlet hot water temperature [°C] | Thermal power consumption [kW] | Distillate production [m³ h⁻¹] | PR |
|------|----------------------------------|--------------------------------|--------------------------------|------|
| 1 | 57 | 137 | 1.9 | 8.9 |
| 2 | 60 | 153 | 2.2 | 9.1 |
| 3 | 63 | 166 | 2.4 | 9.3 |
| 4 | 65 | 191 | 2.7 | 9.0 |
| 5 | 68 | 182 | 2.9 | 10.0 |
| 6 | 70 | 195 | 2.9 | 9.5 |
| 7 | 72 | 203 | 3.0 | 9.4 |
| 8 | 74 | 207 | 3.0 | 9.3 |

- The optimum working conditions when the plant operates at a constant vapor temperature of 31°C in the last evaporator are a hot water inlet temperature in the range of 64–67°C and a feed water flow rate of 8 m³ h⁻¹. The performance ratio at these hot water inlet temperatures was 11 (at 64°C) and 10.4 (at 67°C).
- In the case of operating the plant keeping the temperature difference (vapor temperature difference between the first and last evaporator) at 35°C, the optimum working conditions in terms of maximum efficiency involve a hot water inlet temperature of 68°C, and for maximum distillate production require a temperature of 72°C.

## 2.3  PHOTOVOLTAIC REVERSE OSMOSIS CASE STUDY: KSAR GHILÈNE PLANT

### 2.3.1  *Introduction*

The main objective of this section is to show a data analysis of four years of operation (2009–2012) of the Ksar Ghilène plant, a solar-PV-driven RO desalination plant installed in Tunisia by the Canary Islands Institute of Technology (ITC) within the framework of its strategic research and development lines. This project was funded by the Spanish Agency for International Development Cooperation (Agencia Española de Cooperación Internacional para el Desarrollo, AECID) and the Directorate General of Affairs with Africa (Dirección General de Relaciones con África, DGRA) of the Canary Islands Government (Peñate *et al.*, 2015; Subiela *et al.*, 2009).

The Ksar Ghilène plant is a standalone solar plant for brackish water RO desalination powered by a solar PV field of 10.5 kWp. The PV-driven desalination plant, with a nominal capacity of 50 m³ day⁻¹, is fed by a brackish well. The plant mainly consists of two systems:

- A desalination system with 50 m³ day⁻¹ of nominal capacity (Fig. 2.16). This includes feed-water intake and pre-treatment equipment, which comprise a feed tank, a feed pump, a sand filter, an activated carbon filter and a cartridge filter. Once the raw water has been pre-treated, it is sent to the desalination subsystem. The main components of this subsystem are a high-pressure pump and a membrane module, with three membrane elements connected in series within a single pressure vessel. Finally, there is a product storage tank. Whenever it is full, the desalination plant begins an automatic shutdown procedure.
- A PV system (Fig. 2.17) in which seven PV panels tilted at 40° are connected in parallel, with a total power of 10.5 kWp. This PV field is connected to a battery bank and to an inverter, with a nominal power of 10 kW. A regulator controls the charge state of the batteries and the inverter operation.

Ksar Ghilène village and desalination plant are depicted in Figure 2.18. The following data are continuously recorded in order to control and analyze the plant's performance:

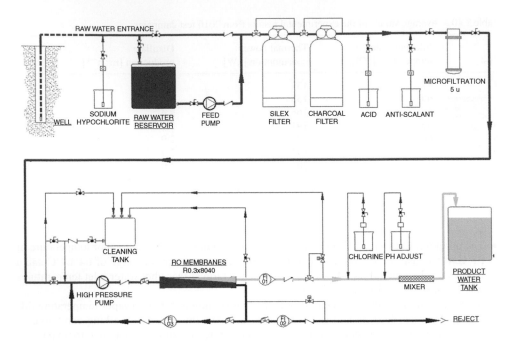

Figure 2.16.    Schematic diagram of desalination plant hydraulics.

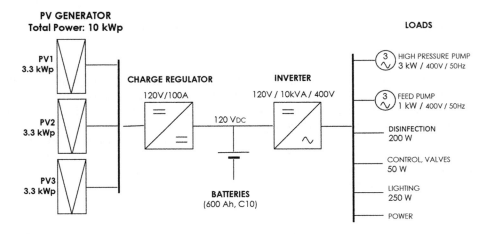

Figure 2.17.    General electrical diagram of the isolated energy system and loads.

- Hourly/daily mean values of voltage and intensities at batteries, inverter, PV field and pumps.
- Temperature of the PV panels.
- Battery data: voltage, density and temperature.
- Main operating parameters of the RO plant components, namely, the pre-treatment unit, membrane module and high pressure pump. These parameters are chloride concentration, conductivity, pH and temperature, together with volume of permeate, concentrate recycling and rejection.
- Morning and afternoon temperature in all rooms of the building as well as ambient temperature.

Section 2.3.2 describes the overall plant performance, by means of monthly and annual results. In addition, a solar PV data analysis is reported in Subsection 2.3.3, where the balance of energy

Figure 2.18.   Ksar Ghilène plant: (a) situation of Ksar Ghilène; (b) standalone PV-RO plant; (c) brackish water RO plant; (d) funding notice at the Ksar Ghilène plant; (e) one of the five freshwater supply points provided by the plant.

production and energy consumption is studied, along with daily and monthly maximum and minimum values. Finally, Subsection 2.3.4 highlights the conclusions to be drawn after several years of operation of the Ksar Ghilène plant.

### 2.3.2   *Desalination plant operation*

Experimental data analysis was performed, with hourly data acquired from the plant startup as of September 2012. The main parameters analyzed in this subsection are product volume and quality, along with operating hours.

Table 2.11 and Table 2.12 show the experimental data obtained in terms of the volume of freshwater produced by the standalone plant, and some average values calculated from them. The year 2009 is not included because there are no data pertaining to operating hours, thus making it difficult to calculate water production rates. Furthermore, the inverter was out of order from

Table 2.11. Monthly freshwater production [m$^3$].

|  | 2010 | 2011 | 2012 |
|---|---|---|---|
| January | 81.06 | 107.92 | 92.73 |
| February | 32.73 | 79.10 | 85.02 |
| March | 92.07 | 141.11 | 121.60 |
| April | 90.09 | 155.61 | 172.27 |
| May | 102.38 | 150.13 | 188.69 |
| June | 102.21 | 178.83 | 179.83 |
| July | 115.75 | 173.14 | 214.37 |
| August | 179.67 | 207.70 | n.a. |
| September | 133.14 | 160.49 | n.a. |
| October | 62.57 | 103.90 | n.a. |
| November | 53.47 | 81.67 | n.a. |
| December | 110.50 | 83.20 | n.a. |

Table 2.12. Overall assessment of 2009–2012 operating period (freshwater production).

|  | 2010 | 2011 | 2012 | 2009–2012 |
|---|---|---|---|---|
| Total production [m$^3$] | 1155.63 | 1622.81 | 1054.51 | 3832.96 |
| Average monthly production [m$^3$ month$^{-1}$] | 96.30 | 135.23 | 150.64 | 123.64 |
| Average daily production [m$^3$ day$^{-1}$] | 3.17 | 4.45 | 4.97 | 4.07 |
| Maximum monthly production [m$^3$ month$^{-1}$] | 179.67 | 207.70 | 214.37 | 214.37 |
| Minimum monthly production [m$^3$ month$^{-1}$] | 32.73 | 79.10 | 85.02 | 32.73 |

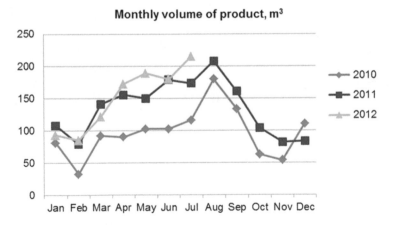

Figure 2.19. Monthly volume of freshwater desalinated.

August 2012 and, therefore, no data have been available since then. In addition, Figure 2.19 represents experimental data from Table 2.11 in order to make the data analysis clearer.

With regard to the operating hours, Table 2.13 and Table 2.14 show experimental data and minimum/maximum values, together with calculated averages. The corresponding graph is presented in Figure 2.20.

The product quality was analyzed by calculating the salt concentration from the conductivity data recorded. The monthly average is derived from the hourly data. Table 2.15 shows the results obtained for the years 2009 and 2010.

Table 2.13.   Monthly operating hours.

|  | 2010 | 2011 |
|---|---|---|
| January | 66.66 | 62.66 |
| February | 43.83 | 43.83 |
| March | 71.32 | 74.49 |
| April | 79.25 | 79.24 |
| May | 69.90 | 75.07 |
| June | 93.23 | 91.23 |
| July | 82.48 | 82.48 |
| August | 91.32 | 98.90 |
| September | 76.98 | 76.98 |
| October | 54.83 | 54.83 |
| November | 44.14 | 44.14 |
| December | 61.58 | 47.22 |

Table 2.14.   Overall assessment of 2009–2012 operating period (operating hours).

|  | 2010 | 2011 | 2012 | 2009–2012 |
|---|---|---|---|---|
| Annual operation [h] | 835.53 | 831.08 | 569.24 | 2235.85 |
| Average monthly operation [h month$^{-1}$] | 69.63 | 69.26 | 81.32 | 72.12 |
| Average daily operation [m$^3$ day$^{-1}$] | 2.29 | 2.28 | 2.69 | 2.37 |
| Maximum monthly operation [h month$^{-1}$] | 93.23 | 98.90 | 109.83 | – |
| Minimum monthly operation [h month$^{-1}$] | 43.83 | 43.83 | 49.58 | – |

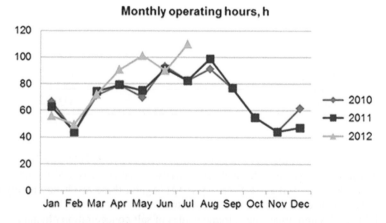

Figure 2.20.   Experimental data of monthly operating hours.

In addition, maximum ranges of product conductivities were analyzed in order to complement information retrieved from the tables and figures above. The data source is the hourly values of product conductivity on operating days. Firstly, the maximum values of product conductivity were analyzed. These are summarized in Table 2.16, expressed in terms of salt concentration. The days on which the maximum values were measured were 15 August 2009 and 10 August 2010 (as well as 20 April 2011). The data are totally consistent with the fact that the raw water temperature is higher in summer than in winter and the RO process increases its productivity with a higher salt passage.

Table 2.15.   Mean values of product quality, expressed in salt concentration [mg L$^{-1}$].

|           | 2009   | 2010   |
|-----------|--------|--------|
| January   | 98.52  | 155.26 |
| February  | 104.38 | 142.14 |
| March     | 114.98 | 180.26 |
| April     | 131.67 | 195.00 |
| May       | 160.91 | 205.52 |
| June      | 198.75 | 235.78 |
| July      | 231.76 | 246.57 |
| August    | 247.88 | 249.60 |
| September | 235.86 | 218.08 |
| October   | 228.81 | 203.05 |
| November  | 199.31 | 173.80 |
| December  | 159.42 | 163.40 |

Table 2.16.   Maximum values of salt concentration of water product [mg L$^{-1}$].

|           | 2009  | 2010  |
|-----------|-------|-------|
| January   | 103   | 164.5 |
| February  | 112   | 158   |
| March     | 130   | 199.5 |
| April     | 143   | 203.5 |
| May       | 187.5 | 218   |
| June      | 213   | 247.5 |
| July      | 247.5 | 257   |
| August    | 257   | 343   |
| September | 252   | 239   |
| October   | 238   | 227   |
| November  | 216.5 | 181   |
| December  | 174   | 182   |
| Annual    | 257   | 343   |

Secondly, the minimum values obtained every month for the years 2009 and 2010 are shown in Table 2.17. The minimum values of product conductivity were recorded on 24 and 25 January 2009 and 24 February 2010 (as well as 16 February 2011). Again, these results are totally consistent with the raw water temperature being lower in winter.

Finally, Table 2.18 summarizes the extreme values of salt concentration obtained during plant operation.

After data recording and analysis, the following observations could be made:

- Water production, summarized in Table 2.11 and Figure 2.19, exhibits maximum values in August and minimums in February for all of the years analyzed.
- In terms of operating hours, shown in Table 2.13 and Figure 2.20, maximum values were obtained in summer, in the month of June in 2010 and August in 2011, while the minimum of operating hours was registered in February in both years.
- In 2009 and 2010, maximum conductivities occurred in August, according to Table 2.16. However, the minimum data was associated with different winter months in these years (see Table 2.17).

Table 2.17.   Minimum values of salt concentration of water product [mg L$^{-1}$].

|  | 2009 | 2010 |
|---|---|---|
| January | 92 | 137 |
| February | 96.5 | 133 |
| March | 101 | 156 |
| April | 121 | 186 |
| May | 148.5 | 167.5 |
| June | 172 | 218.5 |
| July | 170 | 238 |
| August | 243.5 | 238.5 |
| September | 223 | 209.5 |
| October | 224 | 182 |
| November | 180.5 | 167 |
| December | 141.5 | n.a. |
| Annual | 92 | 133 |

Table 2.18.   Extreme values of salt concentration [mg L$^{-1}$] throughout the operating period.

|  | 2009 | 2010 | 2011 |
|---|---|---|---|
| Maximum | 257 | 343 | 197 |
| (date) | (August 15th) | (August 10th) | (April 20th) |
| Minimum | 92 | 133 | 142 |
| (date) | (January 24th & 25th) | (February 24th) | (February 16th) |

- Months with the greatest water production corresponded to those with the largest number of sun hours as the plant operated for longer. Therefore, the water production in summer was higher than that obtained in winter.
- A slight increase of product conductivity occurred in summer, due to the higher raw water temperature and its effect over RO membranes (pure dilation). This is the result of two contrary effects: lower salt rejection but higher productivity. Thus, the greater production of the membranes in summer partially diluted the higher mass of salts that passed through the membranes. In addition, the conductivity increase resulted in an increase of recovery.

### 2.3.3   *Analysis of the PV plant performance*

This section considers the performance of the PV plant that supplies the energy to the brackish desalination system. The energy delivered by the solar PV field is compared to the energy consumed by the desalination plant. When the power supplied by the PV field is insufficient to operate the desalination plant, the control system permits energy consumption from the battery bank. At other times, any surplus energy produced by the solar field is stored in the battery bank. Hourly data records are processed to obtain the total amount of energy both generated and consumed. Table 2.19 and Table 2.20 show the energy production of the PV field, along with the monthly average, maximum and minimum values derived from them. In addition, Figure 2.21 represents monthly energy production. August was the month of maximum production in 2009, 2010 and 2011; February exhibited the minimum production in 2010, 2011 and 2012, while in 2009 the minimum occurred in December.

Table 2.19.  Energy production of the PV field (DC) expressed in monthly averages [kWh].

|           | 2009   | 2010   | 2011   |
|-----------|--------|--------|--------|
| January   | 572.18 | 382.61 | 390.57 |
| February  | 604.44 | 360.81 | 297.94 |
| March     | 693.59 | 414.23 | 444.50 |
| April     | 491.76 | 473.54 | 487.18 |
| May       | 475.98 | 574.49 | 454.96 |
| June      | 660.38 | 521.68 | 520.85 |
| July      | 692.54 | 726.50 | 513.70 |
| August    | 698.43 | 880.23 | 641.11 |
| September | 665.89 | 592.16 | 513.66 |
| October   | 482.04 | 440.99 | 412.93 |
| November  | 420.21 | 435.27 | 331.97 |
| December  | 329.68 | 425.94 | 309.45 |

Table 2.20.  Energy production of the PV field (DC).

|                                                      | 2009    | 2010    | 2011    | 2012    | 2009–2012 |
|------------------------------------------------------|---------|---------|---------|---------|-----------|
| Annual production [kWh]                              | 6787.14 | 6228.46 | 5318.81 | 3488.12 | 21822.53  |
| Maximum monthly production [kWh month$^{-1}$]        | 698.43  | 880.23  | 641.11  | 674.49  | 880.23    |
| Minimum monthly production [kWh month$^{-1}$]        | 329.68  | 360.81  | 297.94  | 339.04  | 297.94    |
| Average monthly production [kWh month$^{-1}$]        | 565.59  | 519.04  | 443.23  | 498.30  | 507.50    |
| Average daily production [kWh day$^{-1}$]            | 18.59   | 17.06   | 14.57   | 16.45   | 16.31     |

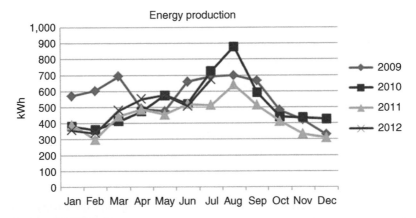

Figure 2.21.  PV field production.

Moreover, Figure 2.22 shows the desalination energy consumption of every month during the plant operation. Months with the highest and lowest values are approximately the same every year, except in 2012.

Since this is an autonomous solar plant, the energy generated is always greater than that consumed. The following figures show, for the respective years, energy produced by the PV system and consumed by the RO unit each month.

Table 2.21.   Energy consumption of the RO desalination plant expressed as monthly averages [kWh][1].

|  | 2009 | 2010 | 2011 |
|---|---|---|---|
| January | 371.54 | 228.64 | 228.11 |
| February | 377.61 | 206.70 | 160.68 |
| March | 391.55 | 231.36 | 256.08 |
| April | 258.80 | 278.99 | 279.15 |
| May | 316.20 | 322.39 | 269.84 |
| June | 370.02 | 324.44 | 325.36 |
| July | 465.06 | 420.67 | 294.52 |
| August | 435.44 | 500.76 | 390.95 |
| September | 386.58 | 320.44 | 390.95 |
| October | 267.65 | 248.91 | 200.86 |
| November | 199.99 | 210.86 | 165.30 |
| December | 215.72 | 241.50 | 178.09 |

[1]AC/electricity losses are included.

Table 2.22.   Energy consumption of the RO desalination plant[1].

|  | 2009 | 2010 | 2011 | 2012 | 2009–2012 |
|---|---|---|---|---|---|
| Annual value [kWh] | 4056.17 | 3535.65 | 3139.89 | 2029.57 | 12761.27 |
| Maximum monthly value [kWh month$^{-1}$] | 465.06 | 500.76 | 390.95 | 391.09 | 500.76 |
| Minimum monthly value [kWh month$^{-1}$] | 199.99 | 206.70 | 160.68 | 181.03 | 160.68 |
| Average monthly value [kWh month$^{-1}$] | 338.01 | 294.64 | 261.66 | 289.94 | 296.77 |
| Average daily value [kWh day$^{-1}$] | 11.11 | 9.69 | 8.60 | 22.55 | 9.76 |

[1]AC/electricity losses are included.

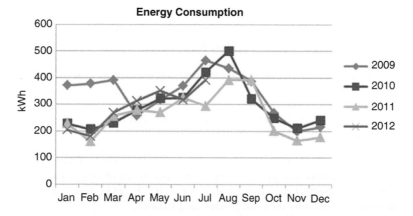

Figure 2.22.   Monthly energy consumed by the RO system (2009–2012) (electricity losses included).

In addition, the days with maximum and minimum energy production in every month and year were identified. The results are depicted in the following figures, which show the years from 2009 to 2012. Extreme values for each year occurred on the following dates:

- 26 June (maximum) and 20 January (minimum) in 2009.
- 5 September (maximum) and 27 January (minimum) in 2010.

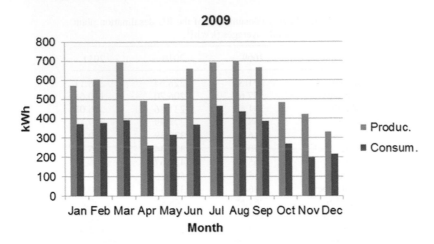

Figure 2.23.    Balance of energy production and consumption in the solar PV plant in 2009.

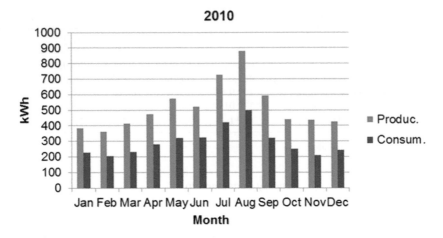

Figure 2.24.    Balance of energy production and consumption in the solar PV plant in 2010.

- 27 August (maximum) and 10 September (minimum) in 2011.
- 15 July (maximum) and 9 March (minimum) in 2012.

Additional detail in relation to energy production ranges is shown in Table 2.23. The specific days and the overall months with maximum and minimum values of energy production, together with the corresponding values, are given from 2009 to 2012. It can be seen that it is only in 2011 and 2012 that the same months show the production extremes.

Turning to the energy consumption of the desalination plant, the following figures summarize the results obtained. The dates on which energy consumption was at a maximum and minimum for each year were:

- 16 September (maximum) and 16 February (minimum) in 2009.
- 27 August (maximum) and 13 March (minimum) in 2010.
- 15 August (maximum) and 18 March (minimum) in 2011.
- 2 July (maximum) and 23 March (minimum) in 2012.

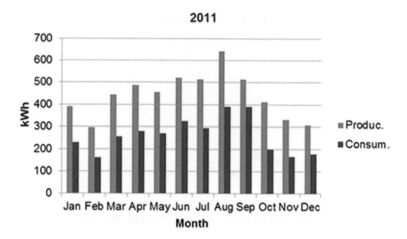

Figure 2.25.   Balance of energy production and consumption in the solar PV plant in 2011.

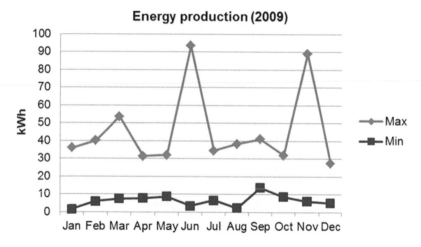

Figure 2.26.   Monthly maximum and minimum energy production in 2009.

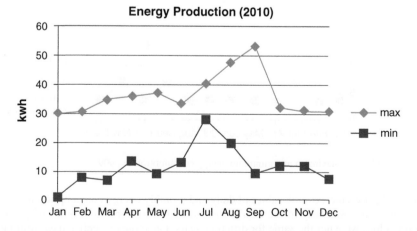

Figure 2.27.   Monthly maximum and minimum energy production in 2010.

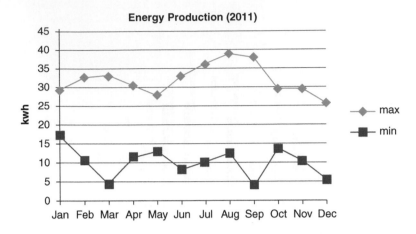

Figure 2.28.    Monthly maximum and minimum energy production in 2011.

Table 2.23.    Energy production ranges [kWh] over the plant operation (2009–2012).

|  | 2009 | 2010 | 2011 | 2012 |
|---|---|---|---|---|
| Maximum | 93.35 | 53.41 | 38.76 | 43.60 |
| (day) | (June 26th) | (September 5th) | (August 27th) | (July 15th) |
| Minimum | 1.63 | 0.92 | 3.90 | 7.72 |
| (day) | (January 20th) | (January 27th) | (September 10th) | (March 9th) |
| Maximum | 698.43 | 880.23 | 641.11 | 674.49 |
| (month) | (August) | (August) | (August) | (July) |
| Minimum | 329.68 | 360.81 | 297.94 | 339.04 |
| (month) | (December) | (February) | (February) | (March) |

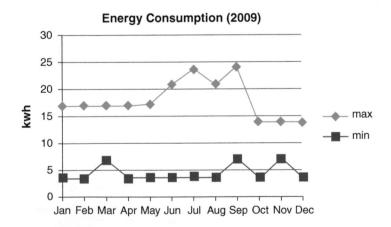

Figure 2.29.    Monthly maximum and minimum energy consumption in 2009.

Table 2.24 shows the extreme values of daily and monthly energy consumption for every year of plant operation. Similarly to the results for energy production, the months of maximum or minimum values were not the same for different years. Once the respective days with minimum and maximum production had been identified for each year, the specific energy consumption was calculated; the results are shown in Table 2.25.

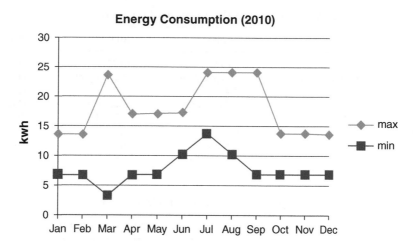

Figure 2.30.   Monthly maximum and minimum energy consumption in 2010.

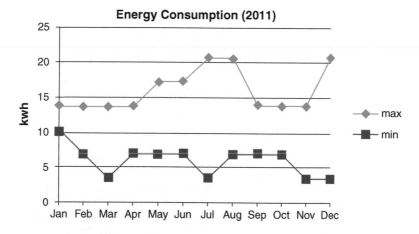

Figure 2.31.   Monthly maximum and minimum energy consumption in 2011.

### 2.3.4   *Operational conclusions*

The most significant conclusions to be drawn from the discussion of these results are as follows:

- From Figure 2.23, Figure 2.24 and Figure 2.25 it can be seen that the PV field delivers more electrical power than is required by the desalination plant. The surplus energy is stored in the battery bank in order to expand the plant's operating hours. The batteries supply the power required when the energy delivered by the PV field is insufficient to operate the desalination plant at nominal capacity, and also when there is no energy production, until the batteries are discharged.
- The maximum and minimum energy generated (see Table 2.23 and Fig. 2.26, Fig. 2.27 and Fig. 2.28) and the maximum and minimum energy consumption (see Table 2.24 and Fig. 2.29, Fig. 2.30 and Fig. 2.31) exhibit similar behavior. This is attributable to the longer sunshine hours in summer resulting in higher energy production and a correspondingly higher number of operating hours and energy consumption.

Table 2.24.   Range of energy consumption [kWh] of the RO desalination plant.

|  | 2009 | 2010 | 2011 | 2012 |
|---|---|---|---|---|
| Maximum (day) | 24.10 (September 16th) | 24.15 (August 27th) | 20.72 (August 16th) | 20.56 (July 2nd) |
| Minimum (day) | 3.43 (February 5th) | 3.47 (March 13th) | 3.47 (March 18th) | 6.85 (March 23rd) |
| Maximum (month) | 465.060 (July) | 500.756 (August) | 390.954 (August) | 391.09 (July) |
| Minimum (month) | 199.985 (November) | 206.701 (February) | 160.678 (February) | 181.03 (February) |

Table 2.25.   Specific energy consumption [kWh m$^{-3}$].

|  | 2009 | 2010 | 2011 | 2012 |
|---|---|---|---|---|
| Maximum | 2.01 | 1.64 | 1.64 | 1.71 |
| Minimum | 2.49 | 1.98 | 2.44 | 2.40 |

- Minimum values for both energy generation (Table 2.20) and water production (Table 2.12) were associated with February, thanks to the lowest number of sunshine hours. By contrast, maximum values occurred in June as that was the month with the highest number of sunshine hours.
- In comparing different years, it is clear that the days showing maximum and minimum energy generation and energy consumption (see Fig. 2.26 through to Fig. 2.31) were not the same. This is mainly attributable to the use of batteries which enable the generation of a specific day to be consumed later. Another explanation might be the automatic plant shutdown when the product tank is full, thus halting water production even though energy generation continues.
- The energy production of the PV field throughout summer is superior, and thus the desalination plant operates for more hours with a consequent increase in permeate production. In summer, the monthly energy consumption of the desalination plant is higher although the specific energy consumption exhibits the best values. As water feed temperature increases in summer, the ability of the membranes to reject salts goes down and hence the product quality decreases. The higher feed water temperature results in higher recovery and, therefore, the increase in desalinated water volume is attributable to both longer operating hours and higher flux on the membrane surface.

## NOMENCLATURE

$M$ — flow rate [m$^3$ h$^{-1}$]
$T$ — temperature [°C]
$Q$ — heat transfer rate [kW]
$STC$ — specific thermal energy consumption [kWh$_{th}$ m$^{-3}$]
$SEC$ — specific electricity consumption [kWh$_e$ m$^{-3}$]
$PR$ — performance ratio

*Subscripts*

h — hot water
v — vapor
d — distillate

in      inlet
out     outlet
fw      feed water
oil     oil
cond    condenser

REFERENCES

Alarcón-Padilla, D. & García-Rodríguez, L. (2007) Application of absorption heat pumps to multi-effect distillation: a case study of solar desalination. *Desalination*, 212, 294–302.

Alarcón-Padilla, D., García-Rodríguez, L. & Blanco-Gálvez, J. (2007) Assessment of an absorption heat pump coupled to a multi-effect distillation unit within AQUASOL project. *Desalination*, 212, 303–310.

Alarcón-Padilla, D., Blanco-Gálvez, J., García-Rodríguez, L., Gernjak, W. & Malato-Rodríguez, S. (2008) First experimental results of a new hybrid solar/gas multi-effect distillation system: the AQUASOL project. *Desalination*, 220, 619–625.

Alarcón-Padilla, D., García-Rodríguez, L. & Blanco-Gálvez, J. (2010) Experimental assessment of connection of an absorption heat pump to a multi-effect distillation unit. *Desalination*, 250, 500–505.

Blanco, J., Alarcón-Padilla, D., Guillén, E. & Wolfgang, G. (2011) The AQUASOL system: solar collector field efficiency and solar-only mode performance. *Journal of Solar Energy Engineering*, 133, 1–6.

Fernández-Izquierdo, P., García-Rodríguez, L., Alarcón-Padilla, D., Palenzuela, P. & Martín-Mateos, I. (2012) Experimental analysis of a multi-effect distillation unit operated out of nominal conditions. *Desalination*, 284, 233–237.

Peñate, B., Subiela, V.J., Vega, F., Castellano, F., Domínguez, F.J. & Millán, V. (2015) Uninterrupted eight-year operation of the autonomous solar photovoltaic reverse osmosis system in Ksar Ghilène (Tunisia). *Desalination and Water Treatment*, 55, 3141–3148.

Subiela, V. de la Fuente, J.A., Piernavieja, G. & Peñate, B. (2009) Canary Islands Institute of Technology (ITC), experiences in desalination with renewable energies (1996–2008). *Desalination and Water Treatment*, 7, 220–235.

Zarza, E. (1991) Solar Thermal Desalination Project, first phase results and second phase description. 1st edn. CIEMAT, Madrid, Spain.

Zarza, E. (1994) Solar Thermal Desalination Project, phase II results and final project report. PSA/CIEMAT, Madrid, Spain.

# CHAPTER 3

## Application of solar nanophotocatalysis in reverse osmosis pretreatment processes

Shaik Feroz

## 3.1 INTRODUCTION

The demand for freshwater is growing rapidly and water scarcity is a major issue in many parts of the world, especially in developing countries (Ghaffour *et al.*, 2015; Goosen *et al.*, 2014). The earth is rich in saline water resources; seawater and brackish water together account for up to 97% of the total water available in the world. Efforts are being made to desalinate them and supply freshwater for human needs. However, desalination is an energy-intensive process and meeting the associated energy requirements from declining conventional sources is a major worldwide challenge. There is now rapid development towards linkage of desalination processes with renewable energy sources such as solar and wind. This will not only reduce the burden on conventional energy sources but will also protect the environment through lower emissions (Brand and Würzburg, 2015).

Reverse osmosis (RO) is a major technique employed to produce freshwater (Ghaffour *et al.*, 2015; Goosen *et al.*, 2014). This process, wherein water from a salt solution is forced through a membrane in the opposite direction by the application of pressure, requires pretreatment to remove substances from the feed water that would interfere with the desalting process. Suspended solids and other particles in the feed water must be removed to reduce fouling of the membranes, which lose their performance with time due to such fouling (Al-Rasheed *et al.*, 2003a). One of the major causes for the loss of RO performance are the substances that become deposited on the membrane surface. Examples are silica, oil, clay, iron, sulfur and humic acids, which may be present in the feed water in a very fine or colloidal form. Even the typical five-micron cartridge filters used in the upstream of a RO system may not completely remove these foulants. Furthermore, surface water contains high-molecular weight organic matter (e.g. humic substances), suspended solids, bacteria and algae, and volatile halogenated carbons. As these substances are traditionally considered hazardous for the membrane, they should be removed from the feed water before they enter membrane systems. Organic acids (i.e. humic substances) build adsorption layers on the membrane surface that reduce membrane flux and rejection. A number of laboratory studies have reported that these biofilms consist of about 50% organic matter, of which about 40% will be organic acids (Al-Rasheed *et al.*, 2003a).

Solar energy is abundant in many parts of world, especially in Middle East countries where desalination plays a major role in the supply of freshwater, and it can be effectively used in reverse osmosis pretreatment processes (Ghaffour *et al.*, 2015). An ultraviolet-driven photocatalytic pretreatment can be employed for the degradation of humic substances (higher molecular weight organics) and microorganisms present in the feed water. The energy generated from the photocatalytic reaction breaks down the humic substances and also kills microorganisms, thereby effectively eliminating the primary source for membrane fouling. Ultraviolet (UV) photo-oxidation uses UV light to activate a catalyst to physically decompose the organic matter into non-toxic components. The degradation of organic compounds in the presence of UV light is a too slow a process (reaction rates are low) but may be considered for large-scale industrial applications. To enhance the reaction rates, several catalysts, such as hydrogen peroxide, ozone,

43

Fe (II) and titanium dioxide, are used in these types of reactions. Titanium dioxide is probably the best choice due to its low cost, low toxicity, resistance to photo-corrosion and catalytic efficiency (Kwak *et al.*, 2001).

Kim *et al.* (2003) have fabricated hybrid polyamide thin-film composite/titanium dioxide ($TiO_2$) membranes possessing photocatalytic properties that reduce biofouling and enhance RO performance. Preliminary feasibility studies were carried out as a part of a project (08-AS-002) sponsored by the Middle East Desalination Research Centre (MEDRC). In addition, detailed examinations were carried out as part of a project (ORG/E1/12003/Grant83) sponsored by the Research Council of Oman. The experimental investigations were carried out using solar nanophotocatalysis in a laboratory-scale batch, as well as in a continuous reactor system, to assess the degradation of pollutants present in seawater. Positive results were reported, with substantial decreases in total organic carbon (TOC). Hence it was demonstrated that solar nanophotocatalysis could be implemented in the pretreatment for the RO desalination process, and could also be applied to any saline or industrial wastewater including oil-produced water, where the trace pollutants are difficult to remove with conventional treatment methods. However, due to the complex nature of seawater and the possible scavenging of oxidative agents by chloride ions, the photocatalytic technique still has challenges to overcome for seawater application.

The aim of the current investigation was to show that the solar nanophotocatalysis technique can be directly employed to effectively degrade pollutants in seawater, and to assess whether this technique can be successfully implemented at the commercial level. A case study was also performed to assess the influence of chloride ions and the scavenging of hydroxyl radicals on TOC levels.

## 3.2   REVERSE OSMOSIS AND FEEDWATER PRETREATMENT PROCESSES

RO membranes are used worldwide for saline water desalination, producing water for drinking and other industrial and agricultural purposes. Seawater RO desalination has become cost-effective when compared to thermal desalination due to recent developments in membrane technology and pressure exchangers (Ghaffour *et al.*, 2015; Goosen *et al.*, 2014). However, challenges remain with respect to fouling of membranes by the organic contents of shallow seawater feeds (Amjad, 1996; Goosen *et al.*, 2014).

A typical RO system consists of a pretreatment system, cartridge filters, RO membrane, and post-treatment system. Pretreatment systems are classified as either conventional or membrane pretreatment. The conventional pretreatment of an RO system involves a clarifier, sand filters and multi-media filters (Amjad, 1996; Goosen *et al.*, 2014). The purpose of pretreatment is to protect the RO membranes from being damaged by foulants and scalants. There is generally a cartridge filter upstream of the high-pressure pump. This filter serves the dual role of protecting the pump from particles that may damage the impellers and protecting the membranes from becoming fouled. Post-treatment includes mineralization, and sterilization with chloramines or UV light. Adjustment of pH or the addition of other chemicals in the post-treatment used will depend on the quality of water required for the end application.

The heart of the system is the RO membrane. All membranes are prone to fouling and therefore great care must be taken in selecting appropriate membranes, designing and building the RO system, and operating the system to avoid membrane fouling. Adequate pretreatment is, therefore, critical in RO plants. Furthermore, the economics of desalination suggest that the greater the percentage of pure water that can be recovered from the feed, the higher the efficiency of the RO process. However, as the concentration of the salts in the brine (i.e. leftover water on the feed stream side of the membrane) increases, the potential for fouling also increases, resulting in the precipitation of scale-forming salts from the brine (Amjad, 1996). The density of a colloidal material may also be increased on the feed side of the membrane during desalination, allowing for coagulation and deposition of colloidal matter from the brine. Concentration polarization further contributes to fouling (Goosen *et al.*, 2004).

Commonly encountered foulants in desalination processes are scales, suspended/colloidal matter and biological material (Amjad, 1996). The major fouling problems in the RO process are due to biofouling and fouling by $CaCO_3$, $Mg(OH)_2$, silica and iron products. Different types of fouling can be distinguished, viz. fouling by particulates (silt), bacteria (biofouling), organics (e.g. oil) and scaling (precipitated inorganic salts). Colloidal impurities, inorganic precipitates, macro-molecules and biological contaminants are the main type of foulants (Characklis, 1981). Pretreatment steps widely employed in seawater RO plants include disinfection, coagulation-flocculation, filtration and anti-scalant dosing. The types and doses of disinfectants, coagulants and/or anti-scalants, and the types of filtration systems used in these plants, are not the same. The quantity of chemical doses and the point of dosing also differ from one plant to another depending on the site-specific characteristics of the feed water. Organics can be successfully removed by filtration or active carbon adsorption. In practice, a combination of different pretreatments often has to be carried out to guarantee a steady process operation. Thus, the pretreatment train often requires a lot more space and attention than the main reverse osmosis system itself (Dudley *et al.*, 1995; Sieburth and Jensen, 1968, 1969).

Experimental studies have reported on coupling photocatalytic oxidation processes and membrane separation for the reuse of dye wastewater (Ou *et al.*, 2015). Prevention of organic fouling and biofouling can be achieved by coating membranes with photocatalysts that are non-toxic (Feroz and Jesil, 2012). Microorganisms are inactivated and organic material from cells is oxidized by photocatalysis. A UV-driven photocatalytic pretreatment technique can be employed, for instance, for the degradation of humic acids and microorganisms.

## 3.3   PHOTOCATALYSIS AND OXIDATION OF POLLUTANTS

Heterogeneous photocatalysis is a rapidly expanding technology for water and air treatment (Ibhadon and Fitzpatrick, 2013). The initial interest began when Fujishima and Honda (1972) discovered the photochemical splitting of water into hydrogen and oxygen with $TiO_2$. Photocatalytic oxidation reactions have the potential to totally degrade organic compounds to carbon dioxide. This provides a clean and energy-saving technology for treatment of polluted water and air.

Most of the semiconductors that have been investigated for the treatment of polluted water are metal oxides (e.g. $TiO_2$, ZnO, $SnO_2$ and $WO_3$) or chalcogenides (CdS, ZnS and CdSe) (Dunlop, 2001). The effectiveness of the process for the photo-oxidation of organics in water depends upon the reduction potentials of the valence band and conduction band. The $TiO_2$ conduction band potential is sufficiently negative for the reduction of $O_2$ and the valence band is sufficiently positive for the oxidation of $OH^-$. Given the reduction potentials of the bands, $SrTiO_3$ ZnO, $WO_3$ and ZnS could also be used for the photocatalytic oxidation of organic pollutants. However, it is usually found that $TiO_2$ is the most efficient semiconductor for the treatment of water containing organic pollutants. ZnO is a suitable alternative to $TiO_2$ and has similar band gap energy. However, it is unstable with respect to incongruous dissolution.

Although ZnO has a higher activity than $TiO_2$, having bandgap energies of 3.436 eV and 3.03 eV respectively, $TiO_2$ is more stable in aqueous media and that is why it is mostly used in water purification processes. Nano-structured $TiO_2$ photocatalysts have three crystal forms: anatase, rutile and brookite. Anatase has a higher photocatalytic activity because the position of oxygen ions on the exposed crystal surface of anatase shows a triangular arrangement, allowing effective absorption of organics, while the position of the titanium ions creates favorable reaction conditions with the absorbed organics. On the other hand, the rutile phase has a wider pore size distribution which increases the photocatalytic activity. To achieve maximum photocatalytic efficiency, the photocatalyst can be formed from a mixture of anatase and rutile (Dunlop, 2001).

Many researchers have reported on the photocatalytic degradation of pollutants in aqueous solution (Ibhadon and Fitzpatrick, 2013). The photocatalytic mechanism of $TiO_2$ is described here and it is similar to other photocatalytic processes. $TiO_2$ is insoluble in water and effectively

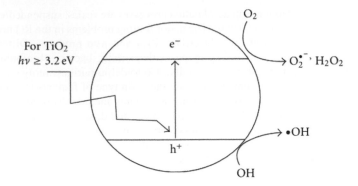

Figure 3.1.   Photocatalytic process on $TiO_2$ particle (Dunlop, 2001).

non-toxic. The photocatalytic process does not require the addition of consumable chemicals and a waste sludge is not produced. $TiO_2$ photocatalysis has been reported to degrade many aqueous chemical pollutants, including pesticides, herbicides, crude oil, surfactants and dyes (Dunlop, 2001; Ibhadon and Fitzpatrick, 2013). Irradiation of $TiO_2$ particles with photons of energy equal to, or greater than, the band gap energy results in the promotion of an electron from the valence band (vb) to the conduction band (cb) of the particle. The outcome of this process is the region of positive charge, termed a hole ($h^+$) in the vb and a free electron ($e^-$) in the cb:

$$TiO_2 + hv \rightarrow TiO_2 \ (e^-cb + h^+vb) \qquad (3.1)$$

At the $TiO_2$ particle surface, the holes can react with surface-bound hydroxyl groups ($OH^-$) and absorbed water molecules to form hydroxyl radicals ($\bullet HO$):

$$h^+ + (vb) + OH^- \rightarrow \bullet HO \qquad (3.2)$$

$$h^+ + (vb) + H_2O \rightarrow \bullet HO + H^+ \qquad (3.3)$$

In the absence of electron acceptors, electron-hole recombination is a possibility. The presence of oxygen prevents recombination by trapping electrons through the formation of superoxide ions according to Equation (3.4). The final product of the reduction may also be hydroxyl radicals and the hydroxyl radical $HO_2$:

$$\text{electron (cb)} + O_2 \rightarrow O_2^- \qquad (3.4)$$

where $O_2^-$ represents the superoxide ion:

$$2O_2^- + 2H^+ \rightarrow 2\bullet OH + O_2 \qquad (3.5)$$

$$2O_2^- + H^+ \rightarrow \bullet HO_2 \qquad (3.6)$$

The photocatalysis mechanism reaction is shown in Figure 3.1 (Dunlop, 2001).

Hydroxyl radicals are known to be powerful, indiscriminate oxidizing agents. During the photocatalytic process the radicals can react with organic compounds and bacterial species adsorbed onto, or very close to, the semiconductor surface resulting in degradation. Similarly, the surface interaction of microorganisms with catalysts used during the photo-disinfection is essential for enhancing the inactivation rate. When the generated radicals make close contact with the microorganisms, the lipopolysaccharide layer of the external cell wall is the initial site attacked by the photo-induced radicals.

Hydroxyl radicals are known to be powerful, indiscriminate oxidizing agents (Chong *et al.*, 2010). During the photocatalytic process the radicals can react with organic compounds and bacterial species adsorbed onto, or very close to, the semiconductor surface resulting in degradation. The surface interaction of microorganisms with a catalyst during the photo-disinfection is essential for enhancing the inactivation rate. The lipopolysaccharide layer of the external cell wall is the initial site attacked by the photo-induced radicals. This is followed by site attack on the peptidoglycan layer, peroxidation of the membrane lipids and eventual oxidation of the membrane proteins. All of these will cause a rapid leakage of potassium ions from the bacterial cells, resulting in direct reduction of cell viability. The decrease in cell viability is usually linked to the peroxidation of polyunsaturated phospholipid components in the cell membrane (i.e. loss of essential cell functions) and eventually leads to cell death. The transfer of a bacterial cell into the close vicinity of the surface-generated radical site is recognized to be the rate-limiting step in the photo-disinfection reaction (Chong *et al.*, 2010).

The UV radiation required for photocatalytic processes may come from an artificial source or the sun. The artificial generation of UV radiation contributes to a large proportion of the operating capital and maintenance costs of a photocatalytic reaction system because of the utility consumption and periodic replacement of the UV lamps (Dunlop, 2001). There is, therefore, a significant economic incentive to develop solar-powered photocatalytic reactors. In addition, the environmental impact induced by the use of solar energy is minimal and this renders the photocatalytic process environmentally attractive. The application of solar-powered photocatalytic reactors to treat water contaminated with organic pollutants holds promise for regions receiving strong sunlight throughout the year, such as the Middle East North Africa (MENA) and the Arabian Gulf regions.

## 3.4   SOLAR PHOTO CATALYTIC PROCESSES AND REMOVAL OF ORGANIC AND INORGANIC COMPOUNDS FROM WATER AND AIR

Over the past three decades research in the photocatalytic field has been extensive, covering the removal of organic and inorganic compounds from contaminated water and air (Mehrjouei *et al.*, 2015). The artificial generation of photons required for the detoxification of polluted water is the primary operational cost in photocatalytic wastewater treatment plants. This suggests that the sun may be employed as an economically and ecologically sensible light source. With a typical UV flux near the surface of the earth of $20–30 \, W \, m^{-2}$, the sun puts out 0.2–0.3 mol photons per $m^2$ per hour in the 300–400 nm range. In principle, these photons are suitable for destroying water pollutants in photocatalytic reactors.

Extensive research work has been performed at various research centers/institutions on the application of solar photocatalysis to the treatment of pollutants in industrial wastewater (Malato *et al.*, 2002; Mehrjouei *et al.*, 2015). Despite its obvious potential for the detoxification of polluted water, there has been very little commercial or industrial use of solar photocatalysis as a technology so far. The published literature shows only a few engineering-scale demonstrations of the solar photocatalytic treatment of polluted groundwater, landfill leachates and industrial wastewater in the US and in Europe. The first engineering-scale field experiments were conducted in 1991 by the National Renewable Energy Laboratory (NREL) and the Sandia National Laboratories at a California Superfund Site, located at Lawrence Livermore National Laboratory (LLNL), treating groundwater contaminated with chlorinated solvents, mainly trichloroethylene (TCE) (Mehos and Turchi, 1993).

Parallel to the work performed in the US under the direction of the NREL and Sandia National Laboratories, several research groups from different European countries, funded by the European Community, have tested the solar detoxification loop with the use of parabolic trough reactors installed at the Plataforma Solar de Almería (PSA) in Spain since 1991 (Malato *et al.*, 2002). Engineering-scale demonstrations of the non-concentrating solar reactor technology were conducted at Tyndall Air Force Base, Florida (US) in 1992 (Goswami *et al.*, 1993), treating groundwater contaminated with fuel, oil and lubricants that were leaking from underground storage tanks. Furthermore, Freudenhammer *et al.* (1997) reported their results from a pilot study using thin-film-fixed bed reactors (TFFBR), which was performed in various Mediterranean countries. Their results showed that biologically pretreated textile wastewater can be cleaned by solar photocatalysis. Likewise, highly polluted olive mill wastewater (OMW) from the olive oil industry, was treated by solar photocatalysis and a solar photo-Fenton process (Gernjak *et al.*, 2004).

Although photocatalysis has been extensively applied to the treatment of industrial wastewater, very few researchers have actually tried this technique for saline water because seawater consists of a complex mixture and there is also the danger of chloride ion interference. In one of the first reports, Al-Rasheed and Cardin (2003b) have shown that humic acids (HA) can be effectively removed from high-saline waters using photocatalysis with titanium dioxide catalysts. The oxidative nature of these reactions was established by the requirements for oxygen, and the optimum conditions have been established (i.e. high temperature, low pH, high oxygen concentration and a $TiO_2$ concentration in the range 2.0–2.5 g $L^{-1}$). Photocatalytic degradation of HA was also observed in low-saline water, where it occurs more rapidly than in seawater. The activation energy was determined in seawater as 17 kJ $mol^{-1}$.

Thin films of $TiO_2$ might be more suitable for use with highly saline waters, in view of the instability of suspended $TiO_2$ in this medium, and the requirement for continuous agitation. The catalytic oxidation process does not conform closely with the Langmuir-Hinshelwood kinetic model, probably owing to competing reactions involving sulfate and bicarbonate ions (Kim *et al.*, 2010). $TiO_2$ (Degussa P25), $TiO_2$ (anatase), $TiO_2$ (rutile), $TiO_2$ (mesoporous) and ZnO dispersions were used as catalysts, employing a medium mercury lamp to study the effectiveness of photocatalytic oxidation of HA in the increasingly important highly saline water. The effect of platinum loading on P25 and ZnO was also investigated and it was found that ZnO with 0.3% platinum loading was the most efficient catalyst. The preferred medium for the degradation of HA using ZnO is alkaline, whereas for $TiO_2$ it is acidic. Kim *et al.* (2010) focused on the photocatalytic membrane reactor treatment of seawater. The turbidity removal was significant and stable without any membrane fouling. Although $TiO_2$-mediated photocatalysis achieved a high degradation of the natural organic matter (NOM) present in freshwater (80% of lake NOM was removed within four hours), unexpectedly, it was found that no seawater organic matter (SOM) was decomposed by the same system. The authors argued that the reason behind this was that the salt level of seawater was altered by electrodialysis.

Kim *et al.* (2010) investigated the effects of seawater sources on photocatalytic microfiltration treatment with respect to the removal of particulate and organic matter, as well as control of membrane permeability. It was observed that photocatalytic degradation of SOM was marginal but it was enhanced significantly when sea salts were removed by electrodialysis. Likewise, Kim *et al.* (2003) designed a hybrid thin-film composite (TFC) membrane consisting of a self-assembly of $TiO_2$ nanoparticles with photocatalytic destructive capability on microorganisms as a novel means to reduce membrane biofouling. The photocatalytic bactericidal effect of the hybrid TFC membrane was examined by determining the survival ratios of the *Escherichia coli* (*E. coli*) cell with and without UV light illumination. The group demonstrated that the photocatalytic bactericidal efficiency was remarkably higher for the hybrid TFC membrane under UV light illumination. In addition, Rahimpour *et al.* (2011) reported on the antibacterial and antifouling properties of $TiO_2$-entrapped nanocomposite polyvinyldine fluoride (PVDF)/sulfonated polyethersulfone membranes. They observed a dramatic increase in anti-bacterial effect on *E. coli*. Shinde *et al.* (2011) studied the photocatalytic activity of seawater under solar light using a photoelectrochemical reactor module consisting of nine photoelectrochemical cells equipped with a spray-deposited $TiO_2$ catalyst and observed complete mineralization of organic compounds.

Very few researchers have attempted the treatment of seawater using solar photocatalysis and hence the work reported in this chapter provides an impetus for the development of technology that can be helpful to the desalination sector, especially in MENA countries.

Solar photocatalytic reactors can be distinguished according to whether the catalyst is applied in suspension or in thin-film form. There are several advantages and disadvantages of choosing

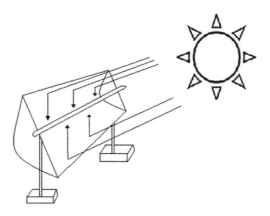

Figure 3.2.  Parabolic trough concentrator system (Thiruvenkatachari *et al.*, 2008).

suspension photocatalysis. The suspension process results in a uniform photocatalyst distribution in the reactor system. It has higher efficiency because of its larger surface area and a low pressure drop because the suspension particles are well-mixed. The process also minimizes catalyst fouling because the catalyst is continuously removed. The main disadvantage is the requirement for a nano filter to separate the catalyst: this will increase the cost of the unit. In comparison, the thin-film method is a continuous operation and does not need a separation step after the reaction takes place. However, it has low efficiency as a result of low light utilization due to the immobilization of the photocatalyst, which results in a smaller surface area for reaction. The key elements affecting the photocatalytic system are the type of catalyst, the light source and the reactor configuration. A concentrated light system (Fig. 3.2) that reflects the solar light onto the photocatalytic reactor via a reflecting surface is preferred because it requires a smaller reactor volume. It also operates at a higher flow rate, has better mass transfer rates and can be operated under cloudy conditions (Thiruvenkatachari *et al.*, 2008). Tubular photo reactors are simple in their operation, can be designed easily, and light can be concentrated by the use of reflectors which will increase the photoreaction.

Thiruvenkatachari *et al.* (2008) showed that the photocatalytic reaction is affected by the dosage of the catalyst, the original concentration of reactants, the illumination time, the intensity of illumination, the pH value and the oxygen flow. It was also found that nano-scale photocatalysts have more activity as they have a larger surface area which reduces the time needed for the carrier to diffuse out of the photocatalyst pores to the photocatalyst surface (Bhattacharya *et al.*, 2013).

The elimination potential of UV irradiation and advanced oxidation processes have been studied for UV/hydrogen peroxide and UV/ozone systems (Sona *et al.*, 2006). It was found that in order to improve the elimination efficiency an oxidant could be added during UV irradiation which absorbs the UV light by itself and reacts with water to form highly reactive hydroxyl radicals. Hydrogen peroxide ($H_2O_2$) and ozone are the commonly used oxidants. It was observed during the experiments that the absorbance efficiency of $H_2O_2$ was dependent on its concentration; the higher the concentration the better the performance of the water treatment system (Sona *et al.*, 2006).

The solar photocatalytic reaction of $TiO_2$ in the presence of inorganic salts results in the deposition of a double layer of the salts on the catalyst surface leading to a reduction in the photocatalyst adsorption and its photocatalytic activity by scavenging of hydroxyl radicals (Guillard *et al.*, 2003). Figure 3.3 shows an illustration of possible scavenging reactions of radicals by $Cl^-$ ions during seawater photocatalysis during which hydroxyl radicals and valence band holes oxidize organic matter (OM). Chloride ions show a negative effect on the photocatalytic oxidation reaction as 50% removal of total organic carbon (TOC) was obtained in the absence of chloride ions created by the addition of excess $AgNO_3$ prior to the photocatalytic reaction of $TiO_2$ in the presence of seawater (Kim *et al.*, 2010; Surolia *et al.*, 2007).

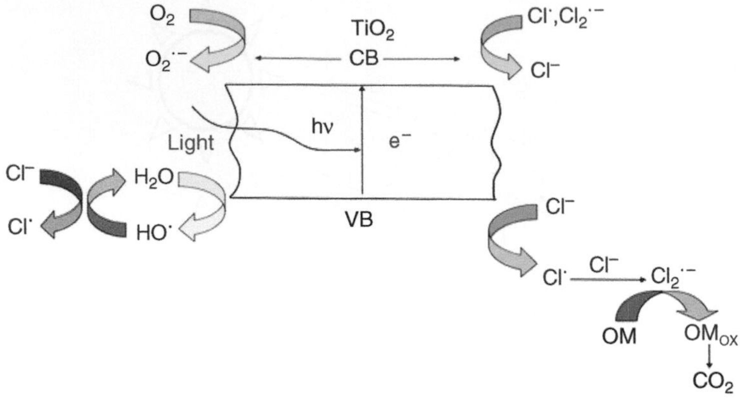

Figure 3.3.   Scavenging of radicals by $Cl^-$ ions during seawater photocatalysis (Kim *et al.*, 2010; Surolia *et al.*, 2007).

Table 3.1.   Seawater composition analysis from case study area.

| Parameter | Value |
|---|---|
| pH | 7.96 |
| Conductivity [mS] | 54.22 |
| Salinity [mg kg$^{-1}$] | 34200 |
| Turbidity [NTU] | 1.60 |
| Dissolved oxygen [mg L$^{-1}$] | 5.38 |
| Total dissolved solids [mg kg$^{-1}$] | 53200 |
| Chemical oxygen demand [mg L$^{-1}$] | 5.0 |
| Biological oxygen demand [mg L$^{-1}$] | 2.0 |
| Total organic carbon [mg L$^{-1}$] | 2.94 |
| *E. coli* | n.d. |

## 3.5   OMANI CASE STUDY OF REVERSE OSMOSIS PRETREATMENT USING SOLAR NANOPHOTOCATALYSIS

A case study was performed to assess the effect of chloride ions and the scavenging of hydroxyl radicals on total organic carbon (TOC), and thus whether solar nanophotocatalysis can be applied for RO treatment of seawater pollutants. Experimental studies were conducted in a batch recirculation reactor system and in a continuous thin-film reactor system. The seawater was taken from the Al Athibha beach area on the coast of the Sultanate of Oman and stored at normal temperature, 25°C. The seawater was collected 1 km offshore because seawater collected near the shore contains up to 50 mg L$^{-1}$ total organic carbon (TOC), whereas water collected from 1 km offshore had TOC in the range 4–7 mg L$^{-1}$. In general, the inlet for any desalination plant is placed 1 km away from the shore, hence the samples used in the present study. The seawater samples were analyzed at the Caledonian College of Engineering solar water research laboratory (Oman). Conductivity, pH, salinity, turbidity, dissolved oxygen (DO), total dissolved solids (TDS), chemical oxygen demand (COD), biological oxygen demand (BOD), TOC and *E. coli* levels were analyzed (Table 3.1). Experimental investigations were carried out with batch as well as continuous reactor systems (Fig. 3.4a through to Fig. 3.4d).

A 2 L glass beaker with magnetic stirrer was used as the batch reactor (Fig. 3.4a). The batch recirculation reactor (Fig. 3.4b) consisted of a 2 L glass tube of diameter 8 cm and length 100 cm fixed to a stand at an angle of 20°C. A small glass tube of diameter 1.2 cm and length 70 cm

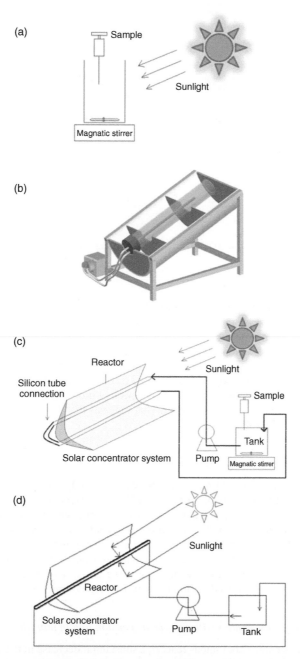

Figure 3.4. (a) Batch reactor system; (b) batch recirculation reactor system; (c) schematic of continuous suspension experimental setup; (d) schematic of continuous thin-film experimental setup (Al Jabri and Feroz, 2015; Cheriyan *et al.*, 2013; Feroz and Jesil, 2012; Feroz *et al.*, 2015).

was fixed inside the glass tube at the exact center and was connected to a peristaltic pump. Glass tubes 60 cm in length with an outer diameter of 2 cm and connected with a peristaltic pump were used for continuous reactor systems with the catalyst in suspension (Fig. 3.4c) or in a thin film (Fig. 3.4d).

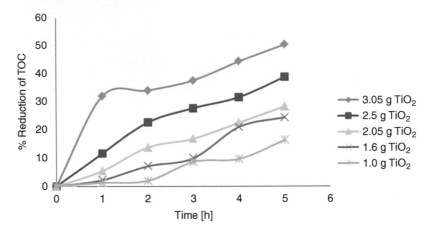

Figure 3.5.   Percentage reduction in TOC at different TiO$_2$ dosages in batch reactor.

### 3.5.1   Batch reactor studies with TiO$_2$

Figure 3.5 shows the percentage decrease in TOC at different dosages of TiO$_2$. It was observed that there was a sharp decrease in TOC over the first two hours of reaction time and thereafter the decline was stable. This may have been due to an increase in the availability of photonic energy. It was also seen that the degradation of pollutants seemed to be on the higher side at higher dosages of TiO$_2$, which may be as a result of availability of additional active sites on the photocatalyst. The amount of TiO$_2$ is directly proportional to the overall photocatalytic reaction rate and a linear dependency exists with TiO$_2$ concentration. A reduction of more than 50% in TOC was observed for 3 g TiO$_2$ dosage within a 5 h reaction time. This suggests that the technique can be applied to the treatment of seawater pollutants.

### 3.5.2   Batch reactor studies with TiO$_2$ recirculation

In the case of a batch reactor with recirculation, when the TiO$_2$ concentration was varied from 0.5 g to 3.5 g, a reduction in TOC of more than 45% was observed at the highest TiO$_2$ dosage (Fig. 3.6). With an increase in concentration of TiO$_2$ P25, the reduction in TOC was enhanced. This was probably due to the availability of more active sites on the catalyst surface. Seawater collected from the foreshore/beach had a high initial concentration of TOC, whereas seawater collected from 1 km offshore had a lower TOC concentration. It was observed that only glass reactors gave accurate results in these experiments; the use of any other material gave erratic results, possibly because of induced carbon content from reactor materials entering the seawater.

### 3.5.3   Continuous reactor studies with TiO$_2$ in suspension and in thin film

With TiO$_2$ in suspension in a continuous reactor, a substantial decrease in the TOC concentration was observed, especially at the lower TiO$_2$ dosage (Fig. 3.7). A sharp reduction in TOC was observed in the first two hours of exposure time and after that it was stable. A reduction in TOC of more than 35% was observed at 0.5 g TiO$_2$ dosage exposed to 5 h of irradiation. With an increase in catalyst dosage the percentage reduction in TOC declined, which may have been as a result of the agglomeration of catalyst particles, thereby inhibiting the generation of hydroxyl radicals.

With TiO$_2$ immobilized in a thin film in a continuous reactor, slightly different results were observed. Polyvinyl alcohol (PVA) was used as a solution thickener and a binder to enhance the chemical bonding of TiO$_2$ to the inner surface of the glass reactor tube. The coated layer of the nanophotocatalyst was stable and strongly bonded to the inner surface of the tube after calcination.

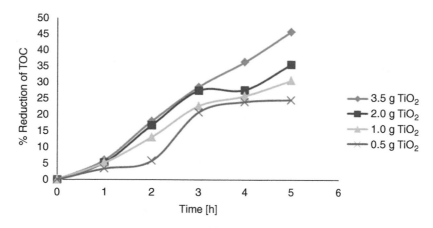

Figure 3.6.   Percentage reduction in TOC at various TiO$_2$ dosages in batch reactor with recirculation.

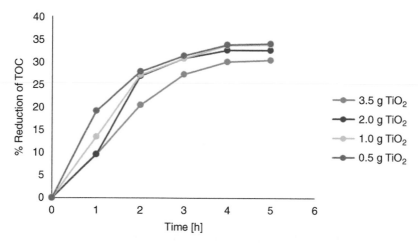

Figure 3.7.   Continuous reactor studies with TiO$_2$ in suspension in 0.002 m$^3$ of seawater, using a peristaltic pumping rate of 170 rpm to a tubular reactor where photocatalytic reaction took place.

It was found that there was a slight increase in TOC (Fig. 3.8), possibly due to the presence of polyvinyl chloride in the thin film or the formation of organic byproducts. The composition of seawater is complex in nature, which may interfere with the photocatalytic process. Though the TOC increased in the initial stages (perhaps due to the presence of PVA), after a certain amount of time it declined again, which shows that even in a thin-film coating the degradation of pollutants takes place. As a possible future study, an inorganic binder may be a better option than PVA to create the thin film.

### 3.5.4   *Effectiveness of combination of photocatalysts and photo-Fenton reagent in reduction of TOC in reactor studies*

Batch experimental studies were carried out with a combination of photocatalysts in a 0.002 m$^3$ reactor system. The combinations used were: 3.0 g TiO$_2$ + 5 mL H$_2$O$_2$; 3.0 g TiO$_2$ + 5 mL H$_2$O$_2$ + 1.0 g Fe$_2$O$_{12}$S$_3$.5H$_2$O; 3.0 g TiO$_2$ + 1.0 g ZnO. The maximum reduction in TOC was observed with the TiO$_2$ + ZnO combination, as shown in Figure 3.9.

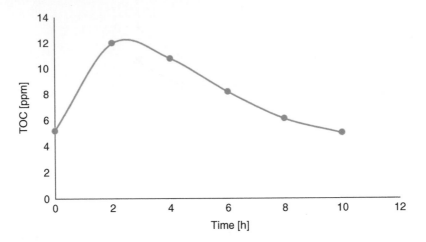

Figure 3.8.   Variation of TOC in continuous reactor studies with TiO$_2$ in thin-film form.

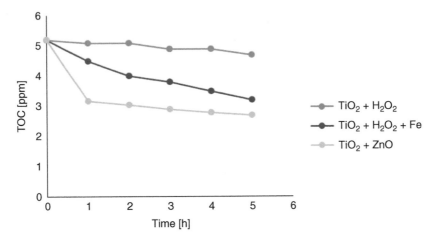

Figure 3.9.   Variation of TOC in batch reactor studies with various combinations of photocatalysts.

## 3.6   CONCLUDING REMARKS

Organic fouling and biofouling of reverse osmosis (RO) membranes requires conventional and/or membrane pretreatment of the feed water. Prevention of biofouling can be achieved by coating the reactor system with a nanophotocatalyst that is non-toxic. Photocatalytic coating is very effective against humic substances and microorganisms. If this technique can be successfully implemented at the commercial level, then solar energy can be utilized in a feasible manner for desalination, especially in RO pretreatment processes. From the case study, it was concluded that although there is interference by chloride ions and scavenging of hydroxyl radicals, substantial decreases in total organic content (TOC) were still observed and hence this technique can be applied to treatment of seawater pollutants. Experimental studies conducted in a batch recirculation reactor system have shown similar reductions in pollutants to those in a normal mixing batch-type reactor system. In a continuous reactor suspension system, a substantial decrease in TOC was observed. However, to apply suspension systems commercially, additional filtration systems are needed to separate nanophotocatalyst particles, which will increase both the capital and the operational costs, and this is one of the challenges that needs to be overcome for commercial exploitation.

## ACKNOWLEDGEMENTS

The author wishes to thank The Research Council of the Sultanate of Oman for financial support (ORG/EI/12/003, Grant Number 83).

## REFERENCES

Al Jabri, H. & Feroz, S. (2015) The effect of combining TiO$_2$ and ZnO in the pretreatment of seawater reverse osmosis process. *International Journal of Environmental Science and Development*, 6(5), 348–351.

Al-Rasheed, R. & Cardin, D.J. (2003a) Photocatalytic degradation of humic acid in saline waters. Part 1. Artificial seawater: influence of TiO$_2$, temperature, pH, and air-flow. *Chemosphere*, 51(9), 925–933.

Al-Rasheed, R. & Cardin, D.J. (2003b) Photocatalytic degradation of humic acid in saline waters: Part 2. Effects of various photocatalytic materials. *Applied Catalysis* A: *General*, 246(1), 39–48.

Amjad, Z. (1996) Scale inhibition in desalination applications: an overview. *Corrosion 96, The NACE International Conference and Exhibition, NACE-96-230, 24–29 March 1996, Denver, CO.*

Bhattacharya, S., Saha, I., Mukhopadhyay, A., Chattopadhyay, D., Ghosh, U.C. & Chatterjee, D. (2013) Role of nanotechnology in water treatment and purification: potential applications and implications. *International Journal of Chemical Science and Technology*, 3(3), 59–64.

Brand, B. & Würzburg, D. (2015) *The Integration of Renewable Energies into the Electricity Systems of North Africa*. Verlag Dr. Kovac, Hamburg.

Characklis, W.G. (1981) Fouling biofilm development: a process analysis. *Biotechnology and Bioengineering*, 23, 1923–1960.

Cheriyan, A.J., Sarkar, J.P., Shaik, F. & Baawain, M.S. (2013) Photocatalytic degradation of organics in municipal treated wastewater in a re-circulation reactor. *Journal of Environmental Protection*, 4, 1449–1452.

Chong, M.N., Jin, B., Chow, C.W. & Saint, C. (2010) Recent developments in photocatalytic water treatment technology: a review. *Water Research*, 44(10), 2997–3027.

Dudley, L.Y., Annunziata, U.A., Robinson, J.S. & Latham, L.J. (1995) Practical studies to investigate microbiological fouling in RO plant. *Proceedings of the IDA World Congress on Desalination and Water Reuse, Abu Dhabi, UAE*. 4, pp. 45–48.

Dunlop, P.S. (2001) *The Photocatalytic Inactivation of Faecal Indicator Organisms in Water*. PhD Thesis, University of Ulster, UK.

Feroz, S. & Jesil, A. (2012) Treatment of organic pollutants by heterogeneous photocatalysis. *Journal of the Institution of Engineers (India)*, Series E, 93(1), 45–48.

Feroz, S., Al Harthy, W., Baawain, B., Al Saadi, S., Varghese, M.J. & Rao, L.N. (2015) Experimental studies for treatment of seawater in a recirculation batch reactor using TiO$_2$ P25 and polyamide. *International Journal of Applied Engineering Research*, 10(10), 26,259–26,266.

Freudenhammer, H., Bahnemann, D., Bousselmi, L., Geissen, S.U., Ghrabi, S.U., Saleh, F., Si-Salah, A., Siemon, U. & Vogelpohl, A. (1997) Detoxification and recycling of wastewater by solar-catalytic treatment. *Water Science and Technology*, 35, 149–156.

Fujishima, A. & Honda, K. (1972) Electrochemical photolysis of water at a semiconductor electrode. *Nature*, 238, 37–38.

Gernjak, W., Maldonado, M.I., Malato, S., Caceres, J., Krutzler, T., Glaser, A. & Bauer, R. (2004) Pilot-plant treatment of olive mill wastewater (OMW) by solar TiO$_2$ photocatalysis and solar photo-Fenton. *Solar Energy*, 77, 567–572.

Ghaffour, N., Bundschuh, J., Mahmoudi, H. & Goosen, M.F. (2015) Renewable energy-driven desalination technologies: a comprehensive review on challenges and potential applications of integrated systems. *Desalination*, 356, 94–114.

Goosen, M.F.A., Sablani, S., Al-Hinai, H., Al-Obeidani, S., Al-Belushi, R. & Jackson, D. (2004) Fouling of reverse osmosis and ultrafiltration membranes: a critical review. *Separation Science and Technology*, 39, 1–37.

Goosen, M.F., Mahmoudi, H. & Ghaffour, N. (2014) Today's and future challenges in applications of renewable energy technologies for desalination. *Critical Reviews in Environmental Science and Technology*, 44(9), 929–999.

Goswami, D.Y., Klausner, J., Mathur, G.D., Martin, A., Schanze, K., Wyness, P. & Marchand, E. (1993) Solar photocatalytic treatment of groundwater at Tyndall AFB: field test results. *Proceedings of the 1993 Annual Conference, American Solar Energy Society, Inc.* pp. 235–239.

Guillard, C., Lachheb, H., Houas, A., Ksibi, M., Elaloui, E. & Herrmann, J.M. (2003) Influence of chemical structure of dyes, of pH and of inorganic salts on their photocatalytic degradation by $TiO_2$ comparison of the efficiency of powder and supported $TiO_2$. *Journal of Photochemistry and Photobiology* A: *Chemistry*, 158(1), 27–36.

Ibhadon, A.O. & Fitzpatrick, P. (2013) Heterogeneous photocatalysis: recent advances and applications. *Catalysts*, 3(1), 189–218.

Kim, M.J., Choo, K.H. & Park, H.S. (2010) Photocatalytic degradation of seawater organic matter using a submerged membrane reactor. *Journal of Photochemistry and Photobiology* A: *Chemistry*, 216(2), 215–220.

Kim, S.H., Kwak, S.Y., Sohn, B.H. & Park, T.H. (2003) Design of $TiO_2$ nanoparticle self-assembled aromatic polyamide thin-film-composite (TFC) membrane as an approach to solve biofouling problem. *Journal of Membrane Science*, 211, 157–165.

Kwak, S.Y., Kim, S.H. & Kim, S.S. (2001) Hybrid organic/inorganic reverse osmosis (RO) membrane for bactericidal anti-fouling. 1. Preparation and characterization of $TiO_2$ nanoparticle self-assembled aromatic polyamide thin-film-composite (TFC) membrane. *Environmental Science and Technology*, 35(11), 2388–2394.

Malato, S., Blanco, J., Vidal, A. & Richter, C. (2002) Photocatalysis with solar energy at a pilot plant scale: an overview. *Applied Catalysis* B: *Environmental*, 37, 1–15.

Mehos, M.S. & Turchi, C.S. (1993) Field testing solar photocatalytic detoxification on TCE-contaminated ground water. *Environmental Progress*, 12, 194–199.

Mehrjouei, M., Müller, S. & Möller, D. (2015) A review on photocatalytic ozonation used for the treatment of water and wastewater. *Chemical Engineering Journal*, 263, 209–219.

Ou, W., Zhang, G., Yuan, X. & Su, P. (2015) Experimental study on coupling photocatalytic oxidation process and membrane separation for the reuse of dye wastewater. *Journal of Water Process Engineering*, 6, 120–128.

Rahimpour, A., Jahanshahi, M., Rajaeian, B. & Rahimnejad, M. (2011) $TiO_2$ entrapped nano-composite PVDF/SPES membranes: preparation, characterization, antifouling and antibacterial properties. *Desalination*, 278(1), 343–353.

Shinde, S.S., Shinde, P.S., Bhosale, C.H. & Rajpure, K.Y. (2011) Zinc oxide mediated heterogenous photocatalytic degradation of organic species under solar radiation. *Journal of Photochemistry and Photobiology* B: *Biology*, 104, 425–433.

Sieburth, J.M. & Jensen, A. (1968) Studies in algal substances in the sea. I. Gelbstoff (humic material) in terrestrial and marine waters. *Journal of Experimental Marine Biology and Ecology*, 2, 174–189.

Sieburth, J.M. & Jensen, A. (1969) Studies in algal substances in the sea. II. The formation of Gelbstoff (humic material) by exudates of phaeophyta. *Journal of Experimental Marine Biology and Ecology*, 3, 275–289.

Sona, M., Baus, C. & Brauch, H.J. (2006) UV irradiation versus combined UV/hydrogen peroxide and UV/ozone treatment for the removal of persistent organic pollutants from water. *International Conference Ozone and UV*, 3rd April 2006, Wasser Berlin, Germany. pp. 69–76.

Surolia, P.K., Tayade, R.J. & Jasra, R.V. (2007) Effect of anions on the photocatalytic activity of Fe (III) salts impregnated $TiO_2$. *Industrial & Engineering Chemistry Research*, 46(19), 6196–6203.

Thiruvenkatachari, R., Vigneswaran, S. & Moon, I.S. (2008) A review on UV/$TiO_2$ photocatalytic oxidation process. *Korean Journal of Chemical Engineering*, 25(1), 64–72.

# CHAPTER 4

## Metal oxide nanophotocatalysts for water purification

Edreese Alsharaeh, Faheem Ahmed, Nishat Arshi & Meshael Alturki

### 4.1 INTRODUCTION

Photocatalysis, which exploits renewable solar energy to trigger chemical reactions via oxidation and reduction, can be classified as a sustainable technology that offers solutions to ecological problems (Kudo and Miseki, 2009). Nanophotocatalysts, for example, can be utilized for water purification and environmental pollution management, such as eliminating residual dye pollutants from wastewater streams (Lee *et al.*, 2016). In a review, Reddy *et al.* (2016) reasoned that technological progress in this research field is dependent on the improvement of solar sensitivity to enhance the efficiency of pollutant decontamination.

The spread of a wide range of contaminants in surface water and groundwater has become a critical issue worldwide due to population growth, rapid industrialization and long-term droughts (Chong *et al.*, 2010; Cundy *et al.*, 2008; Zeng *et al.*, 2011). It is thus essential to be able to better control the harmful effects of pollutants so as to improve the human environment. Toxins persisting in wastewater include heavy metals, inorganic compounds, organic pollutants and many other complex compounds (Fatta *et al.*, 2011; Li *et al.*, 2011; O'Connor, 1996). All of these contaminants that are released in wastewater are harmful to human beings and the natural surroundings. Consequently, the need for efficient methods of pollutant removal has become crucial (Huang *et al.*, 2010; Jiang *et al.*, 2006; Pang *et al.*, 2011a). In an effort to combat the problem of water pollution, for example, rapid and significant progress in wastewater treatment has been made, including photocatalytic oxidation, adsorption/separation processes and bioremediation (Huang *et al.*, 2006; Long *et al.*, 2011; Pang *et al.*, 2011a, 2011b; Zelmanov and Semiat, 2008). However, their applications have been restricted by many factors, such as processing efficiency, operational methods, energy requirements and economics.

Recently, nanomaterials have been suggested as efficient, cost-effective and environmentally friendly alternatives to existing treatment materials from the standpoints of both resource conservation and environmental remediation (Dastjerdi and Montazer, 2010; Dimitrov, 2006; Friedrich *et al.*, 1998). Nanotechnology holds out the promise of immense improvements in manufacturing technologies, electronics, telecommunications, health and even environmental remediation (Gross, 2001; Kim *et al.*, 2005; Moore, 2006). The production and utilization of a diverse array of nanomaterials include structures and devices with sizes ranging from 1 to 100 nm, and the display of unique properties not found in bulk-sized materials (Stone *et al.*, 2010; Wang *et al.*, 2010). Various materials, such as carbon-based (Mauter and Elimelech, 2008; Upadhyayula *et al.*, 2009) and $TiO_2$ nanomaterials (Khan *et al.*, 2002; Shankar *et al.*, 2009), have been widely studied and reviewed. However, there is a need to assess iron oxide-based nanomaterials in greater detail.

This chapter evaluates the important properties of iron oxide nanomaterials. It highlights not only recent developments in their application to wastewater treatment, but also seeks to address the issues which appear to limit their large-scale field application. Primary attention is given to recent progress in the utilization of iron oxide nanomaterials as nanosorbents, followed by critical discussion on their application as photocatalysts. A detailed description of synthesis methods, properties and characterization is also included in the scope of this chapter.

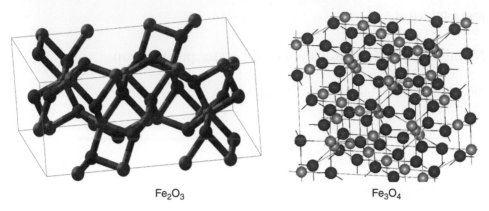

$Fe_2O_3$                                          $Fe_3O_4$

Figure 4.1.    Common structural forms of iron oxide.


## 4.2    IRON OXIDE NANOMATERIALS

Iron oxides exist in many forms in nature. Magnetite ($Fe_3O_4$), maghemite ($\gamma$-$Fe_2O_3$), and haematite ($\alpha$-$Fe_2O_3$) are the most common (Chan and Ellis, 2004; Cornel and Schwertmann, 1996) (Fig. 4.1). In recent years, the synthesis and utilization of iron oxide nanomaterials with novel properties and functions have been widely studied due to their relative size in the nano-range, their high surface area-to-volume ratios and their superparamagnetism (Afkhami *et al.*, 2010; McHenry and Laughlin, 2000; Pan *et al.*, 2010). In particular, their easy synthesis, coating or modification, and the ability to control or manipulate matter on an atomic scale can provide unmatched versatility (Boyer *et al.*, 2010; Dias *et al.*, 2011). In addition, iron oxide nanomaterials with low toxicity, chemical inertness and biocompatibility show tremendous potential in combination with biotechnology (Gupta and Gupta, 2005; Huang *et al.*, 2003; Roco, 2003). They have unique properties, which account for their application as well as the considerable differences among iron oxide bulk materials (Bystrzejewski *et al.*, 2009; Selvan *et al.*, 2010). It was reported that preparation methods and surface coating mediums play a key role in determining the size distribution, morphology, magnetic properties and surface chemistry of nanomaterials (Jeong *et al.*, 2007; Machala *et al.*, 2007). Many researchers have been focusing their efforts on developing chemical and physical methods for the synthesis of magnetic nanoparticles (MNPs) (Dias *et al.*, 2011).

Lately, a variety of synthesis approaches have been developed to produce high-quality nanoparticles (Hassanjani *et al.*, 2011), nano-ovals (Zhong and Cao, 2010), nanobelts (Fan *et al.*, 2011), nanorings (Gotić *et al.*, 2011) and other nanostructures. The three most important published routes for the synthesis of superparamagnetic iron oxide nanoparticles (SPIONs) have been summarized by Mahmoudi *et al.* (2011). Advances in nanomaterial synthesis enable the precise control of surface active sites by manufacturing monodispersed and shape-controlled iron oxide nanomaterials (Bautista *et al.*, 2005; Li and Somorjai, 2010). Emerging methods, such as fungi/protein-mediated biological methods and sonochemical methods, necessitate broader development. Future studies should aim to address different challenges to provide new more efficient and specific magnetic nanomaterials. In addition, the growth of iron oxide nanomaterials to field scale may provide a fruitful area for research.


## 4.3    GRAPHENE OXIDE-BASED NANOCOMPOSITES

Recently, researchers have been exploring the dispersion of nanoparticles on graphene so as to provide new materials with enhanced applications such as catalysis, energy storage and biochemical sensors (Zhang *et al.*, 2015). These new-generation nanoparticles can be beneficial in adding magnetic, optical, electrical and catalytic properties to graphene. Consequently, it can be argued that these metal-nanoparticle graphene-based composites can be very advantageous for many applications. Many researchers have described the use of functionalized graphene with different types of metal nanoparticles. Zhang *et al.* (2015), for example, synthesized ZnO-reduced graphene oxide (rGO) nanocomposites with a uniform dispersion of 10.5 nm on sheets of graphene prepared by a modified Hummers' method. These nanocomposites were tested for their photocatalytic reduction of $CO_2$ into methanol under UV-vis light. Four samples with different percentages of rGO (1%, 5%, 10% and 20%) and one sample of pure ZnO were tested. After irradiation for three hours, ZnO-(10%)rGO yielded 263 $\mu$mol g$^{-1}$ catalyst of methanol, which was five times higher than the 52 $\mu$mol g$^{-1}$ catalyst yield of the pure ZnO nanoparticles and better than all other samples tested. These results show the abundant influence of reduced graphene oxide in enhancing photocatalytic reaction and thus improving the effectiveness of ZnO nanoparticles.

## 4.4 IRON OXIDE NANOMATERIALS IN WASTEWATER TREATMENT

Selection of the best method and material for wastewater treatment is a highly complex task, dependent on a number of factors, such as the quality standards to be met and the efficiency as well as the cost (Huang *et al.*, 2008; Oller *et al.*, 2011). Therefore, the following four aspects must be considered in any decision-making process on wastewater treatment technologies: (i) treatment flexibility and final efficiency; (ii) reuse of treatment agents; (iii) environmental security and friendliness; (iv) cost-effectiveness (Oller *et al.*, 2011; Zhang and Fang, 2010).

Magnetism helps in water purification by influencing the physical properties of contaminants. Adsorption procedures combined with magnetic separation have, for instance, been used extensively in water treatment and environmental cleanup (Ambashta and Sillanpää, 2010; Mahdavian and Mirrahimi, 2010). Furthermore, iron oxide nanomaterials are promising for industrial-scale wastewater treatment, due to their low cost, strong adsorption capacity, easy separation and enhanced stability (Carabante *et al.*, 2009; Fan *et al.*, 2012; Hu *et al.*, 2005). The ability of iron oxide nanomaterials to remove contaminants has been demonstrated in both laboratory- and field-scale trials (Girginova *et al.*, 2010; White *et al.*, 2009).

Current applications of iron oxide nanomaterials in contaminated water treatment can be divided into two groups: (i) technologies which use the materials as a nanosorbent or immobilization carrier for removal efficiency enhancement (referred to here as adsorptive/immobilization technologies), and (ii) those which use iron oxide nanomaterials as photocatalysts to break down or convert contaminants into a less toxic form (i.e. photocatalytic technologies), although it should be noted that many technologies may utilize both processes.

### 4.4.1 *Adsorptive technologies*

#### 4.4.1.1 *Iron oxide as a nanosorbent for heavy metals*
Heavy metal contamination is of great concern because of its toxic effect on plants, animals and human beings, and its tendency for bioaccumulation even at relatively low concentrations. Therefore, effective removal methods for heavy metal ions are extremely urgent and have attracted considerable research and practical interest (Chen *et al.*, 2011; Huang *et al.*, 2006; Pang *et al.*, 2011c). Nowadays, the majority of bench-scale research and field applications of materials for wastewater treatment have focused on magnetic nanomaterials (Iram *et al.*, 2010), carbon nanotubes (Stafiej and Pyrzynska, 2007), activated carbon (Kobya *et al.*, 2005) and zero-valent iron (Ponder *et al.*, 2000). Among these, it appears that iron oxide magnetic nanomaterials, possessing the capability to treat large volumes of wastewater and being convenient for magnetic separation, are the most promising materials for heavy metal treatment (Hu *et al.*, 2010). In a study performed by Nassar (2010), it was found that the maximum adsorption capacity of $Fe_3O_4$ nanoparticles for Pb(II) ions was 36 mg g$^{-1}$, which was much higher than that reported of low-cost

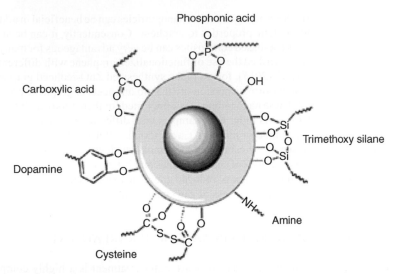

Figure 4.2.    Common chemical moieties for the anchoring of polymers and functional groups at the surface of iron oxide magnetic nanoparticles (adapted from Dias *et al.*, 2011).

adsorbents. The small size of $Fe_3O_4$ nanosorbents was favorable to the diffusion of metal ions from solution onto the active sites of the adsorbent's surface. Nassar (2010) suggested that $Fe_3O_4$ nanosorbents should be employed as effective and economical adsorbents for rapid removal and recovery of metal ions from wastewater effluents. Common chemical moieties for anchoring of polymers and functional groups at the surface of iron oxide magnetic nanoparticles are shown in Figure 4.2.

Laboratory studies have indicated that iron oxide nanomaterials could effectively remove a range of heavy metals, including $Pb^{2+}$, $Hg^{2+}$, $Cd^{2+}$ and $Cu^{2+}$ (Otto *et al.*, 2008). However, iron oxide-based technology for heavy metal adsorption is still at a relatively early stage in terms of wider application. It is recognized that much work is needed to advance knowledge in the area of nanomaterials, and the transfer of technology from laboratory scale to field scale is complex. With increasing emphasis on contaminant removal treatments, more data will become available on performance and cost, which can provide additional information for large-scale industrial applications.

### 4.4.1.2    *Iron oxide as nanosorbents for organic contaminants*

As a well-known separation process, adsorption has been widely applied to remove chemical pollutants from water. It has numerous advantages in terms of cost, flexibility and simplicity of design/operation, and insensitivity to toxic pollutants (Ahmad *et al.*, 2009; Rafatullah *et al.*, 2010; Zeng *et al.*, 2007). Hence, an effective and low-cost adsorbent with high adsorption capacity for organic pollutant removal is desirable. Iron oxide nanomaterials are currently being explored for the efficient treatment of large-volume water samples and fast separation via the application of a strong external magnetic field. Numerous trials have been undertaken to examine the efficiency of organic pollutant removal by using iron oxide nanomaterials (Luo *et al.*, 2011; Zhang *et al.*, 2010; Zhao *et al.*, 2010). For example, hollow $Fe_3O_4$ nanospheres were shown to be an effective sorbent for red dye (with a maximum adsorption capacity of 90 mg g$^{-1}$) (Iram *et al.*, 2010). The saturation magnetization of prepared nanospheres was observed to be 42 emu g$^{-1}$, which was sufficient for magnetic separation (critical value at 16.3 emu g$^{-1}$) (Ma *et al.*, 2005). These studies proved that magnetic nanomaterial technology was a novel, promising and desirable alternative for organic contaminant adsorption.

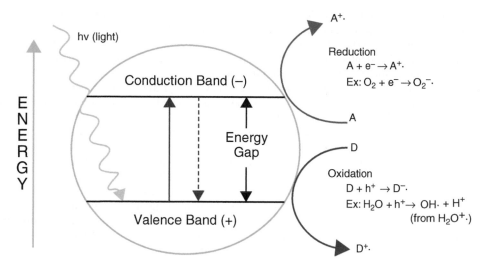

Figure 4.3.   Energy change in photocatalytic reactions.

### 4.4.2   *Photocatalytic technology*

Photocatalysis (Fig. 4.3), one of the advanced physico-chemical technologies applicable in photodegradation of organic pollutants (Akhavan and Azimirad, 2009), has attracted considerable attention. However, several obstacles hinder the wider application of iron oxide nanomaterials to the photocatalysis of toxic compounds. These impediments include: (i) the high expense associated with the separation of materials after the treatment process because of the manpower, time and chemicals used for precipitation, followed by centrifugation or decantation; (ii) restrictions to the kinetics and efficiency of the treatment process as a result of the low quantum yield (Bandara *et al.*, 2007). These limitations should be taken into account in the development of nanomaterial-based technologies. In addition, considerable efforts have been undertaken to enhance photocatalytic activity, such as by decreasing photocatalyst size to increase surface area, combining the photocatalyst with novel metal nanoparticles, and increasing hole concentration through doping (Zhang and Fang, 2010). On the other hand, improved charge separation and inhibition of charge carrier recombination are essential in improving the overall quantum efficiency for interfacial charge transfer (Beydoun *et al.*, 1999; Hu *et al.*, 2009; Watson *et al.*, 2002).

Iron oxide nanomaterials can be good photocatalysts by absorbing visible light. Compared with the commonly applied $TiO_2$, which primarily absorbs UV light at wavelengths of $\sim$380 nm (covering only 5% of the solar spectrum) due to its wide band gap of 3.2 eV, $Fe_2O_3$ has a narrower band gap of 2.2 eV (Akhavan and Azimirad, 2009) and is a fascinating n-type semiconducting material that is a suitable candidate for photodegradation under visible-light conditions. The enhanced photocatalytic performance of iron oxide nanomaterials compared with $TiO_2$ can be attributed to the extensive generation of electron-hole pairs through the narrow band-gap illumination (Bandara *et al.*, 2007).

#### 4.4.2.1   *Preparation of iron oxide/graphene-based nanomaterials*
Iron oxide nanoparticles and iron oxide/graphene nanocomposites were prepared as follows:

- *$\alpha$-$Fe_2O_3$ nanoparticles from microwave-hydrothermal (M-H) process using HMT:* Different concentrations (0.01, 0.02 and 0.05 M) of materials were prepared using ferric nitrate ($Fe(NO_3)_3 \cdot 9H_2O$) or ferric chloride ($FeCl_3 \cdot 6H_2O$). The ferric salt was mixed with similar concentrations of hexamethylenetetramine (HMT; $(CH_2)_6N_4$) in 100 mL of deionized water. The solution was then homogenized by stirring before being heated by microwave irradiation for

20 minutes, in conditions of 100 psi (≈0.689 MPa), 150°C and 100% power. After that, the solution was cooled at room temperature for several minutes and then washed six times by centrifugation at 3000 rpm for two minutes (three times with deionized water and three times with ethyl alcohol diluted to 70%). The sample was then dried at 80°C in an oven overnight. Finally, calcination at 400°C was applied for two hours.

- *α-Fe$_2$O$_3$/graphene nanocomposites from M-H process using HMT:* Different concentrations of commercial graphene (2%, 5%, 10% and 17%) were prepared with 0.05 M ferric nitrate, and a similar procedure followed as for the preparation of α-Fe$_2$O$_3$ nanoparticles described above.
- *α-Fe$_2$O$_3$ nanoparticles from M-H process using PEG:* 0.25 M ferric nitrate was dissolved in 20 mL deionized water and sonicated for 30 minutes to form Solution A. 10 mL of 30% ammonia solution (NH$_3$·H$_2$O) was mixed with 10 mL of polyethylene glycol (PEG) to form Solution B. Solution B was added drop wise to Solution A and stirred for 30 minutes. Next, the mixture was subjected to microwave irradiation for 10 minutes at conditions of 150 psi (≈1.03 MPa), 200°C and 100% power. Finally, the sample was dried overnight in a vacuum oven at 45°C.
- *α-Fe$_2$O$_3$/graphene nanocomposites from M-H process using PEG:* Different concentrations of commercial graphene (2%, 5%, 10% and 17%) were prepared with 0.05 M ferric nitrate, and a similar procedure followed as for the preparation of α-Fe$_2$O$_3$ nanoparticles just described.

The photocatalytic performance of α-Fe$_2$O$_3$ and α-Fe$_2$O$_3$/graphene nanocomposites were evaluated by the degradation of molybdenum blue (MB) in a Pyrex reactor under visible-light irradiation (420 nm) provided by a 500 W halogen lamp with an intensity of ~15000 lux. In each experiment, 20 mg of photocatalyst was dispersed in 100 mL of MB solution with a concentration of $10^{-5}$ M. Before illumination, the suspension was magnetically stirred in the dark for 30 minutes to obtain an adsorption-desorption equilibrium. At given time intervals, 5 mL of aliquot was sampled. Then the supernatant solution was analyzed by monitoring the maximum absorption peak at 665 nm ($\lambda_{max}$ for MB) using a UV-vis spectrophotometer (PerkinElmer LAMBDA 25).

### 4.4.2.2    Properties of α-Fe$_2$O$_3$ nanoparticles

Figure 4.4 and Figure 4.5 show the UV-vis absorption spectra of α-Fe$_2$O$_3$ nanoparticles prepared by the microwave-hydrothermal route using HMT and PEG. All spectra showed a broad absorption band at 425 nm, a characteristic of α-Fe$_2$O$_3$ nanoparticles. This broadening can be caused by many factors, including particle size and shape. All samples showed that as the concentration was increased from 0.01 to 0.05 M, a red shift toward higher wavelengths was observed.

X-ray diffraction (XRD) measurements were performed to determine the crystalline phase of the samples. Figure 4.6 and Figure 4.7 show the XRD patterns of the α-Fe$_2$O$_3$ nanostructures prepared by microwave irradiation. All the diffraction peaks obtained can be successfully indexed to the pure rhombohedral symmetry of Fe$_2$O$_3$ (JCPDS card no. 89-8104), indicating the crystalline haematite phase. No other peaks for any impurities were observed. The strong and sharp diffraction peaks indicated a high degree of crystallization of α-Fe$_2$O$_3$ nanostructures prepared using PEG. However, in the case of samples prepared using HMT, low intensity peaks were seen. The interplanar spacing or d-spacing was calculated to be 0.27 nm, which corresponds to the (104) lattice plane of α-Fe$_2$O$_3$.

To further confirm the crystal phases of the α-Fe$_2$O$_3$ nanostructures, Raman spectroscopy measurements were carried out. Raman spectra of α-Fe$_2$O$_3$ nanoparticles recorded at room temperature in the range from 200–1400 cm$^{-1}$ are shown in Figure 4.8. The α-Fe$_2$O$_3$, crystallized as a corundum-type structure, is the most common iron oxide on earth and seven phonon lines are expected in the Raman spectrum, namely two A1g modes (225 and 498 cm$^{-1}$) and five Eg modes (247, 293, 299, 412 and 613 cm$^{-1}$). It can be clearly seen in the Raman spectra in Figure 4.8 that the peaks appearing in the spectra at 228, 291, 406, and 612 cm$^{-1}$ correspond to the characteristic peaks of α-Fe$_2$O$_3$, that is, the peak located at 228 cm$^{-1}$ corresponds to the A1g mode, and the three peaks at about 291, 406 and 612 cm$^{-1}$ are attributable to the Eg mode. However, some shifts were observed due to differences in size and shape of the nanostructures. These observations imply that

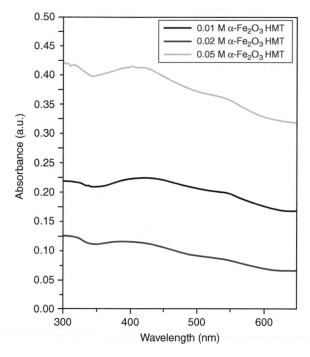

Figure 4.4.   UV-vis spectrum of $\alpha$-Fe$_2$O$_3$ nanoparticles prepared at various concentrations by M-H process using ferric nitrate and HMT.

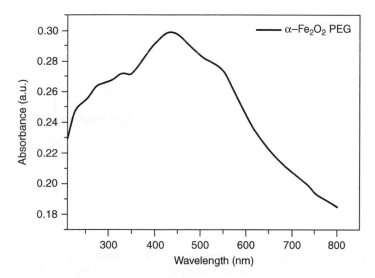

Figure 4.5.   UV-vis spectroscopy of $\alpha$-Fe$_2$O$_3$ nanoparticles prepared by M-H process using ferric nitrate and PEG.

it is feasible to produce highly crystalline $\alpha$-Fe$_2$O$_3$ by our one-step microwave irradiation method. Thus, the Raman spectroscopy measurements are in good agreement with the XRD results, which confirmed that the nanostructures are purely haematite.

The surface morphology of $\alpha$-Fe$_2$O$_3$ nanostructures was investigated by field emission scanning electron microscope (FESEM) analysis. Figure 4.9 depicts the typical morphologies of $\alpha$-Fe$_2$O$_3$

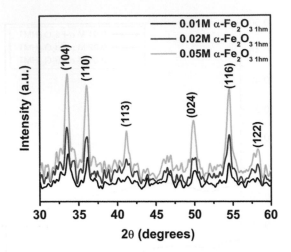

Figure 4.6.    XRD patterns of $\alpha$-Fe$_2$O$_3$ nanoparticles prepared at various concentrations by M-H process using ferric nitrate and HMT.

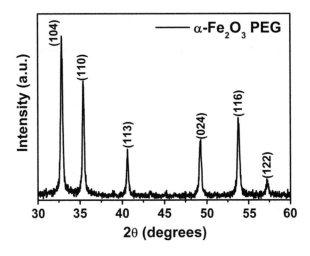

Figure 4.7.    XRD patterns of $\alpha$-Fe$_2$O$_3$ nanoparticles prepared by M-H process using ferric nitrate and PEG.

samples prepared by the microwave-hydrothermal process using HMT. All the nanoparticles exhibit spherical shape, with sizes ranging between 10 and 20 nm. Figure 4.10 shows the transmission electron microscopy (TEM) images of $\alpha$-Fe$_2$O$_3$ nanoparticles generated using PEG, in which fine spherical particles can be clearly seen. The size distribution is uniform, demonstrating the role of PEG as a capping agent.

### 4.4.2.3    $\alpha$-Fe$_2$O$_3$ nanoparticles/graphene nanocomposites from M-H treatment using HMT and PEG

XRD measurements were performed to determine the crystalline phase of the samples. Figure 4.11 shows the XRD patterns of $\alpha$-Fe$_2$O$_3$/graphene nanocomposites prepared by microwave

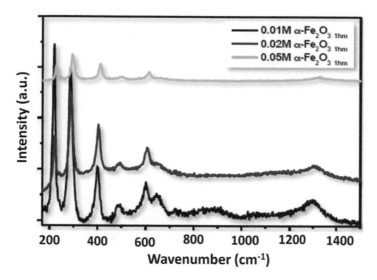

Figure 4.8.    Room temperature Raman spectra of $\alpha$-Fe$_2$O$_3$ nanoparticles prepared at various concentrations by M-H process using ferric nitrate and HMT.

irradiation. All the diffraction peaks obtained can be closely indexed to the pure rhombohedral symmetry of Fe$_2$O$_3$ (JCPDS card no. 89-8104), indicating the crystalline haematite phase. The characteristic peak (002) of graphite at 26.5° disappeared after oxidation, while an additional peak at 11.22° was observed, which corresponds to the (001) diffraction peak of graphene oxide (GO). Furthermore, the d-spacing of the GO was larger than that of graphite. This larger inter-layer distance of GO might be due to the formation of oxygen-containing functional groups, such as hydroxyl, epoxy and carboxyl. Thus, from the XRD pattern of Fe$_2$O$_3$/rGO, it could be inferred that the original graphite powders had been almost completely oxidized. A broad diffraction peak (002) of rGO at about 24.1° was observed in the XRD pattern shown in Figure 4.11. The broadening and shift of the characteristic diffraction peak of graphite from 26.5° to 24.1° was due to the short-range order in stacked stacks. The interlayer spacing of rGO is slightly larger than that of graphite, which results from a small amount of residual oxygen-containing functional groups or other structural defects. No other peaks for any impurities were observed.

The surface morphology of $\alpha$-Fe$_2$O$_3$/graphene nanocomposites was investigated by scanning electron microscopy (SEM) analysis. Figure 4.12 depicts the typical morphologies of $\alpha$-Fe$_2$O$_3$/graphene nanocomposite samples prepared by microwave-hydrothermal processing. In addition, TEM provides further details about the size, morphology and crystallinity of $\alpha$-Fe$_2$O$_3$/graphene. The average particle size of the $\alpha$-Fe$_2$O$_3$ nanoparticles anchored on the graphene sheets was found to be $\sim$8 nm, as shown in Figure 4.13a. Further, the nanosheets had thicknesses of a few nanometers and the nanoparticles were well-dispersed and covered the surface of the nanosheets, indicating successful nanocomposite preparation (see Fig. 4.13b).

The $\alpha$-Fe$_2$O$_3$/graphene photocatalysts have lately been studied for their potential to provide pollution-free environments by degrading air and water contamination through photocatalysis by visible-light absorption. Introducing graphene to $\alpha$-Fe$_2$O$_3$ caused delocalization of the photo-generated electrons through the $\alpha$-network, which led to inhibition of the recombination process and thus improved the photocatalytic performance of the material. Another advantage is that fictionalization of $\alpha$-Fe$_2$O$_3$ nanocomposites on graphene sheets helps in the removal and recycling of compounds when used in water. With graphene forming a sheet under the nanoparticles, it can be argued that the filtration process from water will be easier and more effective. The

Figure 4.9.   SEM images of $\alpha$-Fe$_2$O$_3$ prepared by M-H process using ferric nitrate and HMT.

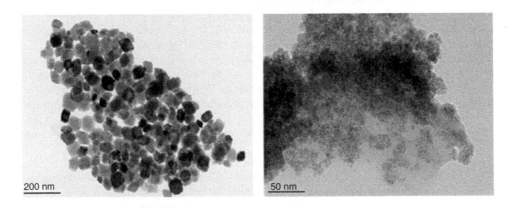

Figure 4.10.   TEM images of $\alpha$-Fe$_2$O$_3$ prepared by M-H process using PEG.

photocatalytic activity of $\alpha$-Fe$_2$O$_3$/graphene nanocomposites was evaluated in degradation of the representative organic dyes methyl orange and MB under visible-light irradiation. Figure 4.14 shows the evolution of MB absorption spectra obtained upon photoirradiation in the presence of $\alpha$-Fe$_2$O$_3$/graphene nanocomposites for different time periods.

For $\alpha$-Fe$_2$O$_3$/graphene sample placed in MB solution, the absorption intensity of both dyes' solution is observed to decrease with irradiation time, indicating that the $\alpha$-Fe$_2$O$_3$/graphene sample demonstrates visible-light photocatalytic properties in the degradation of MB. For 240 minutes 80% degradation of MB was degraded in the presence of $\alpha$-Fe$_2$O$_3$/graphene.

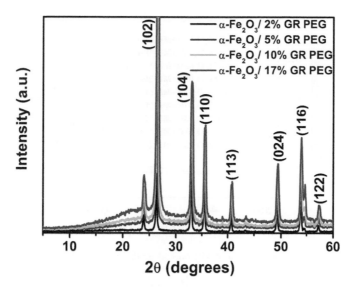

Figure 4.11.   XRD patterns of α-Fe₂O₃/graphene nanocomposites prepared with various concentrations of graphene using PEG.

Figure 4.12.   SEM image of α-Fe₂O₃/(17%)graphene (PEG) nanocomposite.

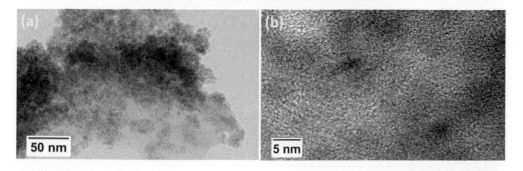

Figure 4.13.   α-Fe₂O₃/(17%)graphene nanocomposite: (a) TEM image; (b) high-resolution TEM image.

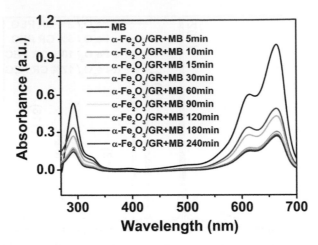

Figure 4.14.   UV-vis absorbance spectra of MB photodegradation in the presence of $\alpha$-Fe$_2$O$_3$/graphene nanocomposites.

## 4.5   CONCLUDING REMARKS

Wastewater treatment and reuse is essential to safeguard human health and to protect environmental ecosystems. Nanomaterials, with their unique physical and chemical properties, have a tremendous potential for contaminant removal. A case study was presented in this chapter showing the successful preparation of $\alpha$-Fe$_2$O$_3$ nanoparticles and their graphene-based nanocomposites using a simple, fast, low-cost, energy-saving and environmentally benign microwave-assisted method. It was demonstrated that the size, shape and diameter of the $\alpha$-Fe$_2$O$_3$ nanostructures can be readily adjusted and optimized by changing the precursors, modifying the solution concentration and employing diverse approaches. Such knowledge allows for the rapid synthesis of $\alpha$-Fe$_2$O$_3$ nanomaterials of specific morphology and size, simply by selecting the appropriate precursor and concentration.

The $\alpha$-Fe$_2$O$_3$ nanoparticles exhibited photocatalytic degradation activity using visible-light irradiation. Their inclusion in graphene-based nanocomposites displayed superior photodegradation properties when compared with $\alpha$-Fe$_2$O$_3$ nanoparticles alone, due to the presence of the graphene, and it was found that the structural, morphological and optical properties of the nanocomposites were very much dependent on the concentration of graphene.

In summary because of their ability to fully utilize visible light, iron oxide nanomaterials are very efficient nanosorbents for heavy metals and organic pollutants. Indeed, it can be argued that employing these nanomaterials to adsorb contaminants is one of their most striking uses. The work described here will contribute to the commercial development of iron oxide-based nanocomposites, where nanostructures having outstanding properties are required. This will aid in the practical application of iron oxide/graphene-based nanocomposites for water treatment and purification.

## REFERENCES

Afkhami, A., Saber-Tehrani, M. & Bagheri, H. (2010) Modified maghemite nanoparticles as an efficient adsorbent for removing some cationic dyes from aqueous solution. *Desalination*, 263(1–3), 240–248.

Ahmad, A., Rafatullah, M., Sulaiman, O., Ibrahim, M.H., Chii, Y.Y. & Siddique, B.M. (2009) Removal of Cu(II) and Pb(II) ions from aqueous solutions by adsorption on sawdust of meranti wood. *Desalination*, 247(1–3), 636–46.

Akhavan, O. & Azimirad, R. (2009) Photocatalytic property of Fe$_2$O$_3$ nanograin chains coated by TiO$_2$ nanolayer in visible light irradiation. *Applied Catalysis* A: *General*, 369(1–2), 77–82.

Ambashta, R.D. & Sillanpää, M. (2010) Water purification using magnetic assistance: a review. *Journal of Hazardous Materials*, 180(1–3), 38–49.

Bandara, J., Klehm, U. & Kiwi, J. (2007) Raschig rings-Fe$_2$O$_3$ composite photocatalyst activate in the degradation of 4-chlorophenol and Orange II under daylight irradiation. *Applied Catalysis* B: *Environmental*, 76(1–2), 73–81.

Bautista, M.C., Bomati-Miguel, O., del Puerto Morales, M., Serna, C.J. & Veintemillas-Verdaguer, S. (2005) Surface characterisation of dextran-coated iron oxide nanoparticles prepared by laser pyrolysis and coprecipitation. *Journal of Magnetism and Magnetic Materials*, 293(1), 20–27.

Beydoun, D., Amal, R., Low, G. & McEvoy, S. (1999) Role of nanoparticles in photocatalysis. *Journal of Nanoparticle Research*, 1(4), 439–458.

Boyer, C., Whittaker, M.R., Bulmus, V., Liu, J.Q. & Davis, T.P. (2010) The design and utility of polymerstabilized iron-oxide nanoparticles for nanomedicine applications. *NPG Asia Materials – Nature*, 2, 23–30.

Bystrzejewski, M., Pyrzyńska, K., Huczko, A. & Lange, H. (2009) Carbon-encapsulated magnetic nanoparticles as separable and mobile sorbents of heavy metal ions from aqueous solutions. *Carbon*, 47(4), 1201–1204.

Carabante, I., Grahn, M., Holmgren, A., Kumpiene, J. & Hedlund, J. (2009) Adsorption of As (V) on iron oxide nanoparticle films studied by in situ ATR-FTIR spectroscopy. *Colloids and Surfaces* A: *Physicochemical and Engineering Aspects*, 346(1–3), 106–113.

Chan, H.B.S. & Ellis, B.L. (2004) Carbon-encapsulated radioactive 99mTc nanoparticles. *Advanced Materials*, 16, 144–149.

Chen, A., Zeng, G., Chen, G., Fan, J., Zou, Z., Li, H., Hu, X. & Long, F. (2011) Simultaneous cadmium removal and 2,4-dichlorophenol degradation from aqueous solutions by *Phanerochaete chrysosporium*. *Applied Microbiology and Biotechnology*, 91(3), 811–821.

Chong, M.N., Jin, B., Chow, C.W.K. & Saint, C. (2010) Recent developments in photocatalytic water treatment technology: a review. *Water Research*, 44(10), 2997–3027.

Cornel, R.M. & Schwertmann, U. (1996) *The Iron Oxides: Structure, Properties, Reactions, Occurrences and Uses*. Wiley-VCH, Weinheim.

Cundy, A.B., Hopkinson, L. & Whitby, R.L.D. (2008) Use of iron-based technologies in contaminated land and groundwater remediation: a review. *Science of the Total Environment*, 400(1–3), 42–51.

Dastjerdi, R. & Montazer, M. (2010) A review on the application of inorganic nano-structured materials in the modification of textiles: focus on anti-microbial properties. *Colloids and Surfaces* B: *Biointerfaces*, 79(1), 5–18.

Dias, A.M.G.C., Hussain, A., Marcos, A.S. & Roque, A.C.A. (2011) A biotechnological perspective on the application of iron oxide magnetic colloids modified with polysaccharides. *Biotechnology Advances*, 29(1), 142–155.

Dimitrov, D. (2006) Interactions of antibody-conjugated nanoparticles with biological surfaces. *Colloids and Surfaces* A: *Physicochemical and Engineering Aspects*, 282, 8–10.

Fan, F.L., Qin, Z., Bai, J., Rong, W.D., Fan, F.Y. & Tian, W. (2012) Rapid removal of uranium from aqueous solutions using magnetic Fe$_3$O$_4$-SiO$_2$ composite particles. *Journal of Environmental Radioactivity*, 2(106), 40–46.

Fan, H.T., Zhang, T., Xu, X.J. & Lv, N. (2011) Fabrication of N-type Fe$_2$O$_3$ and P-type LaFeO$_3$ nanobelts by electrospinning and determination of gas-sensing properties. *Sensors and Actuators* B: *Chemical*, 153(1), 83–88.

Fatta, K.D., Kalavrouziotis, I.K., Koukoulakis, P.H. & Vasquez, M.I. (2011) The risks associated with wastewater reuse and xenobiotics in the agroecological environment. *Science of the Total Environment*, 409(19), 3555–3563.

Friedrich, K.A., Henglein, F., Stimming, U. & Unkauf, W. (1998) Investigation of Pt particles on gold substrates by IR spectroscopy – particle structure and catalytic activity. *Colloids and Surfaces* A: *Physicochemical and Engineering Aspects*, 134(1–2), 193–206.

Girginova, P.I., Daniel-da-Silva, A.L., Lopes, C.B., Figueira, P., Otero, M., Amaral, V.S., Pereira, E. & Trindade, T. (2010) Silica coated magnetite particles for magnetic removal of Hg$^{2+}$ from water. *Journal of Colloid and Interface Science*, 345(2), 234–240.

Gotić, M., Dražić, G. & Musić, S. (2011) Hydrothermal synthesis of $\alpha$-Fe$_2$O$_3$ nanorings with the help of divalent metal cations, Mn$^{2+}$, Cu$^{2+}$, Zn$^{2+}$ and Ni$^{2+}$. *Journal of Molecular Structure*, 993(1–3), 167–176.

Gross, M. (2001) *Travels to the Nanoworld: Miniature Machinery in Nature and Technology*. Plenum Trade, New York.

Gupta, A.K. & Gupta, M. (2005) Synthesis and surface engineering of iron oxide nanoparticles for biomedical applications. *Biomaterials*, 26(18), 3995–4021.

Hassanjani, R.A., Vaezi, M.R., Shokuhfar, A. & Rajabali, Z. (2011) Synthesis of iron oxide nanoparticles via sonochemical method and their characterization. *Particuology*, 9(1), 95–99.

Hu, H.B., Wang, Z.H. & Pan, L. (2010) Synthesis of monodisperse $Fe_3O_4$-silica core-shell microspheres and their application for removal of heavy metal ions from water. *Journal of Alloys and Compounds*, 492(1–2), 656–661.

Hu, J., Chen, G. & Lo, I. (2005) Removal and recovery of Cr (VI) from wastewater by maghemite nanoparticles. *Water Research*, 39(18), 4528–4536.

Hu, X.L., Li, G.S. & Yu, J.C. (2009) Design, fabrication, and modification of nanostructured semiconductor materials for environmental and energy applications. *Langmuir*, 26(5), 3031–3039.

Huang, D.L., Zeng, G.M., Jiang, X.Y., Feng, C.L., Yu, H.Y. & Huang, G.H. (2006) Bioremediation of Pb-contaminated soil by incubating with *Phanerochaete chrysosporium* and straw. *Journal of Hazardous Materials*, 134(1–3), 268–276.

Huang, D.L., Zeng, G.M., Feng, C.L., Hu, S., Jiang, X.Y. & Tang, L. (2008) Degradation of lead-contaminated lignocellulosic waste by *Phanerochaete chrysosporium* and the reduction of lead toxicity. *Environmental Science and Technology*, 42(13), 4946–4951.

Huang, D.L., Zeng, G.M., Feng, C.L., Hu, S., Zhao, M.H. & Lai, C. (2010) Mycelial growth and solid-state fermentation of lignocellulosic waste by white-rot fungus *Phanerochaete chrysosporium* under lead stress. *Chemosphere*, 81(9), 1091–1097.

Huang, S.H., Liao, M.H. & Chen, D.H. (2003) Direct binding and characterization of lipase onto magnetic nanoparticles. *Biotechnology Progress*, 19(3), 1095–1100.

Iram, M., Guo, C., Guan, Y.P., Ishfaq, A. & Liu, H.Z. (2010) Adsorption and magnetic removal of neutral red dye from aqueous solution using $Fe_3O_4$ hollow nanospheres. *Journal of Hazardous Materials*, 181(1–3), 1039–1050.

Jeong, U., Teng, X., Wang, Y., Yang, H. & Xia, Y. (2007) Superparamagnetic colloids: controlled synthesis and niche applications. *Advanced Materials*, 19, 33–60.

Jiang, X.Y., Zeng, G.M., Huang, D.L., Chen, Y., Liu, F. & Huang, G.H.L. (2006) Remediation of pentachlorophenol-contaminated soil by composting with immobilized *Phanerochaete chrysosporium*. *World Journal of Microbiology and Biotechnology*, 22(9), 909–913.

Khan, S.U.M., Al-Shahry, M. & Ingler, W.B. (2002) Efficient photochemical water splitting by a chemically modified n-$TiO_2$. *Science*, 297(5590), 22–43.

Kim, D., El-Shall, H., Dennis, D. & Morey, T. (2005) Interaction of PLGA nanoparticles with human blood constituents. *Colloids and Surfaces* B: *Biointerfaces*, 40(2), 83–91.

Kobya, M., Demirbas, E., Senturk, E. & Ince, M. (2005) Adsorption of heavy metal ions from aqueous solutions by activated carbon prepared from apricot stone. *Bioresource Technology*, 96(13), 1518–1521.

Kudo, A. & Miseki, Y. (2009) Heterogeneous photocatalyst materials for water splitting. *Chemical Society Reviews*, 38(1), 253–278.

Lee, K.M., Lai, C.W., Ngai, K.S. & Juan, J.C. (2016) Recent developments of zinc oxide based photocatalyst in water treatment technology: a review. *Water Research*, 88, 428–448.

Li, X., Zeng, G.M., Huang, J.H., Zhang, D.M., Shi, L.J., He, S.B. & Ruan, M. (2011) Simultaneous removal of cadmium ions and phenol with MEUF using SDS and mixed surfactants. *Desalination*, 276(1), 136–141.

Li, Y. & Somorjai, G.A. (2010) Nanoscale advances in catalysis and energy applications. *Nano Letters*, 10(7), 2289–2295.

Long, F., Gong, J.L., Zeng, G.M., Chen, L., Wang, X.Y. & Deng, J.H. (2011) Removal of phosphate from aqueous solution by magnetic Fe-Zr binary oxide. *Chemical Engineering Journal*, 171, 448–455.

Luo, L.H., Feng, Q.M., Wang, W.Q. & Zhang, B.L. (2011) $Fe_3O_4$/rectorite composite: preparation, characterization and absorption properties from contaminant contained in aqueous solution. *Advanced Materials Research*, 287, 592–598.

Ma, Z.Y., Guan, Y.P., Liu, X.Q. & Liu, H.Z. (2005) Preparation and characterization of micron-sized nonporous magnetic polymer microspheres with immobilized metal affinity ligands by modified suspension polymerization. *Journal of Applied Polymer Science*, 96(6), 2174–2180.

Machala, J., Zboril, R. & Gedanken, A. (2007) Amorphous iron(III) oxides: a review. *Journal of Physical Chemistry* B, 111, 4003–4018.

McHenry, M.E. & Laughlin, D.E. (2000) Nano-scale materials development for future magnetic applications. *Acta Materialia*, 48(1), 223–238.

Mahdavian, A.R. & Mirrahimi, M.A.S. (2010) Efficient separation of heavy metal cations by anchoring polyacrylic acid on superparamagnetic magnetite nanoparticles through surface modification. *Chemical Engineering Journal*, 159(1–3), 264–271.

Mahmoudi, M., Sant, S., Wang, B., Laurent, S. & Sen, T. (2011) Superparamagnetic iron oxide nanoparticles (SPIONs): development, surface modification and applications in chemotherapy. *Advanced Drug Delivery Reviews*, 63, 24–46.

Mauter, M.S. & Elimelech, M. (2008) Environmental applications of carbon-based nanomaterials. *Environmental Science and Technology*, 42(16), 5843–5859.

Moore, M.N. (2006) Do nanoparticles present ecotoxicological risks for the health of the aquatic environment? *Environment International*, 32(8), 967–976.

Nassar, N.N. (2010) Rapid removal and recovery of Pb(II) from wastewater by magnetic nanoadsorbents. *Journal of Hazardous Materials*, 184, 538–546.

O'Connor, G.A. (1996) Organic compounds in sludge-amended soils and their potential for uptake by crop plants. *Science of the Total Environment*, 185(1), 71–81.

Oller, I., Malato, S. & Sánchez-Pérez, J.A. (2011) Combination of advanced oxidation processes and biological treatments for wastewater decontamination: a review. *Science of the Total Environment*, 409(20), 4141–4166.

Otto, M., Floyd, M. & Bajpai, S. (2008) Nanotechnology for site remediation. *Remediation Journal*, 19(1), 99–108.

Pan, B.J., Qiu, H., Pan, B.C., Nie, G.Z., Xiao, L.L. & Lv, L. (2010) Highly efficient removal of heavy metals by polymer-supported nanosized hydrated Fe(III) oxides: behavior and XPS study. *Water Research*, 44(3), 815–824.

Pang, Y., Zeng, G.M., Tang, L., Zhang, Y., Liu, Y.Y. & Lei, X.X. (2011a) PEI-grafted magnetic porous powder for highly effective adsorption of heavy metal ions. *Desalination*, 281, 278–284.

Pang, Y., Zeng, G.M., Tang, L., Zhang, Y., Liu, Y.Y. & Lei, X.X. (2011b) Cr(VI) reduction by *Pseudomonas aeruginosa* immobilized in a polyvinyl alcohol/sodium alginate matrix containing multi-walled carbon nanotubes. *Bioresource Technology*, 102, 10,733–10,736.

Pang, Y., Zeng, G.M., Tang, L., Zhang, Y., Liu, Y.Y. & Lei, X.X. (2011c) Preparation and application of stability enhanced magnetic nanoparticles for rapid removal of Cr(VI). *Chemical Engineering Journal*, 175, 222–227.

Reddy, P.A.K., Reddy, P.V.L., Kwon, E., Kim, K.H., Akter, T. & Kalagara, S. (2016) Recent advances in photocatalytic treatment of pollutants in aqueous media. *Environment International*, 91, 94–103.

Roco, M.C. (2003) Nanotechnology: convergence with modern biology and medicine. *Current Opinion in Biotechnology*, 14(3), 337–346.

Selvan, S.T., Tan, T.T.Y., Yi, D.K. & Jana, N.R. (2010) Functional and multifunctional nanoparticles for bioimaging and biosensing. *Langmuir*, 26(14), 11,631–11,641.

Shankar, K., Basham, J.I., Allam, N.K., Varghese, O.K., Mor, G.K. & Feng, X.J. (2009) Recent advances in the use of $TiO_2$ nanotube and nanowire arrays for oxidative photoelectrochemistry. *Journal of Physical Chemistry* C, 113, 6327–6359.

Stafiej, A. & Pyrzynska, K. (2007) Adsorption of heavy metal ions with carbon nanotubes. *Separation and Purification Technology*, 58(1), 49–52.

Stone, V., Nowack, B., Baun, A., Van Den Brink, N., Von Der Kammer, F. & Dusinska, M. (2010) Nanomaterials for environmental studies: classification, reference material issues, and strategies for physico-chemical characterisation. *Science of the Total Environment*, 408(7), 1745–1754.

Upadhyayula, V.K.K., Deng, S., Mitchell, M.C. & Smith, G.B. (2009) Application of carbon nanotube technology for removal of contaminants in drinking water: a review. *Science of the Total Environment*, 408(1), 1–13.

Wang, L.B., Ma, W., Xu, L.G., Chen, W., Zhu, Y.Y. & Xu, C.L. (2010) Nanoparticle-based environmental sensors. *Materials Science and Engineering* R: *Reports*, 70(3–6), 265–274.

Watson, S., Beydoun, D. & Amal, R. (2002) Synthesis of a novel magnetic photocatalyst by direct deposition of nanosized $TiO_2$ crystals onto a magnetic core. *Journal of Photochemistry and Photobiology* A: *Chemistry*, 148(1–3), 303–313.

White, B.R., Stackhouse, B.T. & Holcombe, J.A. (2009) Magnetic $\gamma$-$Fe_2O_3$ nanoparticles coated with poly-L-cysteine for chelation of As(III), Cu(II), Cd(II), Ni(II), Pb(II) and Zn(II). *Journal of Hazardous Materials*, 161(2–3), 848–853.

Zelmanov, G. & Semiat, R. (2008) Iron(3) oxide-based nanoparticles as catalysts in advanced organic aqueous oxidation. *Water Research*, 42(1–2), 492–498.

Zeng, G.M., Huang, D.L., Huang, G.H., Hu, T.J., Jiang, X.Y. & Feng, C.L. (2007) Composting of lead-contaminated solid waste with inocula of white-rot fungus. *Bioresource Technology*, 98(2), 320–326.

Zeng, G.M., Li, X., Huang, J.H., Zhang, C., Zhou, C.F. & Niu, J. (2011) Micellar-enhanced ultrafiltration of cadmium and methylene blue in synthetic wastewater using SDS. *Journal of Hazardous Materials*, 185(2–3), 1304–3110.

Zhang, L.D. & Fang, M. (2010) Nanomaterials in pollution trace detection and environmental improvement. *Nano Today*, 5(2), 128–142.

Zhang, L., Li, N., Jiu, H., Qi, G. & Huang, Y. (2015) ZnO-reduced graphene oxide nanocomposites as efficient photocatalysts for photocatalytic reduction of $CO_2$. *Ceramics International*, 41(5), 6256–6262.

Zhang, S.X., Niu, H.Y., Hu, Z.J., Cai, Y.Q. & Shi, Y.L. (2010) Preparation of carbon coated $Fe_3O_4$ nanoparticles and their application for solid-phase extraction of polycyclic aromatic hydrocarbons from environmental water samples. *Journal of Chromatography* A, 1217(29), 4757–4764.

Zhao, X.L., Wang, J.M., Wu, F.C., Wang, T., Cai, Y.Q. & Shi, Y.L. (2010) Removal of fluoride from aqueous media by $Fe_3O_4$-$Al(OH)_3$ magnetic nanoparticles. *Journal of Hazardous Materials*, 173, 102–109.

Zhong, J.Y. & Cao, C.B. (2010) Nearly monodisperse hollow $Fe_2O_3$ nanoovals: synthesis, magnetic property and applications in photocatalysis and gas sensors. *Sensors and Actuators* B: *Chemical*, 145(2), 51–56.

# CHAPTER 5

# Wind technology design and reverse osmosis systems for off-grid and grid-connected applications

Eftihia Tzen, Kyriakos Rossis, Jaime González, Pedro Cabrera,
Baltasar Peñate & Vicente Subiela

## 5.1 INTRODUCTION

Wind turbines can be used to supply electricity or mechanical power to desalination plants for the desalination of brackish water or seawater. Within the last century, with the installation of wind farms, wind energy has been mainly used for grid-connected applications with the purpose of selling electricity to the grid. However, in developing countries, and in some areas of developed countries, the use of wind turbines seems to have been more valuable to human life than just an investment for economic benefits. In rural and remote areas, the use of small wind turbines in mini-grids or stand-alone (off-grid) applications is essential for the improvement of inhabitants' lives and their socioeconomic development. Furthermore, the provision of electricity through renewable energy sources can replace the use of diesel generators, kerosene lamps, etc. and reduce environmental impacts, while also providing an affordable and sustainable solution.

The supply of drinking water has always been an important issue in the economic and social development plans of a country. Continuous efforts brought improvements in the conditions of drinking water supplies, in terms of quality and quantity, both in urban and rural areas.

Desalination of water by thermal or membrane processes is a sustainable and reliable solution for the provision of potable or freshwater in areas where there is a requirement. Several applications have been implemented using wind energy to drive reverse osmosis units and a few to drive mechanical vapor compression and electrodialysis units.

Desalination units driven by renewable energy sources (RES), such as those driven by wind energy, guarantee environment-friendly, cost-effective and sustainable production of desalinated water in those regions with severe potable water shortages that, nevertheless, are fortunate to have renewable energy resources. The combination of these technologies is uniquely suited to providing water and electricity in remote areas where the associated infrastructure is currently lacking. This chapter reviews wind technology, its development, and its utilization for the production of fresh or potable water via desalination systems.

## 5.2 WIND ENERGY TECHNOLOGY OVERVIEW

### 5.2.1 *Wind technology history*

From earliest recorded history, people have harnessed the energy of the wind. An example of this is the use of windmills in countries such as Persia, China and Egypt for grinding grain and draining land. In brief, the history of wind energy can be separated into four overlapping time periods (Schaffarczyk, 2014) as follows:

- *600 to 1890 (the Classical period)*. Within this period the exploitation of wind energy was implemented with the use of classic windmills. These were used as mechanical drives and, in most cases, for pumping water and grinding grain. More than 100,000 such windmills were installed in north-western Europe, while six million similar windmills were constructed in USA between 1889 and the advent of the Second World War.

- *1890–1930.* This period saw the start of the development of wind turbines for the generation of electricity. The first automatically operated wind turbine in the world was designed and built in the USA in 1888 by Charles F. Brush (1849–1929). This 12 kW nominal power wind turbine was equipped with 144 cedar blades having a rotating diameter of 17 m. The turbine was used to charge batteries and supply direct current (DC) to lamps and electric motors (Manwell *et al.*, 2009). Brush was one of the founders of the American electrical industry, and his windmills first gave rise to the term "wind turbines". In 1897, Professor Poul la Cour built one of the first DC wind turbines, which electrolyzed water into hydrogen gas for storage purposes, at Askov in Denmark. La Cour also developed a new wind turbine with a power output of 30 kW at 12 m s$^{-1}$ wind speed. Following on from la Cour's experiments, the Danish manufacturers Lykkegaard and Ferritslev (Fyn) developed commercial versions and the resultant La Cour-Lykkegaard wind turbines had generators in the range of 10 to 35 kW nominal power and a maximum rotor diameter of 20 m. The turbines generated direct current that was fed to small DC grids and batteries. As fuel prices had increased significantly, the development of wind technology in Denmark continued during the First World War. In 1925, the Jacobs Wind Electric Company, started on the plains of Montana, USA, began manufacturing wind generators which sold all over the world (Fig. 5.1a). Jacobs Wind was responsible for bringing electricity and a better standard of living to many, improving the technology of wind energy, and pioneering design concepts, some of which are still in use today. Jacobs Wind Electric Co. Inc. is now the oldest renewable energy company in the US.

- *1930–1960.* This period effectively marked the first phase of significant innovation in wind turbine technology. The necessity of electrification in rural areas and the shortage of energy during the Second World War stimulated new developments. The most important of these took place in Denmark, the US and Germany. During the Second World War, the F.L. Smidth Company in Copenhagen developed wind turbines for electricity generation. In 1940, F.L. Smidth developed a two-bladed DC wind turbine (Fig. 5.1b), and two years later it developed a three-bladed machine. Around 1957, J. Juul used Smidth's three-blade concept to build a stall-controlled 200 kW version in Gedser with a rotor diameter of 24 m. The machine had an asynchronous generator and was connected directly to the grid. An important aspect was the development of movable blade tips to control the rotor and to avoid overspeed in situations where the turbine is disconnected from the grid and the load has vanished. This

(a)    (b)

Figure 5.1.    (a) Marcellus Jacobs on a 2.5 kW machine in the 1940s (http://www.jacobswind.net); (b) two-blade wind turbine of 50 kW by F.L. Smidth, 1941 (Hau, 2006).

Gedser wind turbine became the archetypal 'Danish wind turbine' in a generation of very successful wind turbines that followed the 1973 energy crisis. From the 1950s to 1973, besides Denmark, countries such as the US, Germany, France and the UK also contributed to the further improvement of wind technology.

- *1973 onwards.* The period following 1973 is characterized as the second phase of innovation and the time of mass production. In this period the oil crisis and nuclear environmental problems combined with technological advances to ensure a commercial revolution.

Today the development of the so-called "modern wind turbines", such as variable-speed wind turbines with full power-processing capability, enables their use in a variety of circumstances, for example, into a weak or a stiff grid. Modern wind turbines are available in the market in a wide range of nominal power and are classified according to their class (see Section 5.2.2.2 below). Large wind turbines (of more than 2 MW) can be used in both onshore and offshore applications. Mini and small wind turbines are also efficient and can be used for stand-alone or mini-grid applications. The technology is now mature and it appears the most important potential development is an increase in wind turbine capacity.

### 5.2.2   Wind technology description

#### 5.2.2.1   Technical description of modern wind turbines
The principal subsystems of a typical (land-based) horizontal-axis wind turbine (HAWT) include (Fig. 5.2):

- The rotor, which consists of the blades and the supporting hub.
- The drivetrain, which includes the rotating parts of the wind turbine (exclusive of the rotor), and usually consists of shafts, gearbox, coupling, mechanical brake and generator.
- The nacelle and main frame, including the wind turbine housing, bedplate, and yaw system.
- The tower and the foundation.
- The machine controls.
- The balance of the electrical system, including cables, switchgear, transformers, and possibly electronic power converters.

The rotor blade is an important element of a wind turbine as its shape is vital to the capacity yield. The rotor diameter and rated capacity of wind turbines have continually increased over the past 25 years, driven by technology improvements, better design tools, and the need to expand energy capture and reduce energy cost. Rotor diameters have increased from an average of 20 m in 1985 to more than 120 m today (Schaffarczyk, 2014).

The hub of the wind turbine is the component that connects the blades to the main shaft and ultimately to the rest of the drivetrain. Hubs are generally made of steel, either welded or cast. The hub transmits energy and must withstand all the loads generated by the blades. The amount of energy which the wind transfers to the rotor through the blades depends on the density of the air, the rotor area and the wind speed. In a typical wind turbine, the kinetic energy of the wind is converted to rotational motion by the rotor. Wind speed is an important factor for the amount of energy a wind turbine can convert to electricity. This is indicated by the power curve and the terms *cut-in* and *cut-out* speed, typical technical characteristics of a wind turbine.

The power curve describes the relationship between the wind speed and the power generated by the wind turbine. In general, wind turbines are unable to produce electricity at low wind speeds, below 3 m s$^{-1}$, and will attain maximum output at around 12 m s$^{-1}$ before cut-out at about 25 m s$^{-1}$ (RAE, 2014). Simply, as the wind speed increases to the cut-in speed the turbine begins to operate. Above the cut-in speed the level of electrical output power rises rapidly. Typically, with wind speed somewhere between 12 and 17 m s$^{-1}$, the power output reaches the limit that the electrical generator is capable of. This limit is called the rated power output and the wind speed at which it is reached is called the rated output wind speed. In addition, in cases where wind speed is very high the wind turbine shuts down to protect itself from damage (see Table 5.1 and Table 5.2).

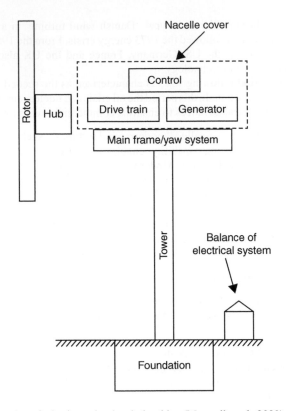

Figure 5.2.   Configuration of a horizontal-axis wind turbine (Manwell *et al.*, 2009).

Table 5.1.   Wind turbine classes -VESTAS wind turbines (Vestas Wind Systems, n.d.).

| Turbine class | IEC I<br>High wind | IEC II<br>Medium wind | IEC III<br>Low wind |
|---|---|---|---|
| Annual average wind speed [m s$^{-1}$] | 10 | 8.5 | 7.5 |
| Extreme 50-year gust [m s$^{-1}$] | 70 | 59.5 | 52.5 |
| Turbulence classes | A 18% | A 18% | A 1.8% |
| | B 16% | B 16% | B 16% |
| Turbine / IEC wind<br>class | IEC I<br>High wind | IEC II<br>Medium wind | IEC III<br>Low wind |
| V90-1.8–2.0 MW | x | x | xx |
| V100-1.8–2MW | x | | |
| V100-2MW | | x | xx |
| V112-3.45 MW | | | x |

x: standard IEC conditions
xx: site dependent

However, cut-out at high wind speed can create problems for the grid system as it occurs more abruptly than cutting in from low wind speeds. Thus, current turbines are being designed to cut out in a more gradual and controlled fashion.

The rotor of a wind turbine turns the main shaft, which transfers the motion into the gearbox. Inside the nacelle, the main (slow) rotating shaft enters the gearbox that greatly increases the

Table 5.2. Technical data of different models of medium wind turbine.

| Technical data | ENERCON E53 | VESTAS V52 | GAMESA G58 | SUZLON S52 |
|---|---|---|---|---|
| Rated power [kW] | 800 | 850 | 850 | 600 |
| Hub height [m] | 60/73 | 36.5/86 | 44/55/65/74 | 75 |
| Tower construction | Conical tubular steel tower | Conical tubular steel tower | Conical tubular steel tower | Lattice tower |
| Cut in wind speed [m s$^{-1}$] | 3 | 4 | 3 | 4 |
| Cut out wind speed [m s$^{-1}$] | 28–34 (with storm control) | 25 | 23 | 25 |
| Rated wind speed [m s$^{-1}$] | 12 | 16 | 12 | 12 |
| Rotor diameter [m] | 52.9 | 52 | 52 | 52 |
| Number of blades | 3 | 3 | 3 | 3 |
| Swept area [m$^2$] | 2198 | 2124 | 2642 | 2124 |
| Blade tip speed [m s$^{-1}$] | No data available | 39.5–85.5 | 44.3–93.5 | 66 |
| Rotor speed, rpm [U/min$^{-1}$] | Variable, 12–28.3 | 14–31.4 | 19.44–30.8 | 24.19 |
| Rotor speed control | Variable | Yes | Variable | Pole-changing |
| Overspeed control | Blade pitch control | Blade pitch control | Blade pitch control | Pitch |
| Generator construction | Synchronous directly driven | Asynchronous | Asynchronous, doubly fed | asynchronous 4-pole |
| Generator cooling system | No data available | Air cooling | Forced convection | Air cooling |
| Grid connection | ENERCON inverter | 850 kW | Directly connected with soft start | IGBT, thyristors |
| Generator voltage [V] | 400 | 690 | 690 | 690/50Hz |
| Gearbox steps | GEARLESS | No data available | 3 | 3 |
| Gear ratio | | 1:62 | 1:61.67 | 1:63.6 |
| Main brake system | Individual brake feathering | Blade pitch control | Blade feathering | Pitch |
| 2nd brake system | Disc brake | Disc brake | Disc brake | Disk brake |
| Turbine class | IEC/NVN IIA | IEC I/II | IEC IIIb/Iia | No data available |
| Distance control | ENERCON SCADA | VESTAS | Gamesa Windnet | SC-PPC Power Plant Controller |

rotational shaft speed. The output (high-speed) shaft is connected to the generator that converts the rotational movement into medium-voltage electricity. The electricity is transferred through heavy electrical cables inside the tower to the transformer, which increases the voltage of the electrical power to the distribution voltage. The electrical power then flows through underground cables to a collection point where the power may be combined with that of other turbines.

Power transformers are important components in any AC power system (Fig. 5.3). Most wind turbine installations include at least one transformer for converting the generated power to the voltage of the local electrical network to which the turbine is connected. In addition, other transformers may be used to obtain voltages of the appropriate level for various auxiliary pieces of equipment at the site (lights, monitoring and control systems, tools, compressors, etc.). Transformers are rated in terms of their apparent power [kVA]. Distribution transformers are typically in the 5–50 kVA range, and may well be larger, depending on the application. Substation transformers are typically between 1000 and 60,000 kVA.

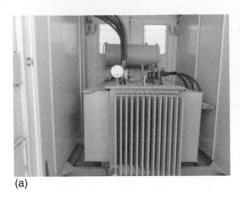

(a)                                                    (b)

Figure 5.3. (a) Medium-voltage/low-voltage substation transformer; (b) cables installation in the low-voltage substation of NEG MICON 750 kW wind turbine at CRES demonstration wind farm, 2015 (Source: Centre for Renewable Energy Sources, CRES, Greece).

The tower of a wind turbine supports the nacelle and the rotor safely under both extreme and fatigue loading, and typically includes the tower's supporting foundation. Wind turbine towers are generally made of steel and can be either tubular or latticed in construction. Most tubular towers have an access door and an internal safety ladder – and/or an elevator in the case of megawatt (MW) wind turbines – to access the nacelle. The principal type of tower design currently in use is the freestanding type, using steel tubes, latticed towers or concrete towers. The tower is usually connected to its supporting foundation by means of a bolted flange connection. For smaller turbines, guyed towers are also used. Tower height is typically one to one-and-a-half times the rotor diameter, but in any case is normally at least 20 m.

The control system for a wind turbine is important in terms of both machine operation and power production. In a wind turbine several controls are used to enable automatic operation, to keep the turbine in alignment with the wind, to engage and disengage the generator, to manage the rotor speed, to protect the turbine from overspeed or damage caused by very strong winds, to sense malfunctions and to warn operators of the need for maintenance or repair. The control system includes the following components:

- Sensors for speed, nacelle direction, position, temperature, current, voltage, etc.
- Controllers for mechanical mechanisms, electrical circuits.
- Power amplifiers, switches, electrical amplifiers, hydraulic pumps and valves.
- Actuators such as motors, pistons, magnets and solenoids.
- Intelligent devices (programmable logic controllers, etc.).

The infrastructure of a wind turbine incorporates the associated civil and electrical works for turbine installation and operation (Fig. 5.4 and Fig. 5.5). The civil works include the construction of roads and drainage, and the foundations of the wind turbine(s) and the meteorological tower ("met mast"), as well as buildings housing electrical switchgear, supervisory control and data acquisition (SCADA) equipment, and potentially spares and maintenance facilities too.

The electrical works are typically composed of the following:

- Equipment at the point of connection (POC), whether owned by the wind farm or by the electricity network operator.
- Underground cable network and/or overhead lines, forming radial "feeder" circuits to strings of wind turbines.
- Electrical switchgear for protection and disconnection of the feeder circuits.
- Transformers and switchgear associated with individual turbines.
- Reactive compensation equipment, if necessary.
- Earth (grounding) electrodes and systems.

(a)                                          (b)

Figure 5.4.   (a), (b) View of foundations for NORDEX N90/2.5MW wind turbines in the region of Peloponnese, Greece, 2012 (Source: CRES).

(a)                                          (b)

Figure 5.5.   (a), (b) Erection of a 20 kW VERGNET wind turbine by CRES at Agios Efstratios island, Greece, 2009 (Source: CRES).

Comprehensive SCADA systems are employed in all commercial wind farms. These are the main systems that control and operate wind turbines and wind farms, and are used for commissioning, operation, troubleshooting and reporting. Moreover, the SCADA system provides the communication network and protocol for information flow between all components of the wind farm. It offers full remote control and supervision of the individual wind turbines as well as the entire wind farm. At its simplest level, the SCADA network connects and controls the wind turbines and enables the collection of production and maintenance data such as provision of real-time monitoring, alarm checking, overview of historical data, etc.

### 5.2.2.2   *Wind turbine classification*

Wind turbines are basically classified by the position of their rotational axis. Thus there is the *horizontal-axis* wind turbine (HAWT) and the *vertical-axis* wind turbine (HAWT). More than two decades of technological progress have resulted in today's wind turbines being state-of-the-art, modern, modular technology-based and rapid to install. Modern wind turbines have improved dramatically in their power rating, efficiency and reliability. Today, the most common design of wind turbine is the HAWT (Fig. 5.6). HAWTs are usually classified according to:

- Rotor orientation (upwind or downwind of the tower).
- Hub design (rigid or teetering hub, hubs for hinged blades).
- Rotor control (pitch control, stall control).
- Number of blades (usually two or three blades).
- Wind alignment (free yaw or active yaw).

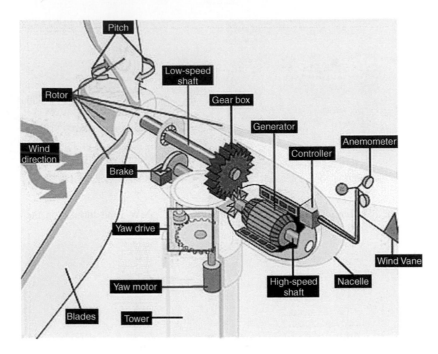

Figure 5.6.    Typical configuration of a horizontal-axis wind turbine (EERE, 2016).

The majority of HAWTs being used today are *upwind* turbines, in which the wind rotors face the wind. The basic advantage of upwind designs is that one avoids the wind shade behind the tower. In terms of the hub design, a *rigid* hub[1] is designed to keep all major parts in a fixed position relative to the main shaft (Fig. 5.7). It must be strong enough to withstand all the loads that can arise from any aerodynamic forces on the blades, as well as dynamically induced loads, such as those due to rotation and yawing. The alternative *teetering* hub is used on nearly all two-bladed wind turbines. This is because a teetering hub can reduce loads due to aerodynamic imbalances or dynamic effects from rotation of the rotor or yawing of the turbine. Teetering hubs are considerably more complex than rigid hubs. As shown in Figure 5.7, a *hinged* hub is in some ways a cross between a rigid and a teetering hub. Basically, it is a rigid hub with 'hinges' for the blades. Examples of this relatively complex hub design include the Lagerwey (Netherlands) small wind turbines, the Lagerwey 18/80 kW and the Lagerwey 30/250 kW.

For rotor control there are two primary strategies: *pitch control* and *stall control* (EWEA, 2009a). The pitch control system is a vital part of the modern wind turbine. This is because the pitch control not only continually regulates the pitch angle of the wind turbine's blade, to enhance the efficiency of wind energy conversion and the stability of power generation, but also serves as a security system in case of high wind speeds or emergency situations. Even in the event of grid power failure, the rotor blades can still be driven into their feathered (edge-on) positions by using either the power of backup batteries or capacitors, or mechanical energy storage devices. Early techniques of active blade pitch control applied hydraulic actuators to control all blades together. However, these pitch control techniques could not completely satisfy all the requirements of blade pitch angle regulation, especially with the increased blade size of MW wind turbines. This

---

[1] The term rigid hub includes hubs in which the blade pitch can be varied, but all other blade motion is not allowed.

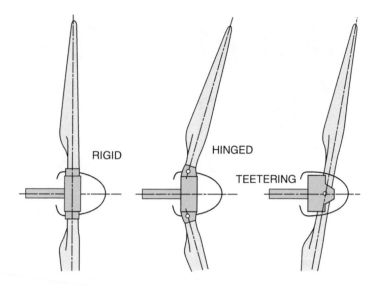

Figure 5.7.   Horizontal-axis wind turbine hub configurations (Gammaenergy, 2016).

is because wind is highly turbulent in flow and wind speed is proportional to height above ground. In today's wind power industry there are primarily two types of blade pitch control systems:

- Hydraulic pitch control.
- Electric pitch control.

A hydraulic pitch control system uses a hydraulic actuator to drive the blade, rotating it with respect to its axial centre line. Electric pitch control systems have been developed as an alternative to hydraulic systems and have higher efficiency (which is usually below 55% in hydraulic systems) and avoid the risk of environmental pollution resulting from spillage or leakage of hydraulic fluid.

Wind turbines are also classified as *fixed* or *variable* pitch, depending on whether the blades embedded in the hub are permanently fixed or not. The inside of the hub houses the hydraulic devices which change the pitch of the blades, if applicable.

Besides pitch control, stall control is another approach for controlling and protecting wind turbines. The concept of stall control is that the power is regulated through stalling (rather than feathering) the blades once the rated speed is achieved. Stall control can be further divided into *passive* and *active* approaches:

- Passive stall control is used in wind turbines in which the blades are bolted to the hub at a fixed installation angle. In a passive stall-regulated wind turbine, the power regulation relies on the aerodynamic features of the blades. In low and moderate wind speeds, the turbine operates near maximum efficiency. At high wind speeds, the turbine is automatically controlled by means of stalled blades, which limit the rotational speed and power output, protecting the turbine from excessive wind speeds.
- The active stall control technique has been developed for large wind turbines. An active stall wind turbine has stalling blades together with a blade pitch control system. Since the blades at high wind speeds are turned towards stall (flat-on), effectively in the opposite direction to that with pitch control systems, this control method is also referred to as negative pitch control.

Most modern wind turbines have three blades. Twin-bladed machines did not prevail, partly because they require higher rotational speed to yield the same energy output. This is a disadvantage both in regard to noise and visual intrusion. One-bladed machines have the same problems as two-bladed ones and they also require a counterweight to be placed on the other side of the hub

(a)

(b)

Figure 5.8.   (a), (b) Replacement of the gearbox of 750 kW NEG MICON wind turbine, at CRES demonstration wind farm, 2007 (Source: CRES).

from the rotor blade in order to balance the rotor. Despite these disadvantages, both single- and double-bladed types of wind turbine are characterized by less weight.

In modern wind turbines, the *active yaw* control is effected by electric motors. The yaw control system usually consists of an electrical motor with a speed-reducing gearbox, a bull gear which is fixed to the tower, a wind vane to gain the information about wind direction, a yaw deck, and a brake to lock the turbine securely in alignment when the required position is reached. For a large wind turbine with high driving loads, the yaw control system may use two or more yaw motors working together to drive a heavy nacelle. An active yaw drive, always used with upwind wind turbines and sometimes with downwind ones, contains one or more yaw motors, each of which drives a pinion gear against a bull gear attached to the yaw bearing. This mechanism is controlled by an automatic yaw control system with its wind direction sensor usually mounted on the nacelle of the wind turbine. Occasionally, yaw brakes are used with this type of design to hold the nacelle in position when it is not yawing. *Free yaw* systems (meaning that they can self-align with the wind) are often used on downwind wind turbines.

Furthermore, according to the drive train condition, wind turbines can be classified into either the *direct drive* or *geared drive* group. To increase the generator rotor rotating speed and gain a higher power output, a regular geared-drive wind turbine typically uses a multi-stage gearbox to take the rotational speed from the low-speed shaft of the blade rotor and transform it into a fast rotation on the high-speed shaft of the generator rotor (Tong, 2010). The advantages of geared generator systems include lower cost and smaller size and weight. However, utilization of a gearbox can significantly lower wind turbine reliability and increase turbine noise levels and mechanical losses. By eliminating the multi-stage gearbox from a generator system, the generator shaft is directly connected to the blade rotor (Fig. 5.8). Thus, this direct-drive concept is superior in terms of energy efficiency, reliability and design simplicity.

Another important parameter that characterizes a wind turbine is the type of electrical machine[2] used. The electrical machines most commonly encountered in wind turbines are those acting as generators, of which the two most common types are *asynchronous* generators (induction generators) and *synchronous* generators. In addition, some smaller turbines use DC generators (Tzen and Mouzakis, 2013).

*Synchronous machines* are used as generators in large central station power plants. In wind turbine applications they are used occasionally on large grid-connected turbines, or in conjunction with power electronic converters in variable-speed wind turbines. A type of synchronous machine

---

[2] In general, generators convert mechanical power to electrical power; motors convert electrical power to mechanical power. Both generators and motors are frequently referred to as electrical machines, because they can usually be run as one or the other.

using permanent magnets is also used in some stand-alone wind turbines. In this case the output is often rectified to DC before the power is delivered to the end load. Synchronous machines may be used as a means of voltage control and a source of reactive power in autonomous AC networks.

The advantages of the synchronous machine, as reported in the literature, are:

- Reactive power control through the excitation circuit.
- Suitable for stand-alone systems.

The disadvantages are:

- High installation and maintenance costs.
- High complexity of grid synchronization circuit.
- Stiffness rotor speed variation: torsional fatigue.

*Asynchronous* or *induction machines* are commonly used for motors in most industrial and commercial applications and are popular because they have a simple and rugged construction. In addition, asynchronous machines are relatively inexpensive, and they may be connected and disconnected from the grid relatively simply. The stator on an asynchronous machine consists of multiple windings, similar to that of a synchronous machine. The rotor in the most common type of induction machine has no windings. Asynchronous machines require an external source of reactive power and an external constant frequency source to control the speed of rotation. For these reasons, they are most commonly connected to a larger electrical network. In these networks, synchronous generators connected to prime movers with speed governors ultimately set the grid frequency and supply the required reactive power. These designs entail a constant or nearly constant rotational speed when the generator is directly connected to a utility network. In cases where the generator is used with power electronic converters, the turbine will be able to operate at variable speed. Furthermore, numerous wind turbines installed in grid-connected applications use squirrel cage induction generators (SQIGs). A SQIG operates within a narrow range of speeds slightly higher than its synchronous speed (a four-pole generator operating in a 60 Hz grid has a synchronous speed of 1800 rpm). The main advantages of this type of induction generator are that it is rugged, of relatively low cost, and easy to connect to an electrical network. An increasingly popular option today is the doubly fed induction generator (DFIG). DFIGs are often used in variable-speed applications (Ekanayake *et al.*, 2003).

A gradually accepted option for utility-scale electrical power generation is the variable-speed wind turbine. There are a number of benefits that such a configuration offers, including the reduction of wear and tear on the wind turbine, and the potential operation of the wind turbine at maximum efficiency over a wide range of wind speeds, yielding increased energy capture. Although there are a large number of potential hardware options for variable-speed operation of wind turbines, power electronic components are used in most variable-speed machines currently being designed. When used with suitable power electronic converters, both synchronous and asynchronous generators of either type can run at variable speed.

The speed of the asynchronous generator will vary with the turning force (moment or torque) applied to it. In practice, the difference between the rotational speed at peak power and at idle is very small, about 1%. It is a very useful mechanical property that the generator will increase or decrease its speed slightly if the torque varies. This means that there will be less wear and tear on the gearbox. The advantages of asynchronous machines are:

- Simple and robust construction.
- Low construction and maintenance costs.
- Massive production.
- High reliability.
- Synchronizing unit not required for grid connection.
- Small capability of speed control by controlling slip.

Some of the disadvantages of asynchronous machines are:

- Need for reactive power from external source.
- No possibility of voltage control.
- Grid connection necessary (cannot be used in stand-alone operation).
- A gearbox is used.

### 5.2.2.3    *Wind turbine capacity*

Though an absolute definition of wind turbines is not available, especially for small sizes, they can be divided into a number of broad categories on the basis of their rated capacities: pico, micro, small, medium, and large wind turbines. Small- and medium-size wind turbines can be used for stand-alone applications, or they can be connected to a utility power grid (RenewableUK, 2011). They can also be combined with a photovoltaic (PV) system or a diesel generating set ("genset") and energy storage systems (e.g. batteries), thus forming "hybrid" systems, which are typically used in remote locations where connection to a utility grid is not available (off-grid areas). Large wind turbines can be used for onshore or offshore applications.

With regard to costs, especially of small wind turbines, a large range of prices have been observed, which are mainly derived from parameters such as the country of origin, construction materials, technology, etc. Medium and large wind turbines also exhibit some differences in their total costs, which are more dependent on installation costs than on the cost of procuring the wind turbine itself. Thus, wind turbines can be categorized as follows:

- Pico wind turbine (PWT) usually refers to wind turbines with capacities from a few Watts to 1.5 kW. The cost of these generators typically ranges from 1500 to 2000 € kW$^{-1}$. In terms of their technical characteristics, as an example, the rotor diameter of a wind turbine of 600 W could be 2 m, and for a 1.5 kW nominal power wind turbine it could be around 3 m. For both wind turbines the height of the mast could range from 7 to 12 m, their cut-in-speed would be around 3.5 m s$^{-1}$ and the rated speed[3] around 11–12 m s$^{-1}$. PWTs are usually used in off-grid applications to cover small loads.
- Micro wind turbine (MWT) generally refers to wind turbines with a nominal power between 1.5 and 10 kW MWTs are especially suitable in locations where the electrical grid is not available. They can be used on a per-structure basis, such as for street lighting, water pumping, and for residents at remote off-grid areas, particularly in developing countries. The rotor diameter of a 6 kW MWT is around 4 m, the mast has a height of 7 to 12 m, the cut-in speed is about 3.5 m s$^{-1}$ and the nominal speed around 12 m s$^{-1}$. For a 10 kW wind turbine the rotor diameter is about 7 m and the mast height could range from 18 to 30 m. The cost of these wind turbines is typically in the range of 2000 to 2500 € kW$^{-1}$.
- Small wind turbine (SWT) usually refers to turbines with output power between 20 and 100 kW (Fig. 5.9). Most SWTs are used for off-grid applications such as residential homes, farms and telecommunications (AWEA, 2012). Over the past decade, there has been a dramatic increase in the number of telecom installations in off-grid regions around the world. The power of these installations relied on diesel fuel. In several countries the diesel gensets for telecom are replaced by renewable power supplies, SWTs or photovoltaics. Moreover, off-grid wind turbines are usually used in conjunction with batteries, diesel gensets and photovoltaic systems to improve the stability of wind-based power supply. However, this can increase the initial cost of a system and make it more complicated. Wind turbines of more than 50 kW nominal power are usually used for on-grid applications. The rotor diameter of a 20 kW wind turbine is around 10 m, the mast height could range from 18 to 30 m, the cut-in speed would be around 4.5 m s$^{-1}$ and the nominal speed around 16 m s$^{-1}$. In general, the cost of SWTs of 50 to 100 kW will range from 2000 to 2500 € kW$^{-1}$.
- Medium wind turbines have a nominal power of more than 100 kW and less than 1 MW (see Table 5.2 for examples). The typical rotor diameter of an 850 kW medium wind turbine is 52 m, the hub height is between 44 and 86 m, the cut-in speed 3.0–4.0 m s$^{-1}$, and the rated wind speed

---

[3] Rated speed: the speed at which the power output reaches its maximum value.

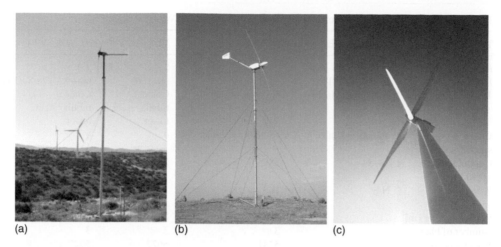

(a)                                    (b)                                    (c)

Figure 5.9.   Various capacities of wind turbines: (a) 900 W Whisper wind turbine installed to provide electricity to a desalination unit at CRES demonstration wind farm in Keratea, Greece, 2004, (b) 20 kW VERGNET wind turbine installed by CRES in Agios Efstratios island, Greece, 2009; (c) 3.3 MW V112 wind turbine in Agios Giorgios island, 2016 (Source: Tzen and Rossis, CRES).

in the range 12–16 m s$^{-1}$. The power output costs of medium wind turbines typically range from 1000 to 1400 € kW$^{-1}$ depending on the country of installation and the size of the wind farm (see Section 5.2.5 below for more information).

- Large wind turbine (LWT) usually refers to wind turbines from 1 up to 10 MW, designed for onshore and offshore applications (see Table 5.3 for examples). A 3 MW wind turbine would typically have a rotor diameter of about 90 m, and a hub height of 65, 80 or 105 m, depending on the wind class[4]. The cut-in speed would be around 3.5 m s$^{-1}$, cut-out speed about 25 m s$^{-1}$ and the rated wind speed would be about 15 m s$^{-1}$.

Turbine classes, and an example of Vestas (Denmark) wind turbine classes, are presented in Table 5.1, while Table 5.2 and Table 5.3 illustrate the main technical characteristics of several types of wind turbines made by significant EU manufacturers.

After several decades of experience with onshore wind technology, offshore wind technology has presently become the focus of the wind power industry. Offshore wind turbines (Fig. 5.10) have been developed faster than onshore ones since the 1990s due to the excellent offshore wind resources available, in terms of wind power intensity and continuity.

A wind turbine installed offshore can generate higher power output and operate for more hours each year compared with the same turbine installed onshore. Due to an absence of obstacles, wind speeds over offshore sea levels are typically 20% higher than those over nearby land. Several structures for the installation of offshore wind turbines (e.g. monopile, gravity-based, etc.) have been studied and implemented for relatively low water depths.

However, more work has to be done on the structures for deeper water structures. For example, this is the primary reason for the absence of offshore wind farms in the deep waters of the Mediterranean (water depths of more than 50 m).

---

[4] IEC classification of turbines: specification of classes of wind turbines (class IA, class IB, etc.) in order to determine which turbine is suitable for the wind conditions of a particular site. Wind classes are mainly defined by the average annual wind speed (measured at the turbine's hub height), the speed of extreme gusts that could occur in a 50-year period, and the degree of turbulence at the site.

Table 5.3.   Technical data of different models of large wind turbine (LWT).

| Technical data | ENERCON E70 | ENERCON E82 | VESTAS V80 | VESTAS V90 |
|---|---|---|---|---|
| Rated power [MW] | 2.5 | 2.0 | 2.0 | 2.0 |
| Hub height [m] | 57/64/85/98/113 | 78–138 | 60–100 | 60/65/67/78/80/100 |
| Tower construction | No data available | No data available | Conical tubular steel tower | Conical tubular steel tower |
| Cut in wind speed [m s$^{-1}$] | no data available | No data available | 4 | 4 |
| Cut out wind speed [m s$^{-1}$] | 28–34 | 28–34 | 25 | 25 |
| Rated wind speed [m s$^{-1}$] | N/S | N/S | 16 | 16 |
| Rotor diameter [m] | 71 | 82 | 80 | 80 |
| Number of blades | 3 | 3 | 3 | 3 |
| Swept area, m$^2$ | 3959 | 5281 | 5027 | 5027 |
| Blade tip speed [m s$^{-1}$] | No data available | No data available | 45.2–80 | 45.2–80 |
| Rotor speed, rpm, [U min$^{-1}$] | 6–21.5 | 6–18 | 10.8–19.1 | 10.8–19.1 |
| Rotor speed control | Variable | Variable | Yes | Yes |
| Overspeed control | Blade pitch control | Blade pitch control | Hydraulic blade pitch | Blade pitch control |
| Generator construction | Directly driven | Directly driven | Asynchronous | Asynchronous |
| Generator cooling system | No data available | No data available | Air cooling | Liquid cooling comb |
| Grid connection | ENERCON converter | ENERCON converter | 2 MW | 2 MW |
| Generator voltage [V] | No data available | No data available | 690 | 690 |
| Gearbox steps | Gearless | gearless | 3 | 3 |
| Gear ratio | – | – | 02:40.5 | 1:100.5 |
| Main brake system | Individual blade pitch control | Individual blade pitch control | Blade feathering | Individual blade pitch |
| 2nd Brake System | Pitch control | Pitch control | Disc brake | Disk brake |
| Technical data | ENERCON E70 | ENERCON E82 | VESTAS V80 | VESTAS V90 |
| Rated power [MW] | 2.5 | 2.0 | 2.0 | 2.0 |

(a)   (b)

Figure 5.10.   View of offshore wind turbines: (a) 325 MW Thornton Bank offshore wind farm, (b) 5 MW offshore wind turbine, in Scotland, 126 m tall, 45 m depth (Source: Senvion SE (formerly REpower Systems SE).

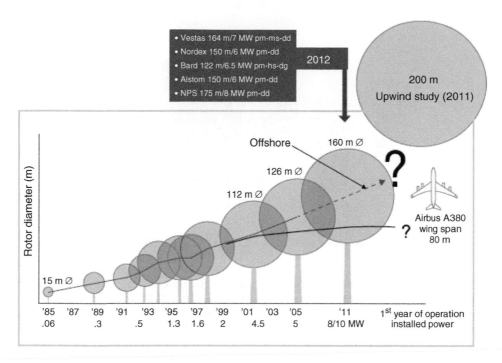

Figure 5.11.   Enlargement trend in modern wind turbines (Schaffarczyk, 2014).

### 5.2.3 *Wind technology progress*

In recent years, megawatt wind turbines have formed the mainstream of the international wind power market (Fig. 5.11). According to the European Wind Energy Association Annual Statistics 2014 (EWEA, 2015), a total of around 129 GW have been installed in the European Union, an increase in installed cumulative capacity of 9.7% compared to the previous year. Germany (39.1 MW) remains the EU country with the largest installed capacity, followed by Spain (22.9 MW), the UK (12.8 MW) and France (9.2 MW). Based on 2014 data from the International Energy Association (IEA, 2015), a total of around 140 MW have been installed in Europe as a whole, while 371 GW of wind plants were operating globally at the end of 2014. Worldwide, China and the US have the greatest capacity share of installed wind with totals for 2014 of 114.59 MW and 65.87 MW respectively. Furthermore, it is estimated that at the end of 2014 the electricity produced by the installed wind power plants in the EU was around 265 TWh, enough to fulfill 9.5% of the EU's electricity consumption. About 1% of EU production is from offshore wind farms.

### 5.2.4 *Wind market share*

According to the available data, EU wind turbine manufacturers have the largest share of the global wind market, followed by China and the US. Figure 5.12 presents the wind industry market share according to the installations in 2014.

### 5.2.5 *Cost of wind energy*

The economics of a wind scheme depend upon technical, resource and cost parameters. The latter two of these vary from country to country. In general, the cost of the equipment represents the largest share of the total investment (around 75%). Thus, according to the International Energy Association (IEA, 2015), the cost of a wind turbine in Austria is estimated at 1180 € kW$^{-1}$ within a total project cost of around 1380 € kW$^{-1}$, while in Spain the cost of a turbine is estimated at approximately 700 € kW$^{-1}$ amid a total project cost of 1100 € kW$^{-1}$.

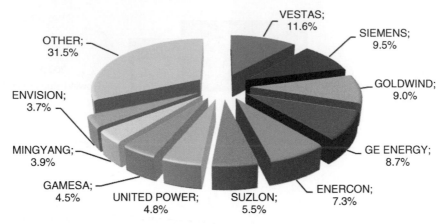

Figure 5.12.    Global wind installation market shares for 2014 (REN21, 2015).

The total installed project cost includes the costs of turbines, civil works, roads, electrical equipment, installation, development and grid connection (EWEA 2009b). Operation and maintenance costs for onshore wind energy are generally estimated to be around 1.2 to 1.5 c€ kWh$^{-1}$ of wind power produced over the total lifetime of a turbine. The levelized cost of electricity[5] (LCOE), in euro cents per kWh of wind-generated power, is calculated as a function of the wind speed (number of full load hours) and the discount rate. The LCOE varies from approximately 8 to 10.5 c€ kWh$^{-1}$ at sites with low average wind speeds, to about 5 to 6.5 c€ kWh$^{-1}$ at windy coastal sites, and to approximately 7.5 c€ kWh$^{-1}$ at a wind site with average wind speeds[6].

Offshore capital costs are much higher than those onshore, mainly due to the larger structures and the complex logistics of installing the towers. For example, offshore wind turbines are generally 20% more expensive and towers and foundations cost more than 2.5 times the price of those for a similar onshore project. On average, investment costs for a new offshore wind farm are expected to be in the range of 2200 to 2500 € kW$^{-1}$ for a nearshore shallow water facility (Rossis, 2015).

In several cases the LCOE from wind is comparable with that from coal-fired generation. In China, given current technology and without considering the cost of long-distance transmission or the resource and environmental benefits of wind power, the cost of wind power is higher than that of coal-fired power by 0.027 € kWh$^{-1}$. If resources and environmental benefits are taken into consideration, the cost of wind power would be almost equal to that of coal-fired power generation.

## 5.3    WIND DESALINATION COMBINATIONS

In recent years there has been prolonged discussion of the environmental impacts of wind turbines and much less on their environmental benefits as they reduce fossil fuel consumption by the power sector. Wind turbines can cooperate with energy-intensive technologies, such as desalination, in an off-grid or on-grid operation, reducing the $CO_2$ footprint of such technologies and creating an almost 100% environmentally friendly technology. The selection of the most suitable technological combination of renewable energy sources with desalination is an important factor in the success of a project (Tzen, 2008).

Wind turbines to drive reverse osmosis (RO) units are the second most-used combination, following that of photovoltaic RO systems (Cipollina et al., 2014). In addition, a small number of

---

[5] The LCOE is the net present value of the unit cost of electricity over the lifetime of a generating asset.
[6] The installed cost of wind turbines is assumed to be 1225€ kW$^{-1}$.

applications exist around the combination of wind technology with mechanical vapor compression (MVC) and electrodialysis (ED/EDR) desalination technology (Tzen, 2009).

The major drawback to the combination of RO desalination processes with wind energy is the fluctuation in power supply generated by the wind turbines. In the following paragraphs an analytical study of the alternatives for energy transfer and the energy fluctuation limitations between the two technologies is presented.

In general, desalination systems were traditionally designed to operate with a constant power input; an unpredictable or non-steady power input forces the desalination plant to operate in sub-optimal conditions. Up until now, only a few stand-alone, battery-less, wind-powered RO units have been developed and studied. To avoid the fluctuations inherent in renewable energies, different energy storage systems may be used. A battery system is typically used for medium-term storage, whereas flywheels and supercapacitors can be used for short-term storage. However, the cost of batteries is still high and affects the final unit cost of water in off-grid (stand-alone) applications. The affordability of the water produced and the sustainability of the renewable energy desalination system are major factors in the success of a project and should be examined with particular attention.

Another important factor, especially in off-grid applications, is the monitoring control and full automation of the system. In both grid and off-grid systems, a special energy management system should be designed. The wind potential and siting issues are major technical parameters for the design and installation of a wind turbine or a wind farm. The capacity of the desalination system should be defined according to the energy availability, the energy and water storage capability and the fresh or potable water requirements. The final decision, with the exception of any technical issues, will be determined according to the available budget.

The decision between off-grid or on-grid application is obvious. Off-grid projects involve remote and rural areas without any access to the electricity grid. On-grid applications involve mostly larger units where the energy requirements of the desalination units can be covered by a wind turbine or a wind farm. This technique is applied in areas where the electricity grid is available. The advantage of these wind desalination applications is that the desalination unit can still have a continuous and dependable operation. Any energy excess from the wind turbine or wind farm can be sold back to the electricity grid.

### 5.3.1   *Wind and MVC combination*

With regards to the wind-MVC combination, only a few applications are known. A pilot plant was installed in 1991 at Borkum island in Germany, where a wind turbine with a nominal power of 45 kW was coupled to a 48 m$^3$ day$^{-1}$ MVC unit. The experience was followed in 1995 by a larger plant at the island of Rügen, where a wind turbine with nominal power of 300 kW was coupled to a 300 m$^3$ day$^{-1}$ MVC unit. In 1999, a 50 m$^3$ day$^{-1}$ wind-MVC unit was installed within the SDAWES project at Gran Canaria in Spain, and in 2008 a wind-MVC plant was installed on a remote Greek island (Symi) within the Operational Program for Competitiveness (OPC) of the Greek Ministry of Development. A significant disadvantage of the combination is the high energy requirement of MVC. The specific energy consumption of MVC units ranges from 9 to 20 kWh m$^{-3}$. Furthermore, in most cases the wind turbine is connected directly to the desalination unit. Due to the intermittent operation of the MVC unit because of the variability of the wind, scaling problems have been anticipated (Subiela *et al.*, 2009).

### 5.3.2   *Wind and ED/EDR combination*

In terms of wind-electrodialysis (ED) combination, little data has been reported. A small number of wind-electrodialysis reversal (EDR) units have been installed in the last twenty years. The best-known wind-EDR system was installed and tested within the SDAWES project (Subiela *et al.*, 2009). This EDR unit had a capacity ranging from 3 to about 8 m$^3$ h$^{-1}$ and an average specific energy consumption of 3.3 kWh m$^{-3}$. The electrical conductivity of the brackish feed water ranged from 2500 to 7500 $\mu S \, cm^{-1}$, while the conductivity of the water produced ranged between 200 and 500 $\mu S \, cm^{-1}$. In the EDR process the energy requirements are directly proportional to the quantity of salts removed, and thus the technology is attractive mainly for the desalination of brackish water. The SDAWES wind-EDR system was tested by the Technological Institute of the Canary Islands (ITC) and the University of Las Palmas in both on-grid and off-grid operation. According to the researchers, the system operation was adequate and flexible, as well as showing smooth adaptation to variations in wind power, even when sudden falls or peaks occurred.

### 5.3.3    *Wind and RO combination*

RO is the most widely used process for seawater desalination, and involves the forced passage of water through a membrane against natural osmotic pressure to accomplish separation of water and ions. The amount of desalinated water that can be obtained ranges between 30% and 75% of the input water volume, depending on the initial water quality (brackish or seawater), the quality of the product needed, and the technology and membranes involved.

During recent decades, the energy requirements for the operation of seawater RO units have been dramatically reduced with the use of energy recovery devices (ERDs). RO plants use ERDs to recover energy from the pressurized reject brine, thereby improving the overall efficiency of the system. ERDs use increases the initial cost of the system but effectively reduces its energy requirement. Current RO plants utilize isobaric pressure exchanger devices, which can reduce the energy consumption of the desalination rack below 3 $kWh \, m^{-3}$. For large- and medium-scale RO plants, there are several pressure exchangers on the market which can achieve specific energy consumptions between 2.5 and 3.5 $kWh \, m^{-3}$. Moreover, RO plants of larger capacity (above 1000 $m^3 \, day^{-1}$) could achieve extremely low energy consumption rates – below 2.5 $kWh \, m^{-3}$ – through the use of ERDs together with the use of the latest generation of membranes and high-efficiency high-pressure pumps. For smaller RO capacities, the ERD opportunities are more limited and the efficiency benefits lower. Nevertheless, for a 20 $m^3 \, day^{-1}$ seawater RO unit the energy requirements could be reduced to less than 3.0 $kWh \, m^{-3}$ with the use of an ERD.

Other advantages of the RO process are the modular design and compactness of the plants, satisfactory performance in all sizes, and easy operation. As regards the matching of wind energy to RO, several units have been designed and tested. Off-grid systems are mainly small in scale. Off-grid wind energy RO systems have also been used in conjunction with other conventional or renewable power sources (e.g. diesel genset, photovoltaic); these are known as hybrid systems. The majority of small-scale installations have so far been developed within research projects. In general, the aim of each project was the evaluation of the combination of the technologies and the development of compact and reliable systems. Table 5.4 presents several European off-grid and on-grid implementations of wind-RO systems (Tzen, 2012).

Within one of these, the Punta Jandía project, a wind-diesel system was installed in an isolated fishing village community on the island of Fuerteventura in the Canarian archipelago (Carta *et al.*, 2003a). The project was implemented with the aim of meeting the complete energy requirements of the community: street lighting and domestic consumption, desalination plant, freezer plant, sewage water purifier, hydro compressor for the supply of potable water and a winch for small vessels. In terms of the quality of the service supplied, the percentage of wind penetration in the system, the fuel savings and the decrease in $CO_2$ emissions, the results obtained led to the conclusion that, from a technical point of view, the developed system supplied all of the service needs of the community on a regular basis, giving an acceptable level of energy quality and a substantial improvement in the quality of the environment.

ITC also participated in the DeReDes project, "Development of Desalination driven by Renewable Energies", carried out during 2006 and 2007. The overall objective of the project was to boost the development of the market in desalination driven by renewable energies in order to find new solutions that reduced the negative effects associated with the energy consumption of desalination processes. During the project, a technical and economic assessment of different technologies was conducted. This was based on different criteria, such as the status of development or its adaptation

Table 5.4. Wind energy-driven reverse osmosis desalination plants.

| Location | RO capacity [m³ h⁻¹] | Power supply | Year of installation |
|---|---|---|---|
| Ile du Planier, France | 0.5 | 4 kW WT | 1982 |
| Island of Suderoog, Germany | 0.25–0.37 | 6 kW WT | 1983 |
| Island of Heligoland, Germany | 40 | 1.2 MW WT, diesel genset | 1988 |
| Punta Jandía, Fuerteventura, Spain | 2.3 | 225 kW WT, 160 KVA diesel genset, flywheel | 1995 |
| Pozo Izquierdo, Spain (SDAWES project) | 8 × 1.0 | 2 × 230 kW WT | 1995 |
| Therasia Island, Greece (APAS RENA project) | 0.2 | 15 kW WT, 440 Ah batteries | 1995/6 |
| Tenerife, Spain (JOULE project) | 2.5–4.5 | 30 kW WT | 1997/8 |
| Syros island, Greece (JOULE project) | 2.5–37.5 | 500 kW WT, stand-alone and grid-connected | 1998 |
| Keratea, Attiki, Greece (PAVET Project) | 0.13 | 900 W WT, 4 kWp PV, batteries | 2001/2 |
| Pozo Izquierdo, Spain, (AEROGEDESA project) | 0.80 | 15 kW WT, 190Ah batteries | 2003/4 |
| Loughborough University, UK | 0.5 | 2.5 kW WT, no batteries | 2001/2 |
| Milos island, Greece (Operational Programme for Competitiveness project) | 2 × 42 | 850 kW WT, grid-connected | 2007 |
| Delft University, The Netherlands | 0.2–0.4 | Windmill, no batteries | 2007/08 |

(a)　　　　　　　　　　　　　　　　　(b)

Figure 5.13. (a) 0.1–0.8 m³ h⁻¹ variable capacity SWRO plant in operation within SODAMEE project; (b) Installation of the 15 kW wind turbine without a battery system (Source: SODAMEE project – ITC-ULPGC, 2015).

to the needs of three scenarios defined with varying water demands and resource availabilities in order to evaluate these technologies and their implementation. It was concluded that a few technologies were sufficiently mature to pass into commercial application (wind-RO and PV-RO), and some others (solar multi-effect distillation and solar membrane distillation) could be viable with further technical development or with subsidies that made them competitive in relation to conventional energy costs. The DeReDes project provided a general view, but because each location has different characteristics it is necessary to evaluate water demands on the basis of quantity and quality, as well as renewable energy potential.

A more recent project implemented by ITC is the SODAMEE project in Pozo Izquierdo in Spain, in which a 0.1–0.8 m³ h⁻¹ (variable capacity) seawater RO unit (SWRO) operates with the use of a 15 kW wind turbine without a battery system and under variable capacity control (Fig. 5.13). The project started in 2009 and ended in December 2015 (Carta *et al.*, 2015).

One of the most important conclusions drawn from the studies undertaken in this project is the feasibility of adapting the consumption of the SWRO desalination plant prototype to widely varying (simulated) power generated by wind turbine (WT). Despite using a variety of time intervals in which it was assumed that the WT output power remained constant, a perfect fit could not be obtained between the theoretical WT-generated power and the power consumed by the SWRO desalination plant. Due to oscillations in the operating parameters and inertia in the desalination system, it was not possible to maintain the permeate recovery rate at the constant target value at all times. Given the slowness of the control system, as well as the response times inherent to the desalination elements, a dynamic regulation device needs to be incorporated to achieve the goal of instantaneous adaptation of the SWRO plant's power consumption to the variable WT-generated power.

### 5.3.3.1   *Off-grid wind reverse osmosis*
In general, a stand-alone RO desalination system using renewable energy sources can be represented in terms of the following blocks or subsystems (Fig. 5.14):

- Electricity generation from renewable energy sources (wind energy in this case).
- Electricity generation from non-renewable energy sources (as backup in the absence of renewable energy sources).
- Renewable energy storage systems (non-massive storage).
- Consumption based fundamentally on seawater desalination.
- Possibility for conventional grid connection (if feasible).

The interconnection of the different blocks that structure the system depends on several factors including the technology used in each block which, in turn, depends among other factors on the energy transfer at each moment and between each subsystem.

Analytically speaking, a typical off-grid battery-bank wind-RO system consists of the following equipment:

- Wind generator.
- Charge controller.
- Battery bank.
- Inverter.
- SWRO unit.

In the case of off-grid operation with a battery bank, the batteries are used for power stability and as an energy supply during periods when wind energy is not sufficient to drive the desalination unit. Charge controllers are used to protect batteries from overcharging, and inverters are used to convert the DC current from the battery output to AC for the load. A diesel genset can also be used as backup to charge the battery bank or to drive the SWRO unit directly. The way in which wind energy is transferred between the different subsystems is of fundamental importance. There are two basic options:

- DC energy transfer.
- AC energy transfer.

With regard to DC energy transfer, one idea that emerged when stand-alone systems in micro-grids were first studied was to simplify the system with DC interconnection of the different subsystems. In this way, the variable to be controlled was the voltage, which in turn controls the flow of current and, consequently, the transfer of active power between subsystems. The option of using a DC machine as generator in the generation subsystem had been discarded because of the problems of voltage control and high cost, as well as the problem of supporting the weight at such a height in the wind turbine. However, the advantage of the DC machine lies in the proportionality

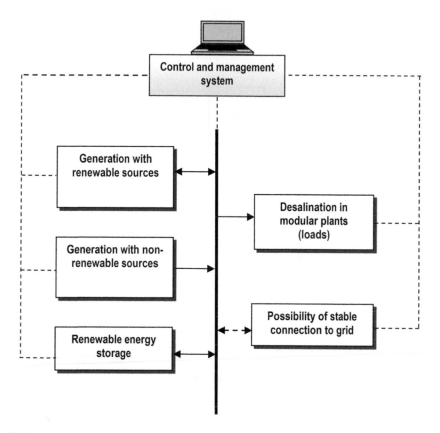

Figure 5.14.   Microgrid subsystems.

between the voltage applied in the armature and its speed in maintaining constant torque. This makes it a valid option in other subsystems, for example, driving the pump in SWRO desalination, or in electrodialysis desalination plants. Figure 5.15 presents a simple wind desalination system in which the water injection pump in the RO unit is DC-driven, with the current being controlled by a DC/DC converter which could even allow the operation of the pump under variable pressure and flow regimes, depending on energy availability. The wind turbine used in this system has an asynchronous machine magnetized by a capacitor bank. Regulation of the active generated power is by stall control. Since there is no control of the reactive power, the system can generally be used only in low-power systems (5 to 15 kW).

The wind turbines that this system uses (for their simplicity, low cost and robustness) have a major instability problem with respect to the voltage and frequency of the output signal, which therefore needs to be rectified. Most of them are usually manufactured without a gearbox. Generally, an energy storage subsystem, such as batteries, is required for circumstances where no power is being generated either as a result of a lack of wind or generator field loss. One valid option for this system is to use a permanent-magnet synchronous generator which, though heavier and more expensive, avoids the need for capacitor banks and all of their associated drawbacks.

An AC-based interconnection between different subsystems requires control of the transfer of both active and reactive powers. Figure 5.16 shows an AC interconnection between two subsystems. The bus voltages of each element, as well as the voltage phase differences, mark the direction and magnitude of the energy transfer, and the expressions that follow show the influence of the electrical parameters on the active and reactive powers transferred between subsystems.

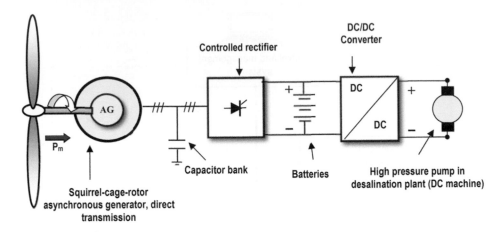

Figure 5.15.    Stand-alone DC desalination system.

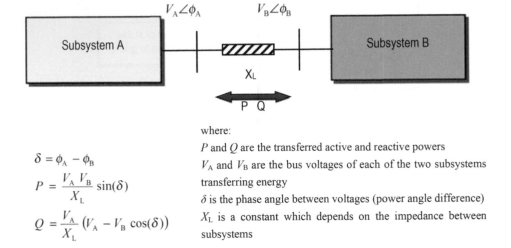

where:

$\delta = \phi_A - \phi_B$

$P = \dfrac{V_A V_B}{X_L} \sin(\delta)$

$Q = \dfrac{V_A}{X_L} \left( V_A - V_B \cos(\delta) \right)$

$P$ and $Q$ are the transferred active and reactive powers
$V_A$ and $V_B$ are the bus voltages of each of the two subsystems
transferring energy
$\delta$ is the phase angle between voltages (power angle difference)
$X_L$ is a constant which depends on the impedance between
subsystems

Figure 5.16.    AC energy transfer between two subsystems.

The flow of active and reactive powers shown in Figure 5.16 and the intervention of the electrical parameters are the key factors in subsystem interconnection. For example, if the power angle, $\delta$, is controlled to maintain constant voltages and frequencies, an energy transfer is obtained of basically active powers; whereas if the power angle is minimized the reactive powers can be controlled simply by varying the magnitudes of the voltages.

The technological state-of-the-art of commercial wind turbines currently suggests growing competition between variable-speed machines and the now-traditional fixed-speed machines, with manufacturers giving particular attention to the high-power end of the wind turbine market and the possibilities offered in the delivery of significant amounts of wind-sourced energy into large-size grids. The use of these types of machine in stand-alone systems is, generally, not contemplated in the actual design of the wind turbine. An analysis of what is required tends to be carried out on a case-by-case basis as it will depend on several factors, including the nature of the electrical system to which the wind turbine will be connected, the loads involved, the energy storage needs, etc.

##### 5.3.3.1.1 *Stand-alone systems with fixed-speed asynchronous wind turbines*

Good results have been obtained with wind-diesel systems for desalination, such as the one installed in Fuerteventura (Canary Islands, Spain) in the 1990s (Calero *et al.*, 1994; Carrillo, 2001; Carta *et al.*, 2001, 2003b; González *et al.*, 1993, 1994). In principle, such a system is ideally suited for off-grid remote areas, without access to potable water and where maintenance needs to be kept to a minimum. Figure 5.17 is a simplified schematic representation of the system in place, designed for a 220 kW wind turbine.

The 220 kW wind turbine has a gearbox transmission system, a two-speed squirrel-cage-rotor asynchronous generator and, therefore, a double-stator winding with a different number of pairs of poles. In this way, better exploitation is made of the wind resource as the wind speed required for connection is low, meaning that the generator can operate with less power but at optimum speed at average wind speeds (about 7 m s$^{-1}$). In addition, a soft electronic starter for grid connection is included in the system.

The synchronous generation subsystem with flywheel has four basic functions:

- To act as a synchronous compensator, supplying the necessary reactive power demanded by the wind turbine and the loads (when disconnected from the diesel machine).
- To maintain the frequency and voltage as stable as possible within certain limits (measured by the flywheel) by supplying or consuming the transitory energy demanded by the loads or supplied by the wind turbine in the event of a sudden gust or drop in wind speed (when disconnected from the diesel machine).
- To operate as diesel generator in parallel with the wind turbine when the latter is unable to supply all the power demanded by the loads.
- To operate as diesel generator supplying all the energy requirements of the loads in the absence of wind.

Clearly, the active powers that are generated and demanded are not in balance at all times, especially given the asynchronous generating condition with large torque variation vs. small variations of slip caused by the wind. Consequently, control of the flywheel and control of blade pitch are insufficient to maintain the stability of the system and a load-dumping system is required to operate in the event of excess power.

This system is limited in terms of the power that can be exploited because of this need to stabilize the system by load dumping in the event of an excess of energy, and also because of the synchronous generation of reactive power required by the asynchronous generation of the wind turbine.

##### 5.3.3.1.2 *Stand-alone systems with variable-speed synchronous wind turbines*

The use in stand-alone systems of fixed-speed wind turbines based on asynchronous generation is limited by the size of the powers that can be transferred between the different subsystems that structure a microgrid. Given the greater control of the energy that is generated, variable-speed wind turbines have been proposed as a more suitable option for renewable energy-based stand-alone microgrid systems.

The pitch control and power electronic systems of variable-speed wind turbines allow a much more stable delivery of power, avoiding the need for load dumping to maintain system stability. Figure 5.18 is a schematic representation of a wind-diesel system with synchronous generation in the wind generation subsystem. In this case, a permanent-magnet synchronous machine is used, which implies no voltage regulation, which will in part depend on the rotational speed.

The advantage of this synchronous machine is that there is no need for magnetizing reactive power, given the use of the permanent magnets, though pitch control is required as well as power electronics for output-power regulation. The power electronics basically comprise unidirectional power converters with a DC link, controlling the output voltage at three points – the rectifier, the DC link, and the PWM (pulse-width modulated) output inverter – through the respective modulation index. At the third of these points, the phase angle and frequency of the output voltage are controlled as well as its magnitude.

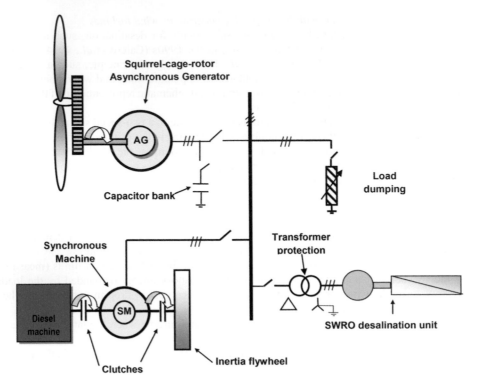

Figure 5.17.   Wind-diesel system for seawater desalination based on a fixed-speed wind turbine.

The consequence of the elimination of a synchronous energy storage system is system instability at transient moments of load connection/disconnection. The inertial flywheel, which does not have to be as large as in the system shown in Figure 5.17, is used to control transient stability and can be used as a reference control of the voltage and frequency limits. In addition, the commercial wind turbine is essentially modeled to behave as a current source and, therefore, it is this synchronous energy system which maintains system voltage and frequency. It can also be used as an uninterruptible power supply via a clutch-coupled diesel motor.

The use of direct-transmission turbines and multi-pole machines is another option, very similar to that shown in Figure 5.18 but with the added advantage of not requiring a wind turbine gearbox. Since the purpose of the system in this case is seawater desalination, the diesel motor can be eliminated and the loads regulated according to the capacity of the system to capture the wind energy available at any given moment, allowing the system to shut down when there is insufficient wind. If the load consumption is in the MW order, then the basic system outlined above is still a feasible option, with the wind turbines able to deliver all of the power they can capture from the wind, and with load regulation according to wind power availability and wind turbine capacity.

Figure 5.19 shows enlargement of the generation block to a MW-sized wind farm and of the load block to various modules for a daily desalinated water production of over 15,000 m$^3$ (Carrillo et al., 2004; Carta et al., 2003b, 2004; Subiela et al., 2004). If, as in this case, an inertial flywheel is used, the diesel machine needs to be replaced with another type of system that allows the start-up from zero of the inertial flywheel-synchronous machine. A speed inverter (frequency converter) for induction machines can be used for this purpose and, given the power control of this type of wind turbine, power from one of the wind turbines can feasibly be diverted to the inverter for use in the start-up by an auxiliary motor, taking the flywheel up to the rated speed. It is advisable in this case to use the DC link of one of the wind turbines whose voltage is kept stable internally with

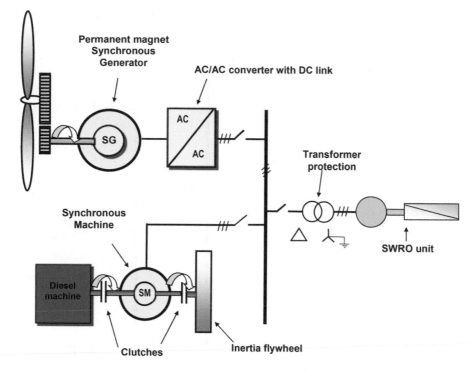

Figure 5.18.   Wind-diesel system based on a variable-speed wind turbine.

a DC/DC converter (generally a boost-type converter), so that through the inverter the flywheel is taken to the synchronous speed of the electrical machine coupled to the same shaft. As the flywheel is on the same shaft as the synchronous machine, the former, once started up, confers a magnetic field on the latter. Thus, a stable electrical grid is created and provides the basis for the connection of more wind turbines and the formation of a wind farm. Staggered connection of the desalination modules now takes place depending on the system's energy availability.

Today, advances in power electronics technology allow replacement of the traditional inertial flywheel storage subsystem with energy storage battery banks. Figure 5.20 shows a large-scale desalination system with battery energy storage. Battery technology has progressed considerably since the 1980s, though its use for large-scale energy storage in electrical grids remains complex and has the drawbacks of high cost and questions of durability. Nonetheless, in stand-alone systems they can be a solution for relatively long periods of time as they can store energy when there is an excess of wind-sourced energy generation and dump it in periods of low-wind energy production.

One example of the investigation of the potential of energy storage systems is the stoRE project (www.store-project.eu), co-funded by the European Union (European Regional Development Fund), the government of Spain (Centro de Desarrollo Tecnológico Industrial), and a consortium headed by the electricity company Endesa (Grupo ENEL) and various industrial partners and research centers. The stoRE project aims to analyze three different energy storage technologies: batteries, flywheels and supercapacitors.

The installations used in the stoRE project are located in the Canary Islands archipelago (Spain), where attempts are being made to overcome problems of dynamic stability caused by feeding into the different island grids energy generated from non-controlled energy sources, as is the case with renewables in general and wind energy in particular. More specifically, a massive energy storage plant has been installed on the island of Gran Canaria using Li-ion batteries (Saft SA,

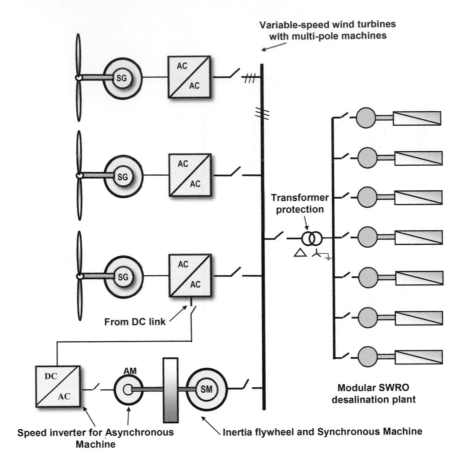

Figure 5.19.    Large-scale desalination system based on variable-speed wind turbines and inertia flywheel.

France) capable of supplying a stand-alone electrical grid with 1 MW for a duration of three hours (1 MW/3 h). On La Gomera island, a 1800/3600 rpm flywheel has been installed, capable of supplying 0.5 MW for a duration of 30 seconds, thereby achieving continuous grid frequency stabilization on the island. Meanwhile, on La Palma island, supercapacitor technology is being used (4 MW/6 s) to avoid transient supply losses due to unforeseen grid disturbances and failures. These latter two installations (La Gomera and La Palma) use energy storage technologies with extremely fast response times, capable of limiting the effect of such transient losses.

Another example is the sodium-sulfur (NaS) battery developed by TEPCO (Tokyo Electric Power Company, Inc.) and NGK Insulators Ltd. NaS batteries have been used in numerous projects and demonstrated their usefulness in mass energy storage in electric power systems.

Their high density of energy storage, efficiency, practically zero maintenance requirements, absence of contaminating gases, noise or vibrations, and a useful life of over 15 years make NaS batteries ideally suited for use in power electronic systems, as well as in large stand-alone or microgrid systems. For example, 20 × 50 kW NaS battery modules, which would be equivalent in size to two standard shipping containers, can store up to 7.2 MWh of, in our case, wind energy. The battery is hermetically sealed and kept at a temperature of around 300°C, operating in conditions in which the active materials at both electrodes are liquid and the electrolyte is solid. At this temperature, both active materials react rapidly with a low internal resistance. NaS batteries allow continuous charge and discharge, making them ideally suited for use in stand-alone energy storage systems.

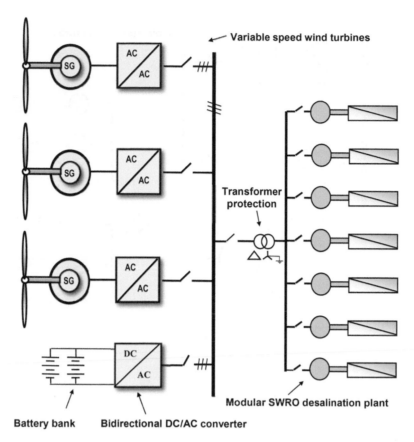

Figure 5.20.  Large-scale desalination system based on variable-speed wind turbines and energy storage in batteries.

Consequently, the operation of the synchronous subsystem can be covered, for large powers, by an NaS battery bank and a reversible converter acting as would a synchronous electrical machine, namely by supplying the system with the necessary reactive power as well as the demanded active power, not only in transient moments of load connection but also uninterruptedly for a time, as would a diesel generator. It should be noted that the flow of active and reactive powers in a microgrid shows the difficulties entailed in the coupling between different (commercial) power electronic devices that are normally designed to be connected, as current sources, at a stable frequency and voltage, that is, to a low impedance grid. As a result, the integration of these electronic devices in the configuration of a microgrid generates serious coupling problems.

In addition to the effects of connection/disconnection, harmonic pollution produced by the converters in the system and in steady-state conditions could also lead to major problems.

The solution lies in the appropriate design, as voltage source, of the bidirectional converter of the battery energy storage bank. In this way, the whole system would operate as a stable reference to which wind generation could be connected, followed by the loads. Otherwise, the installation of another synchronous subsystem would be necessary. Although, theoretically, this could be one of the wind generation output converters, this would not be advisable given the variable nature of the energy source (i.e. the wind).

Consequently, if no commercial converter is suitable for use as reference for the other converters, there remains no choice but to use a synchronous-type subsystem, as in the traditional synchronous machine with inertial flywheel and automatic voltage control. Figure 5.21 shows the resulting configuration.

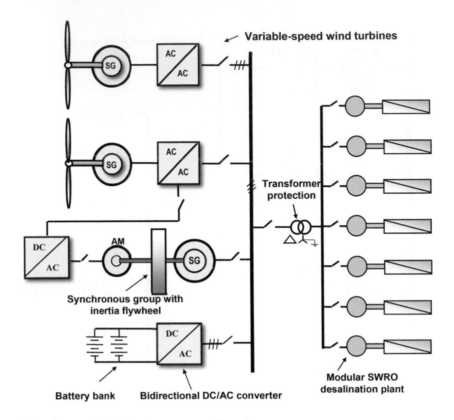

Figure 5.21.    Large-scale desalination system configuration.

In the stand-alone desalination systems that have been described (Carrillo, 2001), the innovation lies basically in managing the loads as opposed to the generation, adapting the consumption of the SWRO plants to the wind energy available at any given moment.

Consequently, the aim is to exploit the maximum amount of wind energy possible for the maximum possible production of desalinated water, regulating the SWRO plants through the connection/disconnection of plants or, in other words, using a type of step control. One potential improvement in load management is for this to be performed continuously, in contrast to the traditional use of such step control.

Continuous control of the consumption of an SWRO plant through regulation of the pressure and flow rate of the seawater fed through has been analytically investigated (Carta et al., 2015). In this approach the consumption is continuously regulated in accordance with the amount of wind energy that can be captured by the wind turbines. As shown in Figure 5.22, wind power is transformed into electrical power (voltage × current), the mechanical power of the high-pressure pump (torque × engine speed), and fluid power (pressure × flow rate). If the wind speed and power-wind speed curve of the wind turbine are known, continuous regulation of the plant's power consumption is possible by varying the flow rate and pressure of the seawater forced through the membranes.

As shown in Figure 5.22, in addition to continuous regulation of the SWRO plant, the traditional flywheel-synchronous machine and batteries have been replaced with a non-massive energy storage system comprised of supercapacitors. The function of the supercapacitors is to reduce the effect of variations in wind power produced by the wind turbine and to assist in continuous regulation of the SWRO plant.

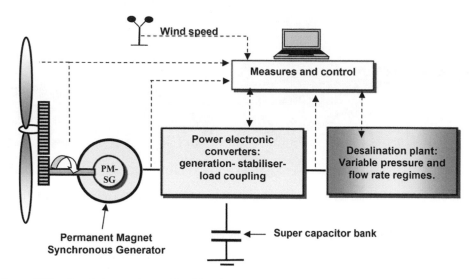

Figure 5.22.   Stand-alone micro-grid with wind power generation and supercapacitors applied to desalination.

### 5.3.3.2   *On-grid wind reverse osmosis*

With regards to on-grid wind-RO applications, there are some successful examples that verify the benefits of this kind of function. In 2008, a wind desalination plant was installed and began its operation on the island of Milos in Greece (Fig. 5.23). The wind-RO plant consists of a V52/850 kW wind turbine able to cover the energy requirements of a 3600 m³ day⁻¹ SWRO plant. The average yearly electricity generation of the wind turbine is estimated at 1800 MWh. A SCADA system is used to balance the operation of the units according to the electrical load of the island and the operating status of the local thermal power plant (Bouzas and Relakis, 2010). The water production is sufficient to meet the water needs of Milos island. The price at which the investor sells the water to the municipality is around $2\,€\,m^{-3}$. Before the installation of the plant, the island covered its water needs with the transportation of water via tankers from the Greek mainland. Water transportation costs for the Cyclades and Dodecanese Aegean islands prior to 2010 ranged from 5 to $8\,€\,m^{-3}$. By 2015 the cost of water transport to several Aegean islands had increased to $10.5\,€\,m^{-3}$.

Another on-grid wind-RO application originated in Australia. In 2006 a SWRO unit began operation in Perth. The plant is designed to operate continuously, drawing water with an input salinity between 35 and $37\,g\,L^{-1}$. With this plant, Western Australia became the first state in the country to use desalination as a major public water source. Located at Kwinana, some 25 km south of the city, the new plant had an initial daily capacity of 140,000 m³ with designed expansion to 250,000 m³ day⁻¹. Electricity for the desalination plant, which has an overall 24 MW power requirement, comes from an 80 MW wind farm, which consists of 48 wind turbines located 30 km east of Cervantes.

The idea of covering the power needs of desalination units with the use of one or more wind turbines even when an electricity network is available is important and beneficial also for islands or remote areas with autonomous power plants, such as the case of the Greek islands. According to a pre-feasibility study that Greece's Centre for Renewable Energy Sources and Saving (CRES) carried out for the Municipality of Syros, the total energy required for 24-hour daily operation of the Syros island RO units amounts to 49,980 kWh, while on an annual basis the total energy required is estimated at around 18,200 MWh (Tzen and Sieros, 2013). Syros island covers a large part of its water needs by the operation of thirteen SWRO units, with a total water production capacity of 8,340 m³ day⁻¹ and total installed power of around 2 MW (December 2013). Taking

Figure 5.23.    View of wind reverse osmosis plant on Milos island, Greece (Source: Bouzas and Relakis, 2010).

Table 5.5.    The gross and net yearly energy production of the Syros wind farm sites.

| Site | Gross production [GWh] | Net production [GWh] |
|---|---|---|
| 1 | 19.157 | 17.3428 |
| 2 | 18.900 | 17.1102 |
| 3 | 18.922 | 17.1301 |
| 4 | 22.007 | 19.9229 |
| 5 | 21.479 | 19.4449 |
| 6 | 14.963 | 13.546 |

the above with electricity pricing of $0.08 \text{€} \text{kWh}^{-1}$, the overall electricity cost for the operation of the RO units on the island amounts to 1.45 million $\text{€} \text{ year}^{-1}$. Within the study, six potential turbine sites were selected with capacity factors ranging from 25% to 50%. Table 5.5 presents the gross and the net yearly energy production of these sites.

Taking into consideration the wind potential of the specific areas and the sites' characteristics, it was concluded that two wind turbines of 2.5 MW each (i.e. a wind farm of 5 MW) could provide the energy required for the island's desalination units. An indicative economic assessment based on the current pricing of energy from RES (assuming the price for the interconnected system was equal to $0.09 \text{€} \text{kWh}^{-1}$) estimated the gross annual revenue from the wind farm at €1.5 million. The cost for the procurement and installation of the two-turbine 5 MW wind farm was estimated at €7 million.

Another illustrative example is the Canary Islands' "Water – Renewable Energy" nexus. This is an initiative of the local agricultural cooperative Soslaires Canarias S.L., which installed a $5000 \text{ m}^3 \text{ day}^{-1}$ SWRO plant associated with a grid-connected 2.64 MW wind farm ($4 \times 660 \text{ kW}$ wind turbines) in Playa de Vargas (east of Gran Canaria island) (Fig 5.24), with a total investment of €5.2 million (wind farm 46%, RO plant 21%). Both installations were commissioned in 2002. The desalination plant is able to produce up to 1.5 million $\text{m}^3 \text{ year}^{-1}$ of water for the irrigation of more than 150 hectares. The water produced is of high quality (salinity slightly over $400 \text{ mg L}^{-1}$) and the plant has excellent specific energy consumption (approximately $2.2 \text{ kWh m}^{-3}$, equivalent to $2.85 \text{ kWh ha}^{-1}$ of irrigated land). The annual electrical energy balance (wind energy production minus energy consumption due to water production) is positive, avoiding the emission of more than 6000 tons of $CO_2$ per year (Flammini et al., 2014).

In 1997, El Hierro was the first of the Canary Islands to modify a sustainable development plan to protect its environmental and cultural richness (Guevara-Stone, 2014). The island was importing and burning 6000 metric tons of diesel per year, emitting 18,700 tons of carbon dioxide. Twenty percent of the electrical energy being generated was consumed by three desalination plants to

(a)                                                          (b)

Figure 5.24.   (a) View of the $5000\,m^3\,day^{-1}$ rack of Soslaires SWRO desalination plant; (b) 2.64 MW (4 × 660 kW wind turbines) on-grid wind farm in Playa de Vargas (east of Gran Canaria island) (Source: J. Lozano, Soslaires Canarias S.L., 2016).

produce water for drinking and irrigation. The government of El Hierro realized that conservation was not enough and the solution was to make the island 100% energy-self-sufficient. The El Hierro Sustainability Plan proposed the idea of making the island a self-sustaining location. Within this framework, the Council of El Hierro, Unelco S.A. and the Canary Islands Institute of Technology (ITC) developed a project called the "El Hierro Hydro-Wind Plant", with the aim of making El Hierro the first island able to supply itself with electrical energy through fully renewable energy sources. Surplus wind-generated energy is used to pump water from a lower to an upper reservoir from where it can be released to generate hydroelectric power when wind generation is insufficient to meet demand.

During the first stage, a technical feasibility study was carried out in which the optimal configuration of wind generators, hydraulic turbines, reservoir volumes and pumping equipment to be installed in the new plant was determined. The project elements are:

- Upper reservoir: located at the La Caldera crater, with a maximum capacity of $380{,}000\,m^3$.
- Lower reservoir: located in the vicinity of the Llanos Blancos thermal power plant and formed via a purpose-built dam, it has a usable capacity of $150{,}000\,m^3$.
- Pumping plant: comprised of 2 × 1500 kW and 6 × 500 kW pump groups, with a total capacity of 6 MW. It has 1500/500 kW inverters.
- Turbine plant: consists of four groups of Pelton turbines of 2830 kW power each, giving a total power of 11.32 MW. The maximum flow during hydroelectric generation is $2.0\,m^3\,s^{-1}$, with a gross head of 655 m.
- Wind farm: consists of a set of five wind turbines (Enercon E-70) of 2.3 MW power each, giving a total power of 11.5 MW.

The island of El Hierro has an installed desalination capacity of $6000\,m^3\,day^{-1}$ which is fed with the electrical energy produced through these RES. The desalinated water is used for human consumption and, for that reason, El Hierro is considered the world's first self-sustaining island in terms of water and energy supply.

## 5.4   WIND DESALINATION MARKET

Various plant manufacturers and firms of consultant engineers, mainly from Germany, have developed, and in some cases also produced and tested, wind-powered desalination technologies.

That said, none of the resulting plants has yet proved itself in the market. In terms of wind-RO, only a few manufacturers of wind turbines have, during the last few years, offered their products in combination with desalination plants in turnkey solutions (Enercon, Vestesen), and occasionally also as hybrid wind-diesel systems in conjunction with a desalination unit (Enercon). Enercon systems have so far only been tested in pilot projects (Norway, Greece) and for simulation purposes (Aurich, Germany). In addition, there are also several companies and consultancies, mostly engineering-based, that advertise various wind desalination plant configurations, although these have not yet been proven in practice.

## 5.5  CONCLUSIONS

As conventional energy costs are expected to increase in the short term, and water availability will decrease due to the consequences associated with climate change, the future of RES-powered desalination is very promising for both environmental and economic reasons. Barriers such as the high cost of small wind turbines and the inefficiency of storage batteries delay the development and market progress of these systems. Hopefully, such environmentally friendly systems should soon be available at competitive prices. More research has to be done on new-generation products – such as small energy recovery devices, new-generation small wind turbines and energy storage systems – and on the development of more advanced control and monitoring systems in order to develop sustainable systems that can provide potable water at reasonable cost.

## REFERENCES

AWEA (2012) 2011 U.S. small wind turbine market report. American Wind Energy Association, Washington, DC.
Bouzas, I. & Relakis, G. (2010) Desalination with the use of wind energy in Milos island. International Technological Applications (ITA) S.A. *PRODES Workshop, Renewable Energy Technologies with Desalination: Technologies Progress-Legislation-Funding Schemes, CRES, 9 September 2010, Athens, Greece.*
Calero, R., González, J. & Carta, A. (1994) Wind system to desalinate seawater on large scale. *European Wind Energy Conference (EWEC '94), 10–14 October 1994, Thessaloniki, Greece.*
Carrillo, C.J. (2001) *Análisis y Simulación de Sistemas Eólicos Aislados [Analysis and Simulation of Isolated Wind Power Systems].* PhD Thesis, University of Vigo, Spain.
Carrillo, C., Feijóo, A.E., Cidras, J. & González, J. (2004) Power fluctuations in an isolated wind plant. *IEEE Transactions on Energy Conversion*, 19, 217–221.
Carta, J.A. & González, J. (2001) Self-sufficient energy supply for isolated communities: wind-diesel systems in the Canary Islands. *Energy Journal (IAEE)*, 22(3), 115–145.
Carta, J.A., González, J. & Gomez, C. (2003a) Operating results of a wind-diesel system which supplies the full energy needs of an isolated village community in the Canary Islands. *Solar Energy*, 74, 53–63.
Carta, J.A., González, J. & Subiela, V. (2003b) Operational analysis of an innovative wind powered reverse osmosis system installed in the Canary Islands. *Solar Energy*, 75, 153–168.
Carta J.A., González, J. & Subiela, V. (2004) The SDAWES project: an ambitious R&D prototype for wind powered desalination. *Desalination*, 161, 33–48.
Carta, J.A., González, J., Cabrera, P. & Subiela, V.J. (2015) Preliminary experimental analysis of a small-scale prototype SWRO desalination plant, designed for continuous adjustment of its energy consumption to the widely varying power generated by a stand-alone wind turbine. *Applied Energy*, 137, 222–239.
Cipollina, A., Tzen, E., Subiela, V., Papapetrou, M., Koschikowski, J., Schwantes, R., Wieghaus, M. & Zaragoza, G. (2014) Renewable energy desalination: performance analysis and operating data of existing RES desalination plants. *Desalination and Water Treatment*, 55(11), 1–21.
EERE (2016) The inside of a wind turbine. Office of Energy Efficiency & Renewable Energy, Washington, DC. Available from: http://energy.gov/eere/wind/inside-wind-turbine-0 [accessed October 2016].

Ekanayake, J.B., Holdsworth, L., Wu, X.G. & Jenkins, N. (2003) Dynamic modelling of doubly fed induction generator wind turbines. *IEEE Transactions on Power Systems*, 8(2), 803–809.

EWEA (2009a) *Wind Energy – The Facts: A Guide to the Technology, Economics and Future of Wind Power*. Earthscan, London.

EWEA. (2009b) The economics of wind energy. European Wind Energy Association, Brussels, Belgium. Available from: http://www.ewea.org/fileadmin/ [accessed October 2016].

EWEA (2015) Wind in power: 2014 European statistics. European Wind Energy Association, Brussels, Belgium.

Flammini, A., Manas, P., Pluschke, L. & Dubois, O. (2014) Walking the nexus talk: assessing the water-energy-food nexus in the context of the Sustainable Energy for All Initiative. Climate, Energy and Tenure Division, Food and Agriculture Organization of the United Nations, Rome, Italy.

Gammaenergy (2016) Wind power: main components. Gammaenergy, Lucca, Italy. Available from: http://en.gammaenergy.it/windpower/main-components.html [accessed October 2016].

González, A., Calero, R., Carta, J.A., Ojeda, L., Gonzalez, J., Perez, A., Torres, J. & Vega, R. (1993) Hybrid wind diesel system for a village in the Canary Islands: development and commissioning. *Proceedings of European Wind Energy Conference EWEC '93, 8–12 March 1993, Lübeck-Travemünde, Germany*. pp. 326–329.

González, A., Cruz, I., Calero, R., Carta, J.A., Ojeda, L., González, A. Nunez A., Perez A, Torres J. & Rocillo M. (1994) Lessons learned during the commissioning and start up of a wind-diesel system with inertial storage in Fuerteventura, Spain. *Proceedings of European Wind Energy Conference EWEC '94, 10–14 October 1994, Thessaloniki, Greece*. pp. 1056–1060.

Guevara-Stone, L. (2014) A high-renewables tomorrow, today: El Hierro, Canary Islands. Rocky Mountain Institute, Boulder, CO. Available from: http://blog.rmi.org/blog_2014_02_13_high_renewables_tomorrow_today_el_hierro_canary_islands [accessed October 2016].

Hau, E. (2006) *Wind Turbines: Fundamentals, Technologies, Application, Economics*. Springer, Berlin.

IEA (2015) Wind Annual Report 2014. International Energy Agency, Paris, France.

Manwell, J.F., McGowan, J.G. & Rogers, A.L. (2009) *Wind Energy Explained: Theory, Design and Application,* 2nd edn. John Wiley, Chichester.

RAE (2014) Wind energy: implications of large-scale deployment on the GB electricity system. Royal Academy of Engineering, London.

RenewableUK (2011) Small wind systems: UK market report. RenewableUK, London. Available from: http://www.renewablesolutionconsultancy.co.uk/images/Downloads/Small_Wind_Systems.pdf [accessed October 2016].

REN21 (2015) RENEWABLES 2015, Global Status Report. Renewable Energy Policy Nework fot the 21st Century, Paris, France. Available from: http://www.ren21.net/wp-content/uploads/2015/07/REN12-GSR2015_Onlinebook_low1.pdf [accessed December 2016].

Rossis, K. (2015) Offshore wind turbine technology and its prospectus. Presented in the workshop of the project Studies for the Strategic Environmental Assessment of the National Program for the Development of Offshore Wind Farms, MIS 375406, 29 September 2015, Athens, Greece.

Schaffarczyk, A. (2014) *Understanding Wind Power Technology: Theory, Deployment and Optimization*. John Wiley, Chichester.

Senvion SE (n.d.) Available from: https://www.senvion.com/global/en/ [accessed December 2016].

Subiela, V.J., Carta, J.A. & González, J. (2004) The SDAWES project: lessons learnt from an innovative project. *Desalination*, 168, 39–47.

Subiela, V.J., de la Fuente, J.A., Piernavieja, G. & Peñate, B. (2009) Canary Islands Institute of Technology (ITC) experiences in desalination with renewable energies (1996–2008). *Desalination and Water Treatment*, 7(1–3), 220–235.

Tong, W. (2010) W*ind Power Generation and Wind Turbine Design*. WIT Press, Southampton.

Tzen, E. (2008) Renewable energy sources for seawater desalination – present status and future prospects. In: Delgado, D.J. & Moreno, P. (eds) *Desalination Research Progress*. NOVA Science, New York, NY. pp. 145–160.

Tzen, E. (2009) Wind and wave energy for reverse osmosis. In: Cipollina, A., Micale, G. & Rizzuti, L. (eds) *Seawater Desalination*. Springer, Berlin. pp. 213–245.

Tzen, E. (2012) Wind energy powered technologies for freshwater production: fundamentals and case studies. In: Bundschuh, J. & Hoinkis, J. (eds) (2012) *Renewable energy applications for freshwater production*. CRC Press, Boca Raton, FL. pp. 161–180.

Tzen, E. & Mouzakis, F. (2013) Capacity building in wind energy & concentrating solar power (WECSP) in Jordan. Europe Aid/129543/C/SER/JO, unpublished report, Centre for Renewable Energy Sources & Saving, CRES, Greece.

Tzen, E. & Sieros, G. (2013) Pre-feasibility study for the covering of desalination power needs of the island of Syros with wind turbines. Unpublished report, CRES, Athens, Greece.

VESTAS wind systems (n.d.). Available from: http: www.vestas.com/en/products/turbines/ [accessed October 2016].

# CHAPTER 6

# Geothermal energy/desalination concepts

Thomas M. Missimer, Noreddine Ghaffour & Kim Choon Ng

## 6.1 INTRODUCTION

Geothermal energy development can be used to produce renewable energy in the form of electricity or directly used to provide hot water for a variety of uses, including desalination of brackish water or seawater (Benjemaa et al., 1999; Chandler, 1982; Chaturvedi et al., 1979; Davies and Orfi, 2014; Dipippo, 2005, 2007; Ghaffour et al., 2014; Ghose, 2004; Goosen et al., 2010; Hammons, 2004; Hiriart, 2008; Kalogirou, 2005; Miller, 2003; Ozgener and Kocer, 2004; Swanberg et al., 1977). An analysis of heat transfer in geothermal-powered desalination was conducted by Bourouni et al. (1999a, 1999b). A great advantage of using a geothermal heat source is that if properly designed, it is truly a renewable energy source that, unlike solar or wind sources, can provide a solution for baseload (non-interruptible) power production (Glassley, 2015). Of all power generation technologies, geothermal facilities (binary and flash) have the highest capacity factors and would deliver a value of 1.0 if "parasitic" electricity usage was not needed to run the high-pressure pumps in the recirculation system.

The natural geothermal gradient allows the harvesting of heat beneath virtually all land areas of the Earth, but development of geothermal energy is most economic when the source areas have high rates of heat flow or where naturally occurring hot springs bring high-temperature water to the Earth's surface. In some development projects, naturally occurring "hot spots" are targeted, where high-temperature water or rock lies close to the land surface (Missimer et al., 2014a). These areas are typically associated with volcanism.

A map of global heat flow shows that many coastal and inland areas that contain sustainable sources of either seawater or brackish water lie coincident with matching areas of high heat flow (Davies, 2013) (Fig. 6.1). The highest heat flow areas are generally associated with the edge of major geologic plate boundaries where active subduction is occurring, such as Japan, New Zealand, Chile, Peru and the western part of the United States. There are also other important areas of high heat flow located along the margins of the Red Sea, North Africa and southern Europe, particularly Italy and the Greek islands. The subsurface water temperatures in southern Europe range from 80–110°C at shallow depths, and in the southern European and Red Sea areas up to 325°C at depths ranging from 70–1400 m (Koroneos and Roumbas, 2011; Missimer et al., 2014a). Hot-rock temperatures are quite high in many areas that do not contain high water-flow rates, with correspondingly high subsurface hydraulic conductivities. Many of these regions have correspondingly arid conditions that currently require desalination of seawater or brackish water to meet current and projected future demands (Ghaffour, 2009).

While the global heat flow areas are well known in terms of geography and geology, detailed heat flow maps of some regions, such as the United States, show a surprisingly large number of "interior" high heat flow areas, not always associated with plate boundaries (Blackwell et al., 2011) (Fig. 6.2). Many areas of the western part of the United States have high heat flows, particularly the area in and around Yellowstone National Park, many areas of New Mexico, Arizona and Nevada, and especially southern coastal California and bordering areas of north-western Mexico (Gutiérrez and Espíndola, 2010). A key issue of note is the number of high heat flow areas in water-poor southern California, which is currently mired in a severe, long-term drought.

**Heat Flow Data-points (mW m^-2)**

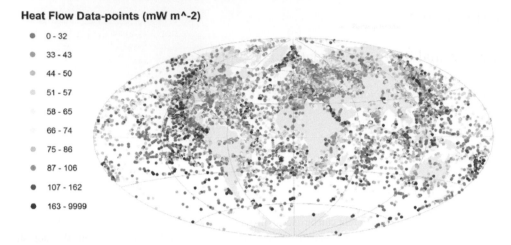

- 0 - 32
- 33 - 43
- 44 - 50
- 51 - 57
- 58 - 65
- 66 - 74
- 75 - 86
- 87 - 106
- 107 - 162
- 163 - 9999

Figure 6.1.   Global map of surface heat flow (Davies, 2013).

**SMU Geothermal Laboratory Heat Flow Map of the Conterminous United States, 2011**

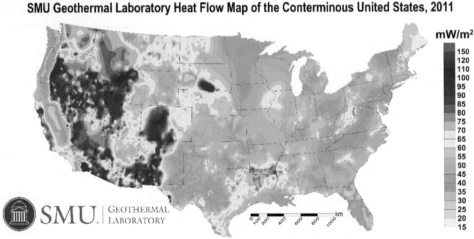

Reference: Blackwell, D.D., Richards, M.C., Frone, Z.S., Batir, J.F., Williams, M.A., Ruzo, A.A., and Dingwall, R.K., 2011, "SMU Geothermal Laboratory Heat Flow Map of the Conterminous United States, 2011". Supported by Google.org. Available at http://www.smu.edu/geothermal.

Figure 6.2.   Surface heat flow in the United States (Blackwell *et al.*, 2011).

The purpose of this chapter is to describe various conventional and emerging desalination methods that can be either directly or indirectly powered by geothermal energy. Some of the systems described are currently operating, while others are proposed, are completely new concepts or are in some stage of development. A key point is that the use of geothermal energy will reduce the cost of desalination and will also reduce the emissions of greenhouse gases (GHG) associated with the use of conventional, non-renewable energy sources.

## 6.2   GEOTHERMAL DESALINATION CONCEPTS

Methods of geothermal desalination can be divided into two general classes: low-enthalpy and high-enthalpy systems (or high- and low-temperature systems). Each classification can be further subdivided into three subtypes, specifically: (i) geothermal desalination systems that use dry steam or naturally hot water directly; (ii) geothermal desalination systems that first generate electricity that is then used to power desalination; (iii) geothermal desalination systems that are combined with other renewable energy sources to work on some type of cyclical basis.

Heat can be harvested from the subsurface by production of hot water or steam, commonly termed wet-rock geothermal energy production (Axelsson, 1991; Ragnarsson, 2003), or by injection of water into the subsurface through wells, pumping it through fractures in the rock, and recovering it as superheated water or steam through recovery wells. This latter geothermal harvesting method is termed hot dry rock recovery or an enhanced geothermal system (Brown *et al.*, 2012; Duchane, 1990; Feng *et al.*, 2012; Glassley, 2015; Goldstein *et al.*, 2009; Smith, 1979, 1983; Wan *et al.*, 2005).

A number of low-enthalpy, and a few high-enthalpy, geothermal desalination systems have been proposed as technically feasible (Awerbach *et al.*, 1976; Borurouni *et al.*, 1999a, 1999b; Bundschuh *et al.*, 2015; Chiam, 2013; Davies and Orfi, 2014; Goosen *et al.*, 2010; Karytsas, 1998; Lindahl, 1973; Ophir, 1982; Sarbatly and Chiam, 2013), and some of these options will be explored within the context of their technical feasibility and, where data are available, some economic information will be provided.

## 6.3 LOW-ENTHALPY DIRECT HOT WATER USE DESALINATION SYSTEMS

Geothermally heated groundwater occurs within permeable shallow and deep aquifers lying close to volcanic sources of heat, and also in other areas where hot rock comes into contact with groundwater. In locations like Iceland, wells can be drilled to various depths to produce water with temperatures ranging from 60 to over 100°C. In fact, some wells that penetrate a hot wet-rock reservoir will produce steam at the land surface as the hot water moves upwards in the well. It is assumed that all of the direct-heating, low-enthalpy desalination methods will not utilize a secondary heat-exchange process.

When deep petroleum reservoirs are developed, the oil production wells tend to produce hot saline water along with condensate. During the lifetime of the reservoir, the relative quantity of saline water becomes greater than the amount of condensate produced. Therefore, oil wells are a source of hot saline water, with a total dissolved solids concentration ranging from about 4000 to 240,000 mg L$^{-1}$, which can be used for desalination in different low-enthalpy systems, and perhaps some high-enthalpy systems too. Abandoned oil wells that produce water have been assessed for use in generating electricity within a low-enthalpy regime (Noorollahi *et al.*, 2015).

Any shallow or deep aquifer containing hot saline water can be used as a source of supply for geothermal desalination by employing a variety of different processes and design concepts. These processes are described below.

### 6.3.1 *Wet-rock harvested seawater or brackish water and adsorption desalination (AD)*

Adsorption desalination (AD) is a low-temperature process that can effectively use brackish water, seawater, or hypersaline water within a temperature range of 55 to 90°C (Chakraborty *et al.*, 2009a, 2009b, 2011b; Ng *et al.*, 2010, 2012). This temperature range lies within the low-enthalpy classification and occurs commonly in natural hot springs or in shallow wet geothermally heated groundwater systems throughout the world. The inflow water enters the process at ambient temperature, but heated water is required for the adsorption process and it flows in a closed loop. AD has the great advantage that it can be applied to a very large range of salinities that occur naturally below the saturation concentration of CaSO$_4$, which is about 235,000 mg L$^{-1}$ of total dissolved solids at 25°C. Therefore, the direct use of AD could be associated with natural springs, hot shallow water, and water produced at oil production wells or separation facilities (after some pretreatment).

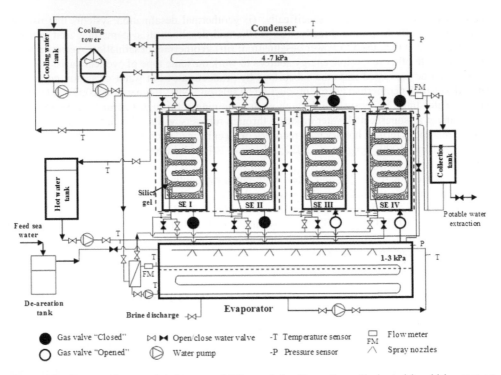

Figure 6.3.    Process diagram showing use of AD coupled with geothermally heated brackish water or seawater.

The designs required to utilize the geothermally heated water could take on a variety of configurations (Fig. 6.3). Hot water of variable salinity could be extracted directly from hot springs, desalted, and the concentrate discharged in some environmentally responsible manner. A shallow geothermally heated aquifer could be tapped using production wells with the concentrate expelled into a deeper part of the aquifer system, reused in some industrial process, or discharged to the sea after dilution or the use of a dispersion nozzle system to minimize marine impacts. If the hot water derived from the groundwater system is treated directly in the desalination process, then only some form of filtration would be necessary in the pretreatment process. In other systems, it may be necessary to convert the heat to energy using some type of heat exchanger.

Deep petroleum production wells tap geologic units that are naturally hot and both the oil and the water produced commonly have temperatures close to 100°C at the well head. The water is commonly separated from the condensate and discharged to waste without utilization of this heat. Some degree of pretreatment would be required to remove as much residual petroleum from the hot saline water as possible, to prevent adverse impacts on the AD treatment process. The concentrate from the desalination process could be re-injected into the petroleum reservoir to assist in maintaining reservoir pressure or to assist in secondary recovery.

### 6.3.2    *Wet-rock harvested seawater or brackish water and membrane distillation (MD)*

Another low-enthalpy desalination technology that can be used to desalinate hot saline groundwater or thermal spring discharges is membrane distillation (MD). MD is a thermally driven process that utilizes a hydrophobic, microporous membrane as a contactor to achieve separation by liquid-vapor equilibrium. Other related membrane processes, such as reverse osmosis (RO), forward osmosis (FO) or electrodialysis (ED), use pressure, concentration or an electrical potential

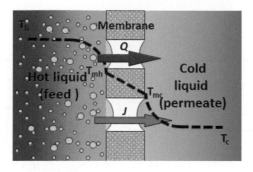

Figure 6.4.  Principle of MD process and temperature profile, using DCMD configuration as an example.

gradient as the driving force, while the MD process force is the partial vapor pressure difference maintained on the two sides of the MD membrane (Fig. 6.4). This process uses hot saline water on one side of a membrane and a cooler circulating fluid on the other side to create diffusion of pure water vapor across the membrane (Al-Obaidani *et al.*, 2008; Alsaadi *et al.*, 2013; El Amali *et al.*, 2004; Bouguecha and Dhahbi, 2003; Francis *et al.*, 2013). The feed solution, after being heated, is brought into contact with the feed side of the membrane which allows only the water vapor to pass through its dry pores so that it condenses on the permeate side (coolant), as shown in Figure 6.4. A temperature difference of just a few degrees, 5–10°C, between the warm and cold streams is potentially enough to produce freshwater (Alkhudhiri *et al.*, 2012; Alsaadi *et al.*, 2013; Curcio and Drioli, 2005; Francis *et al.*, 2013, 2014; Ghaffour *et al.*, 2014; Khayet, 2011). Separation occurs when pure water vapor, with its higher volatility compared to salts, passes through the membrane pores by a convective or diffusive mechanism. This process works at relatively low temperatures, which are well within the range of low-enthalpy geothermally heated water sources. It is also very suitable for treating thermal brines, which are already preheated. Under these operating conditions (low temperatures and ambient pressure), corrosion is not an issue as no metallic materials are used in MD units.

According to the type of the condensing (cold side) design, the MD process can be classified as a direct contact membrane distillation (DCMD), air gap membrane distillation (AGMD), vacuum membrane distillation (VMD) or sweeping gas membrane distillation (SGMD). More recently, other new MD configurations aiming to enhance the flux have been developed, such as water gap MD and material gap MD (Francis *et al.*, 2013; Khalifa, 2015).

MD has the potential to be an efficient and cost-effective separation process that can utilize low-grade waste or low-enthalpy geothermal heat (Bundschuh *et al.*, 2015; Ghaffour *et al.*, 2014; Goosen *et al.*, 2014; Gutiérrez and Espíndola, 2010; Rodriguez *et al.*, 1996; Sarbatly and Chiam, 2013). In order to be cost-effective, with waste heat being available at no-cost, the specific energy consumption needed for pumping can potentially be lowered to 1 kWh m$^{-3}$ (Ghaffour *et al.*, 2014). As it is the case for AD, so one of the main advantages of MD is that the process performance is not highly affected by high feed salinity, as has been proven in bench- and pilot-scale studies (Kui *et al.*, 2011; Zaragoza *et al.*, 2014; Goosen *et al.*, 2014).

### 6.3.3  *Wet-rock harvested seawater or brackish water and humidification-dehumidification desalination*

Another low-enthalpy desalination process that can be powered by geothermally heated water is humidification-dehumidification desalination, as described by Bourouni *et al.* (2001), Ettouney (2005), Al-Enezi *et al.* (2006), Mohamed and El-Minshawy (2009) and Goosen *et al.* (2014). Bourouni *et al.* (2001) also described the operation and performance of various humidification-dehumidification desalination systems around the world. There are a number

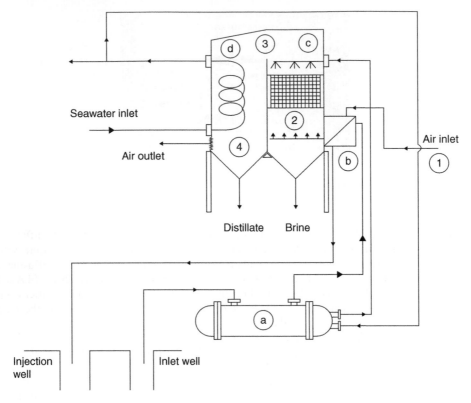

Figure 6.5.    Schematic diagram of a humidification-dehumidification geothermal seawater desalination plant (Mohamed and El-Minshawy, 2009).

of possible configurations for desalination plants using this process, as described by Ettouney (2005). A schematic diagram showing one theoretical geothermal desalination plant is shown in Figure 6.5.

The design created by Mohamed and El-Minshawy (2009) is for desalination of seawater. Air is heated in the air-preheater using a geothermal energy source after passing through the heat exchanger. Seawater is used as the coolant for the condenser. An analysis of desalination production efficiency using this design showed that: (i) a geothermal energy source at between 80 and 100°C is sufficient to produce freshwater from seawater; (ii) the production of freshwater is increased by increasing the ratio of the seawater mass flow rate to the air mass flow rate, with an optimum ratio lying between 1.5 and 2.5; (iii) the freshwater production rate is increased by increasing the geothermal inlet temperature at the optimum ratio of seawater-to-air flow; (iv) increasing the cooling water temperature difference across the condenser produces a higher freshwater production rate at the optimum ratio of seawater mass flow rate to air flow rate (1.5–2.5).

The humidification-dehumidification process appears to be a reasonable seawater or brackish water desalination method where geothermally heated groundwater sources are present. It may also work efficiently at temperatures above 100°C. Based on the research conducted to date, the optimum geographic area in which to apply the technology could be where geothermally heated groundwater is abundant and a source of cool or cold seawater is present. This combination of system temperatures would produce the optimal ratio across the condenser. Coastal areas of California, western Mexico, Peru, and Chile are regions of high potential where this technology could be applied and many small cities in these hyper-arid regions could utilize this technology.

## 6.4   LOW-ENTHALPY HYBRID SYSTEMS: LINKING SOLAR
AND GEOTHERMAL SYSTEMS

Pure solar-powered desalination systems can be successfully operated, but production will be limited solely to daylight hours and reduced production occurs during cloudy or rainy days as well as during dust storms. Geothermal-powered low-enthalpy desalination systems can be linked with solar-powered systems to allow continuous operation without the need for thermal storage (Ghaffour *et al.*, 2014; Missimer *et al.*, 2013). By using solar power during daylight hours, the geothermal reservoir is allowed to "rest", thereby allowing shallower geothermal heat-harvesting wells to be constructed for sustainable operation. The rest period allows the reservoir thermal conductivity to reheat hot dry rock collection systems or to maintain wet geothermal reservoirs based on local or regional groundwater flow (Fan *et al.*, 2007; Gehlin and Hellstrom, 2003; Pillar and Liuzzo-Scorpo, 2013).

### 6.4.1   *Combined-cycle solar- and geothermal-powered AD*

A detailed description of an effective method of linking two sources of renewable energy with AD is given in Missimer *et al.* (2013). AD is a very promising desalination technology that produces freshwater from seawater by sorbing and desorbing water vapor from preheated seawater on a stack of mesoporous silica gel or other appropriate substance (Chakraborty *et al.*, 2009a, 2009b; Ng *et al.*, 2010). The system operates in a vacuum as a batch process. The operating range of temperatures for the heated seawater is 55–100°C, with an optimum temperature of about 80°C. Because the effective operating temperature range is low, AD can be powered by solar, geothermal or waste heat (Fig. 6.3).

A key issue in the scaling up of treatment capacity for an AD desalination system using solar energy as the primary heat source is the issue of an interruptible energy source. Large-capacity systems would be required to have some type of thermal storage system or use a conventional power source for operation when there is no solar or reduced solar power availability. Therefore, development of large-capacity desalination systems, greater than 100,000 $m^3 day^{-1}$, would be challenging using solar energy alone.

One approach to resolving this issue is the development of a hybrid system using a combined cycle of solar and geothermal energy. During daylight hours the system would operate using solar energy with water being heated using a rooftop-mounted heating system (or other configuration). During night-time or other periods of low solar energy input, the system would operate using a geothermal heat system.

The geothermal heating development could be based on either wet-rock or hot dry rock technology, depending upon the location and the local hydrogeological conditions. Missimer *et al.* (2013) outlined a system to be developed in the arid lands adjacent to the Red Sea of Saudi Arabia, so they employed the concept of a hot dry rock (HDR) thermal reservoir using either a dual-well or single-well configuration, depending on the desired capacity of the AD plant (Fig. 6.6). Based on local conditions, they modeled a conceptual two-well HDR geothermal system that could be operated continuously at 80°C for more than eight days without reduction of the operating temperature or breakthrough of the cooler water. The operation would rarely rely solely on the geothermal energy source, thanks to the availability of solar energy during most daylight hours. Cycling to a solar energy source during the day allows the rock material of the geothermal reservoir to reheat and maintains the system in a sustainable manner using wells shallower than those used in continuous heat harvesting.

### 6.4.2   *Combined-cycle solar- and geothermal-powered MD*

Another process design configuration would be to use combined-cycle solar- and geothermal-powered MD. The conceptual operation would be similar to the AD process, running with the same combined-cycle energy source as described in Section 6.4.1. However, the configuration

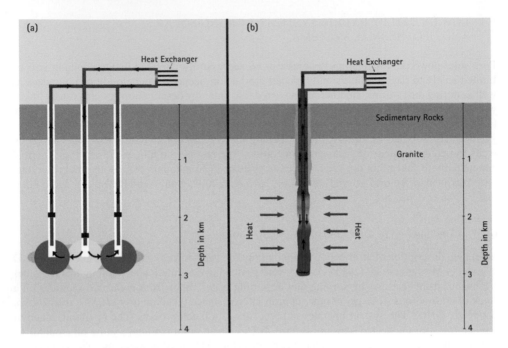

Figure 6.6.    Dual- and single-well heat collection systems in a hot dry rock reservoir (Missimer *et al.*, 2013).

of the process would be somewhat different. The geothermally heated water would enter at the feed side of the process with the opposite side being the cooler water that creates a difference of temperature between the two sides of the hydrophobic membrane and drives the process, allowing water vapor to pass through the membrane pores where it is condensed and collected as distillate, as shown in Figure 6.4. In VMD configuration, the condensation takes place outside the module as the water vapor is extracted using a vacuum pump.

### 6.4.3    *Solar-powered AD or MD with thermal aquifer storage*

Either AD or MD could be operated using solar energy as the primary energy source (Ghaffour *et al.*, 2015), but also utilizing some type of in-ground thermal storage. Aquifer thermal energy storage (ATES) has been used in a variety of configurations to meet long- or short-term needs (Maliva and Missimer, 2010). An aquifer or a subsurface-constructed cylinder or tunnel are ideal for short-term heat storage using water because: (i) an aquifer or suitable underground facility may be available beneath or within a few hundred meters of the desalination facility; (ii) an aquifer, if available, would provide a small footprint; (iii) the rock material surrounding the aquifer or constructed borehole or tunnel is an excellent insulator with a generally low thermal conductance (Schaetzle *et al.*, 1980); (iv) in tight fractured rock materials there would be an insignificant loss of water from the subsurface storage facility; (v) if some form of deep tunnel is used, the stored water may actually gain some additional heat during storage depending on the location and depth of the facility.

The use of a natural aquifer may have limitations at many sites because of the difficulty of maintaining the subsurface plume of heated water due to potential migration away from the storage site. The hydraulic conditions within the target aquifer system would have to meet a set of specific conditions, with an adequate hydraulic conductivity to allow injection and recovery of the heated water and a low potentiometric surface slope to prevent horizontal, down-gradient loss of the heated water. Therefore, the construction of an artificial subsurface heated water storage system would be required at some locations.

The capacity of the thermal storage is dependent on the overall desalination capacity being operated and the desired night-time operation rate. Typically, potable water requirements decline in the evening hours, which may reduce the storage capacity required. In addition, low-enthalpy desalination technologies, such as AD or MD, could operate at slightly lower temperatures during evening hours which could allow a blended inflow of seawater from the source and the hotter stored seawater. Thus there are many possible thermal reservoir design and operation concepts that could be employed.

### 6.4.4  *Co-generation of electricity and desalination using low-enthalpy geothermal systems*

It is possible to combine low-enthalpy electricity generation systems with desalination, depending on the initial temperature of the water and its temperature after the first or second pass through the turbines. Low-enthalpy geothermal electricity generation systems using the organic Rankine cycle (ORC) or Kalina cycle plants (Kalina Power, Australia) operate using a binary-cycle power plant configuration that could be used to power desalination (Stober and Bucher, 2013) (Fig. 6.7c). Sufficient waste heat is available after electricity generation to power low-enthalpy desalination processes, such as ED or MD systems, and in terms of future research, appropriate modeling and design of these systems should be pursued.

### 6.5  HIGH-ENTHALPY GEOTHERMAL DESALINATION USING DRY STEAM OR HOT WATER DIRECTLY

High-enthalpy geothermal-powered desalination systems may have a number of configurations depending on the heat source: dry steam or hot water. A dry-steam system could be used to directly heat seawater inflow after possible filtration to remove any particulates, which is then directed to other parts of the thermal process to deliver the desired heat source (Fig. 6.7). A secondary heat exchange system may not be necessary depending on the temperature of the dry steam developed. Thus, direct use does not produce electricity before entering the desalination process or processes. The cooled steam would be vented to ambient surroundings after heat utilization. Use of dry steam from a geothermal source would replace the steam normally produced by a boiler, which is sent into the thermal distillation processes (Fig. 6.8).

The use of naturally heated water in a high-enthalpy system has a number of complexities depending on the temperature and quality of the water or wet steam produced. For systems producing water with a temperature above 170°C, it is recommended that the water be kept under pressure in order to prevent it flashing to steam. Any hardness or other chemical species within the raw water would be likely to precipitate within the piping system downstream of the flashpoint (Stober and Bucher, 2013). If a closed-loop system doublet is used to obtain the heat and inject the cooled water, a heat exchanger will be required to convey the harvested heat to the incoming seawater and to critical parts of the desalination process. Geochemical modeling of the system would be required during design to ensure that mineral precipitation will not occur anywhere within the piping system and especially within the injection well, where it would be very difficult and expensive to remove (if it can be removed at all, e.g. silicate precipitation).

Because of the rather rare occurrence of a geothermal reservoir that produces abundant dry steam at the shoreline of an ocean, the desalination processes described herein would utilize hot water with a temperature range between 170 and around 300°C and heat exchangers would probably be required.

### 6.5.1  *Wet-rock-heated water for multi-stage flash distillation*

Wet geothermal systems could be used to provide water for direct entry into a multi-stage flash (MSF) distillation system, provided that the water has a sufficiently high temperature to meet the

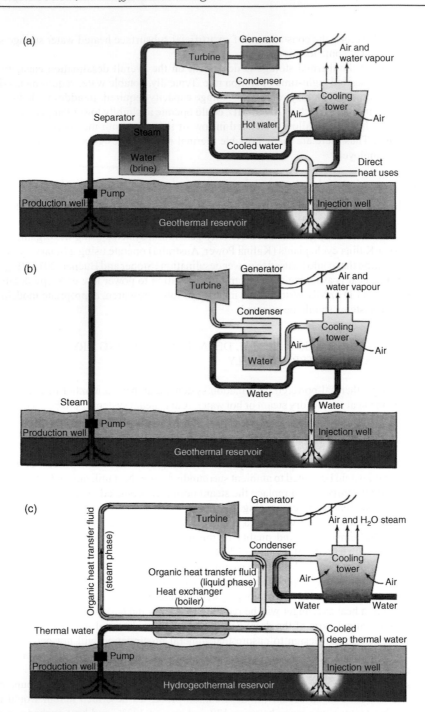

Figure 6.7.    Process diagrams for geothermal power plants using (a) flash steam; (b) dry steam; (c) binary cycle (Stober and Bucher, 2013).

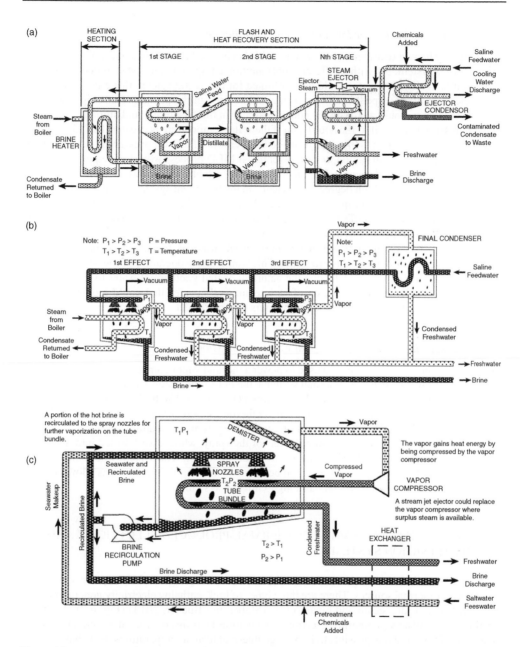

Figure 6.8.   Dry-steam-driven, high-enthalpy desalination for: (a) MSF; (b) MED; (c) TVC. The dry steam enters the system at the steam entry points marked (Buros, 1990).

feed requirements for the first stage (*TBT*: top brine temperature), and has low scaling potential (silica in this case). The very hot water would typically enter the system within the brine heater and mix with the seawater feed used for condensation (Fig. 6.8a). Since water produced by hot springs can be close to 100°C at the surface, and from deep wells can be up to 300°C, a wet geothermal system could be used as a direct heat source in the MSF process as described. Another possible design would be to add a flash chamber before the brine heater in which superheated water would flash to steam and enter the MSF process in a conventional manner.

### 6.5.2    Wet-rock-heated water for multiple-effect distillation

Geothermally heated seawater or freshwater could also be used to operate a multiple-effect distil-lation (MED) system, but with a different configuration compared to the MSF process described in Section 6.5.1. The hot water could be substituted for the heating steam that enters the effects, and cooler seawater pumped from the ocean could be used in the primary feed (Fig. 6.8b). The superheated water would mix with the cooler seawater in the first effect and, initially, would use a lower pressure to allow it to flash to steam. If the water temperature is greater than 250°C, there may be no need to operate the first stage at a lower pressure. If a lower pressure is required in the first stage, a design modification to the conventional MED process would be required. Because no fuel would be required to create the incoming steam, establishing a lower pressure in the first effect would still create a high degree of efficiency in the treatment system. The number of effects within the geothermal design might be reduced. However, if the temperature of the water from the geothermal source was sufficiently high, then the number of effects could be increased, taking into account the scaling control at higher temperatures.

In the event that the natural wet geothermal system water is fresh, consideration could be given to allowing the water to flash to steam. This would allow operation of the MED system in a conventional configuration. Allowing the naturally heated geothermal water to flash to steam could cause considerable precipitation of various mineral species that could require cleansing and, perhaps, filtering of the steam before it enters the first effect.

### 6.5.3    Wet-rock-heated water for thermal vapor compression

Geothermally heated water could be used to power a thermal vapor compression (TVC) desalina-tion system. In wet-recovery geothermal systems, the hot saline water would be pumped into the system to be sprayed over the heat exchange tubing (Fig. 6.8c). The higher heat required within the tubing could be provided using recovered geothermal steam or hotter water extracted from a greater depth in the geothermally heated groundwater system. For example, seawater extracted from a saline spring could be the source water for the spray, while a well drilled into the source aquifer could deliver the higher heat for the inside of the tubing. The system would then operate in a conventional manner. The system could be operated at lower temperatures as a low-enthalpy system, but would not be as efficient as a high-enthalpy system.

### 6.6    HIGH-ENTHALPY HYBRID GEOTHERMAL-PRODUCED ELECTRICITY AND DESALINATION SYSTEMS

Some geothermal electricity generation facilities operate using high-enthalpy systems to harvest heat from geothermal sources. There are three types of high-enthalpy electricity generation facil-ities: flash steam, dry steam and binary cycle (Stober and Bucher, 2013) (Fig. 6.7). All of the existing high-enthalpy geothermal electricity generation facilities operate at a feed temperature of at least 175°C, and it is possible to harvest geothermal heat at temperatures well above 300°C.

It is relatively easy to link thermal desalination processes into any of these three electricity generation schemes. Within the flash plant configuration (Fig. 6.7a), hot water (commonly saline) can be obtained directly at the separator and steam can be collected from the exhaust side of the turbine. In most flash plants the waste steam is directed to a condenser where it is cooled and ultimately discharged to a cooling tower where the hot air and water vapor are conveyed to the ambient surroundings. Therefore, two streams of heat can be used in thermal-based processes with temperatures of the steam or water at or above 100°C.

The configuration linking a thermal desalination process with a dry-steam geothermal electric generation plant is less complex. The steam would be taken from the turbine exhaust stream and used similarly to all of the current hybrid electricity generation/MSF or MED plants described by Awerbach (2007). Many such systems currently operate in the Middle East with the waste steam produced from oil-fired power plants being used to drive either MSF or MED processes (co-generation).

The binary-cycle geothermal power plant configuration may yield the most flexibility in terms of linking a single high-enthalpy desalination process or multiple high- and low-enthalpy processes to geothermal electricity generation. The heat exchanger could be configured to allow distribution of steam to various parts of a desalination facility to allow correct delivery of heat within the overall system, especially when multiple high- and low-enthalpy desalination processes are linked to harvest and utilize as much latent heat as possible with minimal venting to ambient surroundings.

### 6.6.1   *Hot dry rock heat reservoirs: methods of sustainable heat harvesting*

Since geothermal hot water reservoirs or dry steam reservoirs are geologically rare, hot dry rock systems can be developed nearly anywhere beneath the land area of the Earth with naturally high heat flow areas. The harvesting of energy from a hot dry rock reservoir requires a very sophisticated degree of geological assessment and engineering (Harlow and Pracht, 1972). Great care must be used in the engineering design to prevent unanticipated side effects of the geothermal energy recovery system, such as induced seismicity (Majer *et al.*, 2007). The heat harvesting method is commonly described as an "enhanced" geothermal system (EGS) development, which has three primary components: (i) an injection well; (ii) a fractured rock heat reservoir; (iii) a recovery well (Fig. 6.9).

Early hot dry rock geothermal system analyses were focused on the use of naturally fractured hot rock reservoirs. This requires that the rock material has been fractured and contains sufficient permeability to allow injected distilled water or another fluid to move through the rock, become heated and discharge into a recovery well. Location of natural systems that meet the design criteria without the need to use a very high injection pressure severely limits the number of potential geographic locations at which hot dry rock geothermal energy systems could be successfully developed. Furthermore, if high injection pressures are required, increased pore pressures within the naturally fractured rocks could connect to dormant or active faults, thereby inducing seismicity sufficient to create earthquakes of varying magnitude. Induced earthquakes have occurred at several geothermal energy production systems, including The Geysers in California, the Cooper Basin test site in Australia, the Soultz-sous-Forêts site in France, and the Deep Heat Mining project in Basel, Switzerland (Majer *et al.*, 2007). Ultimately, the Basel project was abandoned because of public concern over the earthquakes created.

To limit the development of induced seismicity, the natural fractures of the hot rock reservoir can be enhanced by chemical means (e.g. acidification) or by the use of hydraulic fracturing. In fact, based on the rock type, the heat capture system can be fully engineered using hydraulic fracturing techniques. The key design challenge is to create sufficient permeability between the injection and collection wells to allow water or another fluid used to gather heat while passing through the reservoir at the correct rate to obtain the desired temperature without depleting the heat flow within the reservoir.

Development of a sustainable heat capture system requires that the thermal conductivity of the reservoir effectively replaces the heat being removed by the circulating fluid. It is desirable to reach an equilibrium condition at a temperature at or above the desired recovery temperature. Pre-design information gathering about the subsurface thermal reservoir is required. The thermal conductivity and the geothermal gradient are measured to determine the necessary depth of the wells used for injection and recovery, the desired degree of hydraulic fracturing, and determination of the necessary distance between injection and recovery wells to ensure that there is sufficient time to allow for heat exchange to occur at the intended recovery temperature. The pumping rate and pressure of the injected fluid are other very important considerations. Once the appropriate data have been collected, a three-dimensional heat flow model should be used to design the system (Faust and Mercer, 1979; Kolditz, 1995).

The testing, engineering and construction of hot dry rock geothermal facilities to generate electricity is rather expensive and carries some degree of extra risk compared to conventional

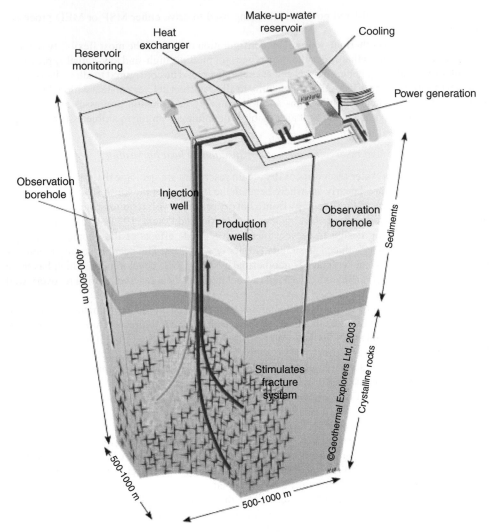

Figure 6.9.   Configuration of a binary- or multiple-well hot dry rock heat-harvesting system (from www.hotrockenergy.com).

generation facilities fuelled by hydrocarbons. However, the addition of a coupled source of revenue, such as freshwater obtained from linked desalination, makes hot dry rock geothermal energy development more economically viable overall. It is therefore very important to create the most efficient integrated electricity generation/desalination system possible to maximize the use of the latent heat acquired.

### 6.6.2   *Linking geothermal electricity generation with MSF or MED systems*

Depending upon the temperature of the steam after exhausting through the electricity generation turbines, either a one- or two-cycle system can be used to power either an MSF or MED system (Gunnarsson *et al.*, 1992; Hamed *et al.*, 2006; Ko *et al.*, 1979; Mamed *et al.*, 2002). The configuration of the steam within the process train is similar to that used in hybrid

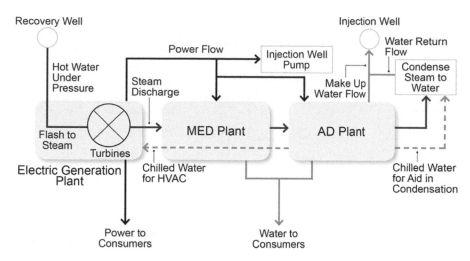

Figure 6.10.    Schematic diagram of linked geothermal electricity generation, MED and AD (Missimer *et al.*, 2016).

hydrocarbon-powered electricity generation plants linked with desalination (Awerbach, 2007). There are many of these hybrid systems currently operating in the Middle East region with most being electricity generation/MSF hybrids. At least one large-capacity electricity generation/MED hybrid exists in the United Arab Emirates. Process diagrams for some of these systems are shown in Figure 6.8 (Buros, 1990).

### 6.6.3    *Linking electricity generation and multiple desalination processes*

The novel concept of linking multiple desalination processes with geothermal-powered electricity generation was suggested by Missimer *et al.* (2014a, 2016), based on the linkage between MED and AD systems suggested by Thu *et al.* (2003, 2014). The concept involves the conservation of latent heat by first using the superheated steam to generate electricity via the flash steam process with one or two cycles, and then directing the cooler steam to act as heat source for a MED process, with use of the remaining steam to heat seawater for an AD system (Fig. 6.10). The AD process would provide desalinated water and a chilled water stream (Chakraborty *et al.*, 2011a), which could be used in the condensation process within the MED, for condensation of the cool steam for reuse in the heat capture process, and for operation of the heating, ventilation and air-conditioning (HVAC) system within the overall facility. Side-streams of the steam were designed to create greater efficiency within the overall set of linked processes (Fig. 6.11). The original design used a recovered steam temperature of 190°C at the entrance to the turbine. Greater efficiency could, perhaps, be obtained by increasing the temperature and running the wet steam through a vapor separator before allowing it to flash prior to the electricity generation turbine. MD could also be linked into the system using a separate steam offtake. However, the use of the AD process as a major component within the system yields greater operational flexibility, especially in the use of the chilled water stream.

The concept of conservation of latent heat within the overall combined geothermal electricity generation and desalination system offers a large number of research opportunities to innovate and develop a dynamic system model to test various design configurations and operating temperatures. A life-cycle cost analysis could be linked to the dynamic model to assess how to develop the most cost-effective combined system with the goal of reducing both electricity and desalination costs.

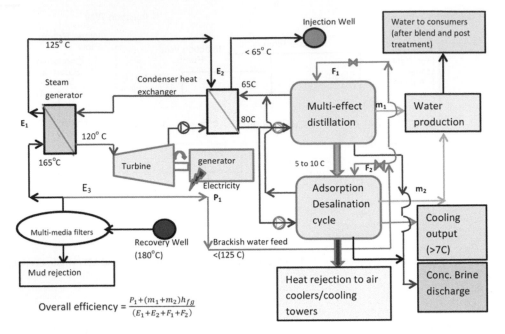

Figure 6.11.   Hybrid electricity generation/MED/AD system with steam side-streams and design details (Missimer *et al.*, 2016).

### 6.6.4   *The geothermal electricity generation and desalination "campus" concept*

The concept of creating an electricity generation/desalination "campus" was first suggested by Missimer and Maliva (2010). Previously, Wright and Missimer (1999) had suggested linking electricity generation using aquifer storage and recovery (ASR) with water treatment, and Al-Katheeri and Agaschichev (2008) specifically linked co-generation facilities with SWRO and ASR. The concept of linking electricity generation with thermal and membrane seawater desalination and including an ASR system would improve operational efficiency (Ghaffour *et al.*, 2013a) (Fig. 6.12). Missimer *et al.* (2014b) filed a patent describing operation of an electricity generation/desalination campus using geothermal energy and incorporating ASR. The system also includes multiple thermal desalination processes using the concept of latent heat conservation, beginning with a high-enthalpy system and ending with low-enthalpy processes. SWRO desalination was added, with some of the electricity generated at the front end of the geothermal recovery process used to operate it. The steam temperature proposed for operation of the system was 180°C, for recovery at the well head and entry into the generation turbine.

There is another concept that should be considered within the context of high-enthalpy systems that could be jointly applied to electricity generation and the SWRO process. In the event that a dry-steam recovery facility was to be operated at a steam temperature of about 300°C, the pressure in the geothermal reservoir would be about 8.5 MPa (Fig. 6.13). Since a SWRO process operates at a pressure of about 6.0–7.0 MPa, a side-stream of the pressurized dry steam could be directed to some form of mechanical work exchanger to create the pressure to operate the SWRO process while the primary steam line would be used to generate electricity. No electricity-driven high-pressure pumps would need to be operated. Therefore, the geothermal system could be used to power all of the desalination processes operated on-site with use of a minimal quantity of electricity for the SWRO process, perhaps only for the high service pumps used to place the permeate into the distribution system, for intake and chemical-dosing pumps, and for internal lighting.

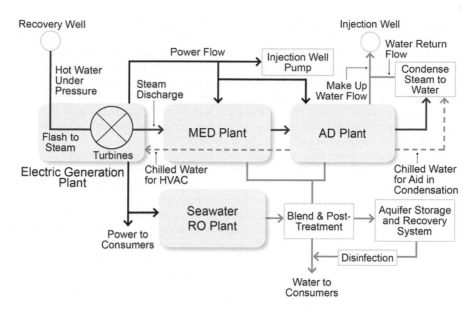

Figure 6.12. Conceptual design for the geothermal generation/desalination campus (Missimer *et al.*, 2016).

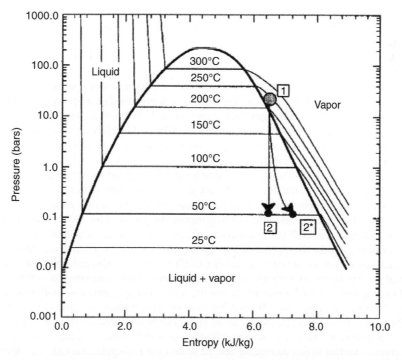

Figure 6.13. Pressure-entropy relationship for a dry-steam generating facility showing the pressure required to link SWRO with the thermal desalination process (adapted from Glassley, 2015).

Since dry-steam recovery systems are rare within the overall context of geothermal energy recovery sites, the use of a hot dry rock system with a higher steam recovery temperature and pressure could be considered for the direct mechanical driving of the SWRO process. This type of system would require the removal of particulates within the steam for both electricity generation and ahead of the mechanical work exchangers. The vapor component of the steam would also have to be removed. Some inefficiency would occur, therefore, and may require that the system initially be operated at 300°C (at 8.5 MPa) or slightly higher to produce the constant 6.0–7.0 MPa required to operate the SWRO process. This is a fundamentally new concept and would require proper modeling and assessment to finalize the engineering design.

ASR is a key element within the generation/desalination campus concept. In this case, the ASR system would be used to store treated water during times of excess treatment capacity. The treated water could be injected into the aquifer for recovery during peak demand or, if necessary, could be used to provide water security wherein the system storage would be increased (Ghaffour *et al.*, 2013a; Missimer *et al.*, 2012). Given that it both generates electricity and desalinates seawater, the system could also be operated in various modes to meet the two differing demand patterns. In the Middle East and North Africa (MENA) region, the treated water demand is fairly constant despite some seasonal changes, but the electricity demand rises sharply during the hot summer season (Missimer *et al.*, 2012). The flexibility of a generation/desalination campus with ASR would allow the geothermal energy recovered to be diverted to increased electricity generation and decreased water treatment during the peak power requirement period, with additional potable water being extracted from storage to meet demand during this period. Many different potential configurations and capacities could be designed into the campus system to cover the demand patterns of virtually any location in the world.

## 6.7    LIFE-CYCLE COST ANALYSIS COMBINED WITH GREENHOUSE GAS EMISSIONS

Several general cost analyses and a few life-cycle economic analyses have been performed to evaluate geothermal electricity systems (GEA, 2005; Harrison *et al.*, 1990; Sullivan *et al.*, 2010), but few life-cycle cost analyses have been performed on any of the high-enthalpy hybrid systems described in this chapter (Hamed *et al.*, 2006).

The economic analysis conducted by Sullivan *et al.* (2010) produced a substantial comparison between geothermal and other electricity generation systems based on a life-cycle analysis which included the impacts of greenhouse gas (GHG) emissions. This report demonstrated that enhanced geothermal systems are one of the lowest emitters of GHGs. The elimination of the hydrocarbon fuel cycle produces fewer economic and environmental impacts in comparison to nearly any of the other electricity generation methods, and the total energy consumption per kWh output for geothermal generation systems (EGS, hydrothermal-binary and hydrothermal-flash) are very favorable when compared to the other electricity generating technologies (Fig. 6.14).

Since a significant proportion of desalination cost is for the energy required to either heat the feed water for thermal desalination or to generate the pressure for SWRO, any reduction in energy cost will reduce overall treatment cost and cost to the consumer (Ghaffour *et al.*, 2013b). Therefore, linking desalination processes to geothermal electricity generation reduces the cost of the processes as well as the cost impacts of GHG generation. Overall, the cost to develop a geothermal energy reservoir with electricity generation is high compared to construction of a standard hydrocarbon-fuelled power plant, but the long-term operating cost will be significantly lower given no fuel cycling costs. By linking in desalination processes, the development costs for geothermal reservoir development can be spread over two product streams, electricity and freshwater production, thereby providing effective cost reduction, and the reduction in GHG impacts creates another major economic benefit within the life-cycle analysis. The development of detailed life-cycle cost analyses for linked geothermal-powered electricity generation and desalinated water production should be a primary research objective for researchers in this field.

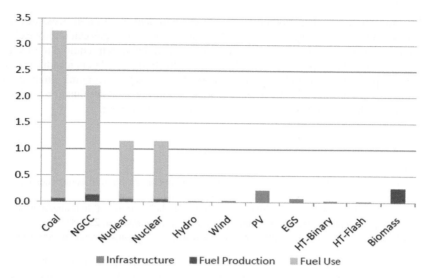

Figure 6.14.  Total energy consumption [kWh] per kWh output modeled for different electricity-generating technologies (Sullivan *et al.*, 2010).

## 6.8  SUMMARY AND CONCLUSIONS

Geothermal energy is the only renewable energy source that can be developed nearly anywhere on the Earth. It is not only renewable, if designed properly, but provides a baseload energy source that produces 24 hours a day, seven days a week. Geothermal energy systems used to generate electrical power have the highest capacity factors among all generation technologies (93–95%, with coal at 60% and solar at 22–27%). Geothermal systems also produce very low greenhouse gas emissions and, despite relatively high development costs, produce very favorable economic benefits in a comparative life-cycle cost analysis.

A key aspect of geothermal energy development is the location of harvesting facilities at optimal positions where there is high natural heat flow close to the land surface, so-called "hot spot" areas. Another key issue is the design of heat collection systems that operate in a balanced fashion, allowing replenishment of the heat extracted in order to maintain the system in a sustainable manner. The heat reservoir must be designed to minimize the impacts of extraction, particularly induced seismicity. This can be controlled by either operating an injection and recovery system at lower pressures or locating the system away from large population centers that would be impacted by seismic vibration. Another important economic element is the co-location of geothermal energy development with other processes to allow multiple products to be generated, such as electricity and desalinated water, or perhaps direct heating of houses and/or industrial uses of the heat.

Desalination of brackish water or seawater using any of the different geothermal-powered systems described herein represents an environmentally friendly method of producing an abundant supply of freshwater. Linking desalination with electricity generation provides two products from the development of a single geothermal reservoir and may significantly improve the overall economics of geothermal energy development.

Geothermal desalination systems can be linked to both low- and high-enthalpy desalination processes. Desalination systems can, therefore, be developed at a variety of capacities to meet local need, particularly where hot water or steam is emergent at the surface (e.g. some Greek Islands and parts of Italy). In hot, arid regions, low-enthalpy desalination systems can be developed with combined-cycle solar and geothermal energy sources to create innovative hybrids that meet local needs.

Finally, innovation in which geothermal energy development, multiple-process desalination, and aquifer storage and recovery are linked within a single site or campus can improve efficiency within an operational mode. There is still more innovation and research required to link as many useful processes together as possible to maximize the utilization of the latent heat harvested, with as little loss as possible to the ambient surroundings. New life-cycle economic analyses need to be performed to demonstrate the effectiveness of geothermal-powered desalination, including the reduction of greenhouse gas emissions within an economic framework.

## REFERENCES

Al-Enezi, G., Ettouney, H. & Fawzy, N. (2006) Low temperature humidification dehumidification desalination process. *Energy Conversion and Management*, 47, 470–484.

Al-Katheeri, E. & Agaschichev, S.P. (2008) Feasibility of the concept of hybridization of existing co-generation plant with reverse osmosis and aquifer storage. *Desalination*, 222, 87–95.

Alkhudhiri, A., Darwish, N. & Hilal, N. (2012) Membrane distillation: a comprehensive review. *Desalination*, 287, 2–18.

Al-Obaidani, S., Curcio, E., Macedonio, F., Profio, G.D., Al-Hinai, H. & Drioli, H. (2008) Potential of membrane distillation in seawater desalination: thermal efficiency, sensitivity study and cost estimation. *Journal of Membrane Science*, 323, 85–98.

Alsaadi, A., Ghaffour, N., Li, J.D., Gray, S., Francis, L., Maab, H., Nunes, S. & Amy, G. (2013) Modeling of air-gap membrane distillation process: a theoretical and experimental study. *Journal of Membrane Science*, 445, 53–65.

Awerbach, L. (2007) Hybrid systems and technology. In: Wilf, M. (ed.) *The Guidebook to Membrane Desalination Technology: Reverse Osmosis, Nanofiltration and Hybrid Systems Process, Design, Application and Economics*. Balaban, L'Aquila. pp. 395–453.

Awerbach, L., Rogers, A.N. & Fernelius, W.A. (1976) Geothermal desalination. In: Delyannis, A. & Delyannis, E. (eds) *Proceedings of the Fifth International Symposium on Fresh Water from the Sea, 16–20 May 1976, Alghero, Italy*. European Federation of Chemical Engineering, Athens.

Axelsson, C. (1991) Reservoir engineering studies of small low-temperature hydrothermal systems in Iceland. *Proceedings, Sixteenth Workshop on Geothermal Reservoir Engineering, 23–25 January 1991, Stanford, CA*.

Benjemaa, F., Houcine, I. & Chahbani, M.H. (1999) Potential of renewable energy development of water desalination in Tunisia. *Renewable Energy*, 18(3), 331–347.

Blackwell, D., Richards, M., Frone, Z., Batir, J., Ruzo, A., Dingwall, R. & Williams, M. (2011) Temperature at depth maps for the conterminous US and geothermal resource estimates. *Geothermal Resources Council Transactions*, 35, 1545–1550.

Bouguecha, S. & Dhahbi, M. (2003) Fluidized bed crystallizer and air gap membrane distillation as a solution to geothermal water desalination. *Desalination*, 152(1–3), 237–244.

Bourouni, I., Martin, R. & Tadrist, L. (1999a) Analysis of heat transfer and evaporation in geothermal desalination units. *Desalination*, 122(2–3), 301–303.

Bourouni, I., Martin, R., Tadrist, L.M. & Chaibi, M.T. (1999b) Heat transfer and evaporation in geothermal desalination units. *Applied Energy*, 64(1–4), 129–147.

Bourouni, I., Chaibib, M. & Tadrist, L. (2001) Water desalination by humidification and dehumidification of air: state of the art. *Desalination*, 137, 167–176.

Brown, D.W., Duchane, D.V., Heiken, G. & Hrison, V.T. (2012) *Mining the Earth's Heat: Hot Dry Rock Geothermal Energy*. Springer, New York.

Bundschuh, J., Ghaffour, N., Mahmoudi, H., Goosen, M., Mushtaq, S. & Hoinkis, J. (2015) Cheap low-enthalpy geothermal heat for freshwater production: innovative applications using thermal desalination processes. *Renewable and Sustainable Energy Reviews*, 43, 196–206.

Buros, O.K. (1990) The ABCs of desalting. International Desalination Association, Topsfield, MA.

Chakraborty, A., Saha, B.B., Ng, K.C., Koyama, S. & Srinivasan, K. (2009a) Theoretical insight of physical adsorption for single component adsorption-adsorbate system. I. Thermodynamic property surfaces. *Langmuir*, 25(4), 2204–2211.

Chakraborty, A., Saha, B.B., Ng, K.C., Koyama, S. & Srinivasan, K. (2009b) Theoretical insight of physical adsorption for single component adsorption-adsorbate system. II. The Henry region. *Langmuir*, 25(13), 7359–7367.

Chakraborty, A., Leong, K.C., Thu, K., Saha, B.B. & Ng, K.C. (2011a) Theoretical insight of adsorption cooling. *Applied Physics Letters*, 98(22), 221910.

Chakraborty, A., Thu, K. & Ng, K.C. (2011b) Advanced adsorption cooling cum desalination cycle: a thermodynamic framework. *ASME 2011 International Mechanical Engineering Congress and Exposition, Volume 4: Energy Systems Analysis, Thermodynamics and Sustainability; Combustion Science and Engineering; Nanoengineering for Energy, Parts A and B, Denver, Colorado, USA, November 11–17, 2011.* American Society of Mechanical Engineers, New York. pp. 605–610.

Chandler, P.B. (1982) Coastal zone geothermal desalination. *Technical Proceedings: 10th Annual Conference and Trade Fair of the Water Supply Improvement Association, July 25–29, 1982, Honolulu, Hawaii.* Water Supply Improvement Association, Topsfield, MA.

Chaturvedi, L., Keyes, C.G., Swanberg, C.A. & Gupta, Y.P. (1979) Use of geothermal energy for desalination in New Mexico. New Mexico Energy Institute Report No. NMEI-42, Las Cruces, NM.

Curcio, E. & Drioli, E. (2005) Membrane distillation and related operations – a review. *Separation and Purification Reviews*, 34, 35–86.

Davies, J.H. (2013) Global map of solid earth surface heat flow. *Geochemistry, Geophysics, Geosystems*, 14(10), 4608–4622.

Davies, P.A. & Orfi, J. (2014) Self-powered desalination of geothermal saline groundwater: technical feasibility. *Water*, 6, 3409–3432.

Dipippo, R. (2005) *Geothermal Power Plants: Principles, Applications and Case Studies.* Elsevier Science, Oxford.

Dipippo, R. (2007) *Geothermal Power Plants: Principles, Applications, Case Studies and Environmental Impact.* 2nd edn., Butterworth-Heinemann, London.

Duchane, D.V. (1990) Hot dry rock: a realistic energy option. *Geothermal Resources Council Bulletin*, 19(3), 83–88.

El Amali, A., Bouguecha, S. & Maalej, M. (2004) Experimental study of air gap and direct contact membrane distillation configurations: application to geothermal and seawater desalination. *Desalination*, 168(1–3), 357–358.

Ettouney, H. (2005) Design and analysis of humidification dehumidification desalination process. *Desalination*, 183, 341–352.

Fan, R., Jiang, Y., Yao, Y., Shiming, D. & Ma, Z. (2007) A study on the performance of a geothermal heat exchanger under coupled heat conduction and groundwater advection. *Energy*, 32(11), 2199–2209.

Faust, C.R. & Mercer, J.W. (1979) Geothermal reservoir simulation. 1. Mathematical models for liquid and vapor-dominated hydrothermal systems. *Water Resources Research*, 15(1), 23–50.

Feng, Z., Zhao, Y., Zhou, A. & Zhang, N. (2012) Development program of hot dry rock geothermal resource in the Yangbajing Basin of China. *Renewable Energy*, 39, 490–495.

Francis, L., Ghaffour, N., Alsaadi, A. & Amy, G.L. (2013) Material gap membrane distillation: a new design for water vapor flux enhancement. *Journal of Membrane Science*, 448, 240–247.

Francis, L., Ghaffour, N., Alsaadi, A., Nunes, S.P. & Amy, G.L. (2014) Performance evaluation of the DCMD desalination process under bench scale and large scale module operating conditions. *Journal of Membrane Science*, 455, 103–112.

GEA (2005) Factors affecting costs of geothermal power development. Geothermal Energy Association, Washington, DC. Available from: http://geo-energy.org/reports/Factors%20Affecting%20Cost%20of%20Geothermal%20Power%20Development%20-%20August%202005.pdf [accessed November 2016].

Gehlin, S. & Hellstrom, G. (2003) Influence on thermal test response by groundwater flow in vertical fractures in hard rock. *Renewable Energy*, 28, 2221–2235.

Ghaffour, N. (2009) The challenge of capacity building strategies and perspectives for desalination for sustainable water use in MENA. *Desalination & Water Treatment*, 5, 48–53.

Ghaffour, N., Missimer, T.M. & Amy, G.L. (2013a) Combined desalination, water reuse and aquifer storage and recovery to meet water supply demands in the GCC/MENA region. *Desalination & Water Treatment*, 51, 38–43.

Ghaffour, N., Missimer, T.M. & Amy, G.L. (2013b) Technical review and evaluation of the economics of water desalination: current and future challenges for better water supply sustainability. *Desalination*, 309, 197–207.

Ghaffour, N., Lattemann, S., Missimer, T.M., Ng, K.C., Sinha, S. & Amy, G. (2014) Renewable energy-driven innovative energy-efficient desalination technologies. *Applied Energy*, 136, 1155–1165.

Ghaffour, N., Bundschuh, J., Mahmoudi, H. & Goosen, M.F.A. (2015) Renewable energy-driven desalination technologies: a comprehensive review on challenges and potential applications of integrated systems. *Desalination*, 356, 94–114.

Ghose, M.K. (2004) Environmentally sustainable supplies of energy with specific reference to geothermal energy. *Energy Sources*, 26, 531–539.

Glassley, W.E. (2015) *Geothermal energy: renewable energy and the environment*, 2nd edn. CRC Press, Boca Raton, FL.

Goldstein, B.A., Hill, A.J., Long, A., Budd, A.R., Holgate, F. & Malavazos, M. (2009) Hot dry rock energy plays in Australia. *Proceedings, Thirty-fourth Workshop on Geothermal Reservoir Engineering, 9–11 February 2009, Stanford University, Stanford, CA.*

Goosen, M., Mahmoudi, H. & Ghaffour, N. (2010) Desalination using geothermal energy. *Energies*, 3(8), 1423–1442.

Goosen, M., Mahmoudi, H. & Ghaffour, N. (2014) Today's and future challenges in applications of renewable energy technologies for desalination. *Critical Reviews in Environmental Science and Technology*, 44, 929–999.

Gunnarsson, A., Steingrimsson, B., Gunnlaugsson, E., Magnusson, J. & Maack, R. (1992) Nesjavellir geothermal co-generation power plant. *Geothermics*, 21, 559–583.

Gutiérrez, H. & Espíndola, S. (2010) Using low enthalpy geothermal resources to desalinate seawater and electricity production on desert areas in Mexico. *GHC Bulletin*, August 2010, 19–24.

Hamed, O.A., Al-Washmi, H.A. & Al-Otaibi, H.A. (2006) Thermoeconomic analysis of a power/water cogeneration plant. *Energy*, 31, 2699–2709.

Hammons, T.J. (2004) Geothermal power generation worldwide: global perspective, technology, field experience, research and development. *Electric Power Components and Systems*, 32, 520–553.

Harlow, F.H. & Pracht, W.E. (1972) A theoretical study of geothermal energy extraction. *Journal of Geophysical Research*, 77, 7038–7038.

Harrison, R., Doherty, P. & Coulson, I. (1990) HDR cost modeling. In: Baria, R. (ed) *Hot dry rock geothermal energy, Proceedings of the Camborne School of Mines International Conference, 27–30 June 1989, Redruth, Cornwall, UK*. Robertson Scientific Publications, London. pp. 245–261.

Hiriart, G. (2008) Geothermal energy for desalination of seawater. *33rd International Geological Congress Abstracts 33, Oslo, Norway.*

Kalogirou, S. (2005) Seawater desalination using renewable energy sources. *Progress in Energy Combustion Science*, 31, 242–281.

Karytsas, C. (1998) Low-enthalpy geothermal seawater desalination plants. *Bulletin of the Geothermal Resources Council*, 4, 111–115.

Khalifa, A.E. (2015) Water and air gap membrane distillation for water desalination – an experimental comparative study. *Separation and Purification Technology*, 141, 276–284.

Khayet, M. (2011) Membranes and theoretical modeling of membrane distillation: a review. *Advances in Colloid and Interface Science*, 164, 56–88.

Ko, A., Guy, D.B. & Cabibbo, S.V. (1979) Geothermal desalination and power generation. *Geothermal Resources Council Transactions*, 3, 341–344.

Kolditz, O. (1995) Modelling flow and heat transfer in fractured rocks: conceptual model of a 3-D deterministic fracture network. *Geothermics*, 24(3), 451–470.

Koroneos, C. & Roumbas, G. (2011) Geothermal waters heat integration for the desalination of seawater. *Desalination and Water Treatment*, 37, 1–8.

Kui, Z., Heinzl, W., Bollen, F., Lange, G., van Gendt, G. & Hoong, C.F. (2011) Demonstrating solar-driven membrane distillation using novel Memsys vacuum multi-effect-membrane-distillation (V-MEMD) process. *Proceedings of Singapore International Water Week, Suntec City, Singapore*. pp. 23–25.

Lindahl, B. (1973) Industrial and other applications of geothermal energy, except power production and district heating. In: Armstead, H.C.H. (ed) *Geothermal Energy*. UNESCO, Paris. pp. 135–148.

Majer, E.L., Baria, R., Stark, M., Oates, S., Bommer, J., Smith, B. & Asanuma, H. (2007) Induced seismicity associated with Enhanced Geothermal Systems. *Geothermics*, 36, 185–222.

Maliva, R.G. & Missimer, T.M. (2010) Aquifer storage and recovery and managed aquifer recharge using wells: planning, hydrogeology, design, and operation. *Methods in Water Resources Evaluation Series* No. 2. Schlumberger, Sugar Land, TX.

Mamed, O.A. (2005) Overview of hybrid systems – current status and future prospects. *Desalination*, 186, 2699–2709.

Miller, J.E. (2003) Review of water resources and desalination technologies. Report 2003-0800. Sandia National Laboratories, Albuquerque, NM.

Missimer, T.M. & Maliva, R.G. (2010) Efficiency improvement of co-located electric power and seawater desalination plants using aquifer storage and recovery technology. *Proceedings of Singapore Water Week 2010, Sustainable Cities: Clean and Affordable Water, 28 June–2 July 2010, Singapore.*

Missimer, T.M, Sinha, S. & Ghaffour, N. (2012) Strategic aquifer storage and recovery of desalinated water to achieve water security in the GCC/MENA region. *International Journal of Environment and Sustainability*, 1(3), 89–100.

Missimer, T.M., Kim, Y.-D., Rachman, R. & Ng, K.C. (2013) Sustainable renewable energy seawater desalination using combined-cycle solar and geothermal heat sources. *Desalination and Water Treatment*, 51, 1161–1170.

Missimer, T.M, Mai, M. & Ghaffour, N. (2014a) A new assessment of combined geothermal electric generation and desalination in western Saudi Arabia: targeted hot spot development. *Desalination and Water Treatment*, 55(11), 3056–3063.

Missimer, T.M., Ng, K.C., Thu, K. & Kim, Y. (2014b) Systems and methods for integrated geothermal electricity generation and water desalination. United States Provisional Patent Application 6197081.

Missimer, T.M., Ng, K.C., Thuw, K. & Shahzad, M.W. (2016) Geothermal electricity generation and desalination: an integrated process design to conserve latent heat with operational improvements. *Desalination and Water Treatment*, 57(48–49), 23,110–23,118.

Mohamed, A.M.I. & El-Minshawy, N.A.S. (2009) Humification-dehumification desalination system driven by geothermal energy. *Desalination*, 249(2), 602–608.

Ng, K.C., Thu, K., Hideharu, Y., Saha, B.B., Chakraborty, A. & Al-Ghasham, T.Y. (2010) Apparatus and method for improved desalination. Patent PCT/SG2009/000223.

Ng, K.C., Thu, K., Saha, B.B. & Chakraborty, A. (2012) Study on waste heat-driven adsorption cooling cum desalination cycle. *International Journal of Refrigeration*, 35, 685–693.

Noorollahi, Y., Pourarshad, M., Jalilinasrabady, S. & Yousefi, H. (2015) Numerical simulation of power production from abandoned oil wells in Ahwaz oil field in southern Iran. *Geothermics*, 55, 16–23.

Ophir, A. (1982) Desalination plant using low grade geothermal energy. *Desalination*, 40(1–2), 125–132.

Ozgener, O. & Kocer, G. (2004) Geothermal heating applications. *Energy Sources*, 26, 353–360.

Piller, M. & Liuzzo-Scorpo, A. (2013) Numerical investigation of forced convection from vertical boreholes. *Geothermics*, 45, 41–56.

Ragnarsson, A. (2003) Utilization of geothermal energy in Iceland. *Proceedings of the International Geothermal Conference, Sept. 2003, Reykjavik, Iceland*. S10 Paper 123, pp. 39–45.

Rodríguez, G., Rodríguez, M., Perez, J. & Veza, J. (1996) A systematic approach to desalination powered by solar, wind and geothermal energy sources. *Proceedings of the Mediterranean Conference on Renewable Energy Sources for Water Production, 10–12 June 1996, Santorini, Greece*. European Commission. pp. 20–25.

Sarbatly, R. & Chiam, C.K. (2013) Evaluation of geothermal energy in desalination by vacuum membrane distillation. *Applied Energy*, 112, 737–746.

Schaetzle, W.J., Brett, C.E., Grubbs, D.M. & Seppanen, M.S. (1980) *Thermal Energy Storage in Aquifers*. Pergamon Press, New York.

Smith, M.C. (1979) The future of hot dry rock geothermal energy systems. Publication 79-PVP-35, American Society of Mechanical Engineers, New York.

Smith, M.C. (1983) A history of hot dry rock geothermal energy systems. *Journal of Volcanology and Geothermal Research*, 15, 1–20.

Stober, I. & Bucher, K. (2013) *Geothermal Energy: From Theoretical Models to Exploration and Development*. Springer, Heidelberg, Germany.

Sullivan, J.L., Clark, C.E., Han, J. & Wang, M. (2010) Life-cycle analysis results of geothermal systems in comparison to other power outputs. Report ANL/ESD/10-5, U.S. Department of Energy, Oak Ridge, TN.

Swanberg, C.A., Morgan, P., Stoyer, C.H. & Witcher, J.C. (1977) An appraisal study of the geothermal resources of Arizona and adjacent areas in New Mexico and Utah and their value for desalination and other uses. Report No. NMEI-6-1, New Mexico Energy Institute, Las Cruces, NM.

Thu, K., Kim, Y.D., Amy, G., Chun, W.G. & Ng, K.C. (2003) A hybrid multi-effect distillation and adsorption cycle. *Applied Energy*, 104, 810–821.

Thu, K., Kim, Y.D., Amy, G., Chun, W.G. & Ng, K.C. (2014) A synergetic hybridization of adsorption cycle with multi-effect distillation (MED). *Applied Thermal Engineering*, 62(1), 245–255.

Wan, Z.J., Yhao, Y.S. & Kang, J.R. (2005) Forecast and evaluation of hot dry rock geothermal resource in China. *Renewable Energy*, 30(12), 1831–1846.

Wright, R.R. & Missimer, T.M. (1999) Power optimization in membrane plants using aquifer storage and recovery. *Proceedings, International Desalination Association World Congress on Desalination and Water Reuse, San Diego, CA*. pp. 361–376.

Zaragoza, G., Ruiz-Aguirre, A. & Guillén-Burrieza, E. (2014) Efficiency in the use of solar thermal energy of small membrane desalination systems for decentralized water production. *Applied Energy*, 130, 491–499.

# CHAPTER 7

## Fuel cells as an energy source for desalination applications

Nadimul H. Faisal, Rehan Ahmed, Sheikh Z. Islam, Mamdud Hossain,
Mattheus F.A. Goosen & Sai P. Katikaneni

### 7.1 INTRODUCTION

Nowadays, from academia to industry, there is a renewed interest in fuel cell technology, with a primary focus in the area of electrochemistry and catalysis science. This interest is due to environmental legislation around $CO_2$ and other greenhouse gas emissions (Tamiotti, 2009) that demands the use of high-efficiency energy production systems. Such systems also have great potential in the area of desalination technology (Al-Hallaj et al., 2004; Kenet and Belmar, 2003; Singh, 2008; Wang et al., 2011). Fuel cells are characterized by high operational efficiency, which results in decreased fuel consumption and a lower environmental impact. A fuel cell is a device that converts the chemical energy of a fuel directly into electricity through electrochemical reactions, with significant waste heat (e.g. solid oxide fuel cell in Fig. 7.1). The first fuel cell was fabricated in the 1830s, and slow but steady progress has been made toward their commercialization ever since.

Ensuring access to clean and fresh (i.e. potable) water is among the major challenges faced by the world's growing population (Wali, 2014). A recent report, published by the World Resources Institute (Maddocks et al., 2015), puts the Middle East high in the ranking of the world's most water-stressed regions. According to this report, 14 of the 33 most water-stressed countries by 2040 will be in the Middle East, nine of which are expected to be extremely stressed: Bahrain, Kuwait, Palestine, Qatar, United Arab Emirates, Israel, Saudi Arabia, Oman and Lebanon. In general, rapidly growing populations will drive increased consumption by people, agriculture and industry. However, it is not clear where all this water will come from. Water desalination will, therefore, play a dominant role in resolving water scarcity, and solid oxide fuel cell (SOFC) technology is one of the key technological advances that can enable sustainable water desalination.

As described by Ghaffour et al. (2015), Goosen et al. (2014), Khawaji et al. (2008) and Jones (2013), desalination processes require external energy. A variety of desalination technologies have been developed over the years on the basis of thermal distillation, membrane separation, freezing and electrodialysis. The two most important technologies are based on the multi-stage flash (MSF) and reverse osmosis (RO) technologies. It is anticipated that these processes, which include membrane distillation (MD), will be dominant and competitive in the future (Jones, 2013; Khawaji et al., 2008). In 1999, approximately 78% of the world's seawater desalination capacity consisted of MSF-based plants, while RO-based plants represented only 10%. However, there has been a gradual increase in RO seawater desalination, primarily due to its lower cost and simplicity (Khawaji et al., 2008). A variety of energy sources, such as wind turbines, solar thermal systems, photovoltaics (PV), biogas plants and fuel cells, are being widely used for desalination, but some of these only provide intermittent power (Chittu and Jeyaprabha, 2013). As a solution to this problem, the intermittent energy excess may be used to split water in an electrolyzer and then used to produce hydrogen to run a fuel cell that can provide a backup energy source for desalination purposes (Kenet and Belmar, 2003).

The development of low-cost fuel cell materials with high durability and lower operating temperatures is the key technical challenge facing a range of fuel cell technologies. The future of

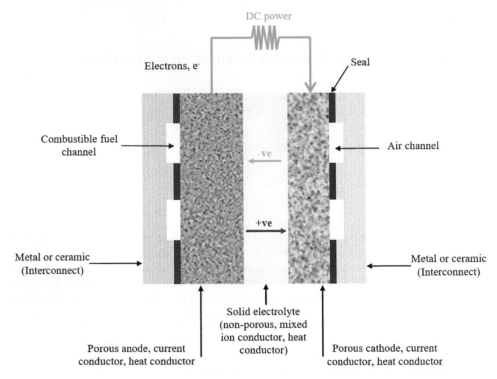

Figure 7.1.    Scheme for SOFC. Fuel is fed to the anode (negative electrode) and an oxidant is fed to the cathode (positive electrode). Electrochemical oxidation and reduction reactions take place at the electrodes and produce electric current. In a traditional SOFC, oxygen (from air) is reduced by a porous cathode to produce oxide ions ($-$ve) which migrate through a solid electrolyte to the porous anode and react with the fuel ($H_2$, CO or $CH_4$) to form $H_2O$ and/or $CO_2$. The electrolyte may conduct either oxygen (SOFC) or, in a proton-conducting SOFC, hydrogen ions (PC-SOFC). In a PC-SOFC, the reaction product (water vapor) is evolved at the cathode side instead of at the anode (fuel side) as in the case of oxide-ion-conducting SOFCs. Thus, no dilution of the fuel takes place in a PC-SOFC, resulting in significantly higher efficiency (adapted from Faisal et al., 2015).

this technology depends upon the development of new materials (i.e. electrodes and electrolytes) which can be used to manufacture fuel cells in a cost-effective manner. Furthermore, fuel cells present a number of inherent challenges. Low mechanical strength, slow start-up time (i.e. measured in minutes) and serious anode deterioration represent some of these challenges. The high operating temperatures (800–1000°C) place additional durability demands on fuel cell materials. The development of low-cost materials with high durability at lower operating temperatures is the key technical challenge facing fuel cell technology. Reducing the operating temperature to an intermediate range can lower cost but it also reduces reaction rate (Shao et al., 2005). The purpose of this section is to summarize the fuel cell systems (Table 7.1) that are actively involved in power generation (i.e. converting the chemical energy of fuels such as natural gas or oil into electrical power).

As shown in Figure 7.1, SOFCs can take many different forms but they all share the same basic structure. There will be a solid oxide electrolyte, sandwiched between a cathode and anode. The cathode is typically a thin, porous layer, where oxygen reduction will occur. The electrolyte is a dense ceramic layer which is ionically conductive but electrically insulating. The electrolyte does not characteristically become ionically active until it is at a temperature between 850 K and

Table 7.1.   Fuel types (some data compiled from Al-Hallaj *et al.* (2004), Singh (2008) and Al-Hallaj and Kiszynski (2011)).

| # | Fuel cell type | Operating temperature [°C] | Power ratio [heat:electric] | Energy grade | Electrodes/electrolytes |
|---|---|---|---|---|---|
| 1 | Solid oxide fuel cell (SOFC) | 650–1000 | 0.67:1 | High | Ceramic; charge carrier: $O^-$ |
| 2 | Molten carbonate fuel cell (MCFC) | 300–700 | 1:1 | High | Immobilized liquid molten carbonate; charge carrier: $CO_3^-$ |
| 3 | Phosphoric acid fuel cell (PAFC) | 80–200 | 1.27:1 | Moderate | Immobilized liquid phosphoric acid; charge carrier: $H^+$ (proton) |
| 4 | Proton-exchange membrane fuel cell (PEMFC) | 80–120 | – | Moderate | Ion-exchange membrane; charge carrier: $H^+$ (proton) |
| 5 | Direct methanol fuel cell (DMFC) | 50–120 | – | Moderate | Polymer electrolyte; charge carrier: $H^+$ (proton) |
| 6 | Alkaline fuel cell (AFC) | 50–90 | – | Low | Mobilized or immobilized potassium hydroxide; charge carrier: $H^+$ (proton) |
| 7 | Microbial fuel cell (MFC) | 20–40 | – | Low | Chemical that transfers electrons from the bacteria in the cell to the anode |

1250 K. The ceramic anode layer is typically a porous layer to ensure maximum contact between the fuel and the anode surface. Once the cell is at an optimum temperature, spare electrons in the cathode layer reduce oxygen atoms from either the air or an oxygen supply to give oxygen ions. These ions diffuse through the ionically conductive electrolyte to the anode, where they can oxidize a fuel. This reaction produces a combination of water and $CO_2$ (dependent on the fuel), as well as electrons. Since the electrolyte is electrically insulating, the electrons flow back to the cathode through an external circuit, where they can do work on a load. This cycle will repeat as long as sufficient fuel and oxygen can be supplied. As summarized by Singh (2008), it is possible to construct a fuel cell of high efficiency (as high as 80 to 90%). However, in current practice, because of irreversible losses (over-potentials), the efficiency of a fuel cell system is 40 to 45%, based on the net calorific value (NCV) of the fuel. However, efficiencies of 80% have been achieved for fuel cell power plants with cogeneration (i.e. combined heat and power systems), and hybrid fuel cell-gas turbine cycles have efficiencies approaching 70%.

Novel works on the integration of fuel cells with desalination systems are summarized in this chapter. The underlying motivation for such system integration is that the exhaust from a power plant (e.g. fuel cell) contains a considerable amount of thermal energy, which may be harnessed for desalination units. Water is an important resource for all and it is essential for agricultural and industrial growth, as well as for populations who require safe drinking water. It is known that 97% of the entire world's water can be found in oceans, with 2% in glaciers and ice caps, and the rest in lakes, rivers and underground aquifers (Al-Shayji, 1998). Available freshwater accounts for less than 0.5% of the earth's total water supply (Khawaji, 2008). Natural resources alone cannot satisfy the growing demand for low-salinity water for industrial development either, together with the increasing worldwide demand for supplies of potable water.

In the Arabian Gulf countries, most power plants now use cogeneration, producing electrical power and heat for water desalination processes (Al-Hallaj *et al.*, 2004). For a given fuel input, the production of water in a cogeneration system is associated with a reduction in electrical power generated. Although desalination costs have decreased markedly in the last few decades, cost remains a primary factor in selecting a desalination technique for drinking water production. Some reduction in desalination costs may be realized from improvements in plant design, fabrication techniques, heat exchange materials, plant automation and scale control techniques (Al-Hallaj *et al.*, 2004). Energy costs in distillation plants (steam and electricity) represent at least 40 to 50% of the cost of the water produced (Lunghi *et al.*, 2001). The minimum cost of obtaining water from seawater desalination occurs when power and desalination are combined in a single "dual-purpose facility", which simultaneously produces electricity and water (Al-Hallaj *et al.*, 2004).

High-temperature fuel cells produce electricity and high-temperature exhaust gases, which can be used as a heat source for desalination applications (Al-Hallaj *et al.*, 2004; Jones, 2013). The exhaust gas temperature depends on the fuel cell type and may range from 20°C to 1000°C. This chapter outlines how fuel cell technologies can be used for desalination purposes. It also describes the current status and future prospects of combined/hybrid technologies.

## 7.2   ENERGY CONSUMPTION IN DESALINATION

According to Jones (2013) and the WateReuse Association (2011), a typical RO desalination plant requires 8.7–9.7 kWh of energy per thousand gallons of water processed. This supports water pretreatment and the actual seawater reverse osmosis process itself, which accounts for the largest proportion of energy consumed during the process. Post-treatment conditioning and pretreatment make up 1–2% of the energy consumed in RO desalination. As summarized by Jones (2013), RO desalination has proved effective in supplying water to regions that lack potable water. However, this process requires significant amounts of energy to produce clean drinking water.

Despite success in the commercial realm, desalination is an expensive and energy-intensive process. One way to reduce the energy expense associated with desalination is through the use of alternative cogeneration systems (Al-Hallaj *et al.*, 2004; Jones, 2013). This type of system would combine alternative energy technologies that use low-grade waste heat with a desalination system for heat and electrical energy production. However, with some processes, such as RO (discussed later in the chapter), requiring chemical preprocessing of feed streams and a large amount of equipment maintenance, the time and money required to maintain a desalination plant, even with the use of alternative technologies, makes such desalination unfeasible for places with small economies.

In an effort to improve the energy efficiency of desalination processes, researchers have been designing, simulating, developing and testing desalination plants that combine water treatment and energy production in a single location. The idea behind these combination plants is that the energy efficiency of the desalination process can be increased by utilizing the waste heat. This is an important aspect because of environmental legislation for $CO_2$ and other greenhouse gas emissions, according to the United Nations Environment Program and the World Trade Organization (Tamiotti, 2009).

## 7.3   INTEGRATED AND HYBRID SYSTEMS: FUEL CELLS WITH DESALINATION

Integration refers to a fuel cell being connected as an energy source for a desalinator, where a heat exchanger transfers waste heat from the fuel cell to the desalinator for the purpose of water desalination (Ghalavand *et al.*, 2015). The heat exchangers that play a key role in this active integration may be in the form of a tube bundle or coil that surrounds the fuel cell and/or its heated water output, and may be made of copper tubing or other heat-transfer-promoting material. Other possible forms of heat exchanger may include a concentric counterflow thin-film heat exchanger or a plate-type exchanger. Several case studies have been reported, describing fuel cell integration with desalination using reverse osmosis (RO), thermal multi-stage flash (MSF), membrane distillation (MD), multi-effect distillation (MED), multi-effect boiling (MEB) and mechanical vapor compression (MVC) units (Ghalavand *et al.*, 2015). In such examples, the underlying motivation for system integration is that the exhaust gas from a power plant

Table 7.2.    Desalination types and energy grade sources temperature range.

| # | Desalination type | Operating temperature or energy grade [°C] | Energy demands for desalination (Al-Hallaj and Kiszynski, 2011) | Reference |
|---|---|---|---|---|
| 1 | Reverse osmosis (RO) | 4–40 | 4–5 kWh m$^{-3}$ (electric) 4–5 kWh m$^{-3}$ (heat) | Al-Hallaj and Kiszynski (2011) |
| 2 | Multi-stage flash (MSF) | 25–110 | 4 kWh m$^{-3}$ (electric) 14 kWh m$^{-3}$ (heat) | Wu *et al.* (2012) |
| 3 | Membrane distillation (MD) | 20–85 | – | Gryta (2012) |
| 4 | Multi-effect distillation (MED) | 70–80 | 4 kWh m$^{-3}$ (electric) 11 kWh m$^{-3}$ (heat) | Bataineh (2016) |
| 5 | Multi-effect boiling (MEB) | Up to 70 | – | Darwish *et al.* (2006) |
| 6 | Mechanical vapor compression (MVC) | Up to 100 | – | Ghalavand *et al.* (2015) |

(i.e. fuel cell) contains considerable amounts of thermal energy, which may be effectively utilized in desalination units. Thus, the high operating temperature of SOFCs or molten carbonate fuel cells (MCFCs) (see Table 7.1) provides an opportunity to use the waste heat produced in desalination units (see operating temperature ranges listed in Table 7.2). Similarly, proton-exchange membrane fuel cells (PEMFC), which generally operate at lower temperatures (about 80–120°C), are also promising for desalination purposes (Kenet and Belmar, 2003). It was also reported that fuels cells in which phosphoric acid is the electrolyte (PAFCs), and which generate high waste heat of up to 180°C or more, are nevertheless considered less efficient or practical than the lower-temperature PEMFC technology. Although only a few case studies are presented in the following section, the concept of integrating fuel cells with desalination units should not be not limited to RO, MSF and MD mechanisms. Similar benefits can accrue with other desalination technologies (e.g. MED, MEB and MVC).

Various fuel cells release a considerable amount of waste energy (Table 7.1) during operation in the form of hot water or steam (Al-Hallaj *et al.*, 2004). This waste energy can be captured by cogeneration systems to improve overall hybrid fuel cell/desalination (HFCD) system efficiency. It is important to recognize that HFCD systems can be more efficient than separate fuel cell desalination systems, and the use of excess electricity generated by fuel cells for water desalination can eliminate the need for electricity storage in batteries during off-peak hours (Al-Hallaj and Kiszynski, 2011). Al-Hallaj and Kiszynski (2011) have also suggested that in HFCD system design, the demand for electricity should be afforded greater significance than the demand for water, as variations in water demand are never as sharp as those in electricity demand and water storage is easier and cheaper than electricity storage. Hence, HFCD systems must be designed according to the peak electricity demand. A general schematic for integrating fuel cells with desalination systems is shown in Figure 7.2.

The maximum power outputs from fuel cells can be calculated using a simple heat transfer analytical method. The difference between the fuel cell temperature and the desalination temperature will determine the power ($Q$) that can be exchanged between the cooling air and brine streams, as analyzed by Al-Hallaj and Kiszynski (2011):

$$Q = \dot{m}C(T_{\text{FuelCell}} - T_{\text{Desalination}}) \tag{7.1}$$

where $T_{\text{FuelCell}}$ is the fuel cell temperature, $T_{\text{Desalination}}$ is the top brine temperature ($TBT$) of the desalination process, $C$ is the lower specific heat capacity between the water and the air, and $\dot{m}$ is the mass flow rate of the cooling air. These two temperatures and the mass flow rate of the cooling

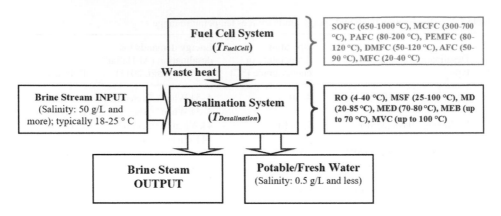

Figure 7.2.    General schematic for integrating fuel cells with desalination systems.

air are dictated by what type of fuel cell and what type of desalination unit are used (refer to Table 7.1 and Table 7.2). From Table 7.1, a 1 MW PAFC plant will provide 1.27 MW of thermal power (operating temperature between 80 and 200°C). Therefore, the mass flow rate can be calculated from Equation (7.2):

$$Q = \dot{m}h(T) \tag{7.2}$$

where $h(T)$ is enthalpy as a function of temperature (Al-Hallaj and Kiszynski, 2011).

### 7.3.1    *Integration with reverse osmosis desalination units*

Reverse osmosis (RO) is a pressure-driven process commonly used for water distillation where electricity is used to drive the feed pump, thereby increasing the feed pressure, which ranges between 50 and 80 atm (≈5.07–8.11 MPa) (Bataineh, 2016; Jones, 2013; Khawaji *et al.*, 2008; Lee *et al.*, 2011). In this process, an aqueous solution is pressurized above the osmotic pressure of the solvent. The pressurized solution is then sent through the center of a semi-permeable, spiral-wound membrane, where the solvent passes through the membrane pores to the product collection compartment at the center of the membrane module, leaving the solute behind. Membranes used for RO are typically made from cellulose acetate (e.g. Fig. 7.3a). It is important that the pores of the membrane only be large enough to allow the solvent molecules to pass through. Furthermore, the solution must be pretreated before coming into contact with the membrane. Pretreatment is needed to prevent membrane fouling – surface build-up and pore clogging, which can decrease the usable life of the membrane. In addition, the spiral-wound design of the membrane only allows one-way flow, which makes it impossible to clean the membrane module of unwanted solids and residues. Typical energy consumption in seawater RO plants operating at 40–45% product water recovery and with energy recovery from the high-pressure reject stream is about 3–4 kWh m$^{-3}$ of potable water produced (Singh, 2008).

RO can work with a low grade of energy source (4–40°C). Al-Hallaj *et al.* (2004) presented a novel fuel cell/desalination system (Fig. 7.3b) where two types of hybrid system were presented (i.e. fuel cell/reverse osmosis and fuel cell/thermal desalination). It was shown that by preheating the feed water to the RO system using the exhaust gas from the fuel cell, the energy demand of the desalination system reduced by 8%.

### 7.3.2    *Integration with thermal multi-stage flash desalination units*

As shown in Figure 7.4a, multi-stage flash (MSF) desalination is a thermal process consisting of a set of *n* stages, each of which operates at a lower pressure and temperature than the preceding

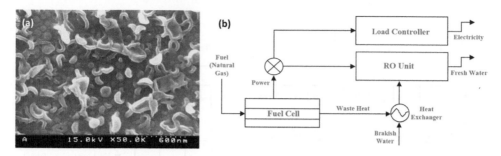

Figure 7.3. (a) Field emission scanning electron micrographs of RO membrane surface: *m*-phenylenediamine (MPD) in aqueous phase and trimesoyl chloride (TMC) in organic phase (reproduced with permission from Kwak and Ihm, 1999); (b) schematic representation of a hybrid system combining fuel cell and reverse osmosis unit (adapted from Al-Hallaj, 2004).

one (Jones, 2013; Tarifa and Scenna, 2001). In this process, a high-salinity aqueous solution is fed into the system at stage *n* through the condensing pipes. From stage *n* to stage 1, the solution simultaneously cools the newly distilled water vapor on the outer surface of the cooling pipes as it is heated by the latent heat released from the condensing vapor. Once the solution exits stage 1, it enters a brine heater where its temperature is raised to the saturated temperature of water at the maximum operating pressure of the plant. The heated solution then re-enters stage 1. Its pressure is immediately reduced, causing a portion of the heated solution to flash into vapor. The vapor travels through a wire mesh, which removes any remaining salt, and condenses on the set of cooling pipes within that stage. This process is repeated until the solution reaches stage *n* (Jones, 2013; Kalogirou, 2005; Khawaji *et al.*, 2008; Tarifa and Scenna, 2001).

MSF can work with a wide grade of energy source (25–100°C). In the configuration shown in Figure 7.4b, Al-Hallaj *et al.* (2004) demonstrated that such a hybrid fuel cell/thermal desalination system is 5.6% more efficient than a fuel cell and thermal desalination system operating separately. In this case, the exhaust from the fuel cell replaces some of the steam feed requirement met by a gas turbine and is, therefore, expected to enable increased power generation from the steam turbine while improving the overall efficiency of the hybrid system.

### 7.3.3   *Integration with membrane distillation desalination units*

Membrane distillation (MD) is a thermally driven process; a low-temperature, low-pressure process that operates at near atmospheric pressure (Alkhudhiri *et al.*, 2012; Jones, 2013; Khawaji *et al.*, 2008). It is a process in which an aqueous solution is heated and placed in direct contact with a microporous (about 10 nm to 1 μm), hydrophobic membrane. The membranes used are typically made of polypropylene, polyethylene, polyvinylidene fluoride or polytetrafluoroethylene (PTFE) (Jones, 2013). The material properties of the membrane prevent liquid from entering the membrane pores, and this creates a liquid/vapor interface at the entrance to the membrane pores. The opposite side of the membrane (i.e. the permeate collection side) is at a lower temperature. The temperature difference between the two surfaces of the membrane creates a vapor pressure drop within the membrane pores, which allows volatile materials in the feed to diffuse through the pores as the liquid feed comes into contact with the solid membrane. The volatile materials are collected on the cooler side of the membrane. As shown in Figure 7.5, there are four types of MD system: air-gap membrane distillation (AGMD), direct-contact membrane distillation (DCMD), vacuum membrane distillation (VMD) and sweeping-gas membrane distillation (SGMD).

MD can work with a low-grade energy source (20–85°C). As revealed by Jones (2013), it is expected that the heat discarded by the fuel cell will be sufficient for the amount of heat

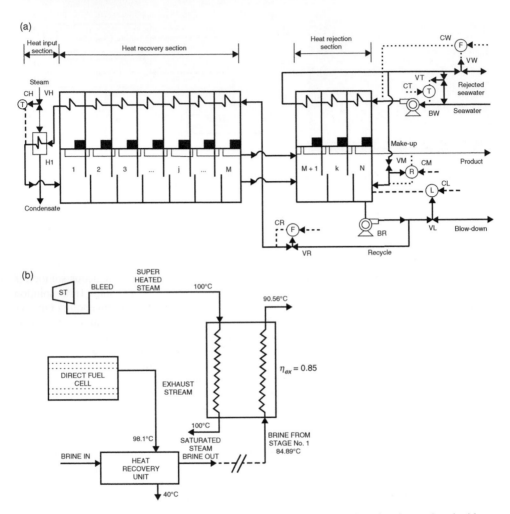

Figure 7.4.    Multi-stage flash (MSF) process: (a) desalination system (adapted and reproduced with permission from Tarifa and Scenna, 2001); (b) integration with molten carbonate fuel cell (MCFC) (adapted and reproduced with permission from Al-Hallaj *et al.*, 2004).

required for direct-contact membrane distillation (DCMD), as the temperature range for this process (20–80°C) is within the range of operating temperatures for a Nafion® membrane PEMFC. PEM stands for proton-exchange membrane (or polymer electrolyte membrane) and PEMFCs use a proton-conductive polymer membrane as electrolyte. Sometimes PEMFCs are also called polymer-membrane fuel cells, or just membrane fuel cells. Jones (2013) also expected that the power produced by a combination of a PEMFC and a steam turbine that is part of a steam generation system can have the capacity to provide all of the electrical power needed for both steam generation and pumping.

### 7.3.4    *Integration with multi-effect distillation units*

It is recognized that multi-effect distillation (MED) is the most efficient distillation process and that reverse osmosis (RO) offers the lowest energy consumption (Bataineh, 2016). In the MED thermal process (Fig. 7.6), the column pressures are adjusted such that the cooling (i.e. thermal energy removal) of one column functions as the heating source (thermal energy input) for another

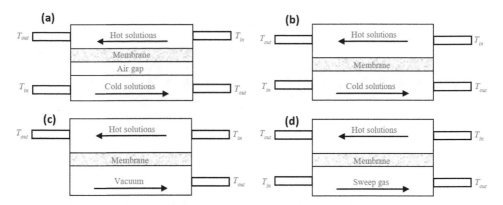

Figure 7.5. Membrane distillation (MD) systems: (a) air-gap membrane distillation (AGMD); (b) direct-contact membrane distillation (DCMD); (c) vacuum membrane distillation (VMD); (d) sweeping-gas membrane distillation (SGMD) (adapted from Alkhudhiri *et al.*, 2012).

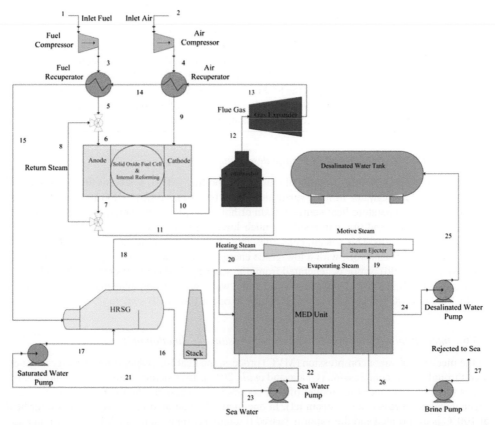

Figure 7.6. Integration of multi-effect distillation (MED) process with SOFC-GT (reproduced with permission from Meratizaman *et al.*, 2016).

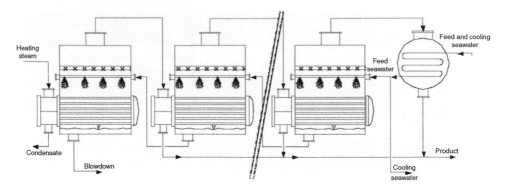

Figure 7.7. Multi-effect boiling (MEB) process: backward-feed multi-effect desalting system using horizontal tube evaporators (reproduced with permission from Darwish *et al.*, 2006).

column (Al-Shammiri and Safar, 1999; Bataineh, 2016; Ghalavand *et al.*, 2015). In the MED process, the seawater enters the first effect and its temperature is raised to its boiling point. Both the water feed and heating vapor to the evaporators flow in the same direction. The remaining water is pumped to the second effect, where it is once more applied to a tube bundle. This process continues for several effects, typically from 4 to 21 in a big plant. An increase in flashing stages can increase the internal energy recovery.

MED can work with a low-grade energy source (70–80°C) (Bataineh, 2016), and the thermal nature of the MED process means that the integration of a desalination unit with a high-temperature power cycle, such as a gas turbine (GT) in combination with a SOFC, gives an electrical efficiency of about 60% (Meratizaman *et al.*, 2014, 2016). Meratizaman *et al.* (2014) presented such a MED unit as a solution for heat recovery in the SOFC 300–1000 kW range of a SOFC-GT power cycle. The exhaust heat of the SOFC-GT power cycle was used in a heat-recovery steam generator to produce the motive steam required for the desalination unit.

### 7.3.5    *Integration with multi-effect boiling desalination units*

As shown in Figure 7.7, the conventional multi-effect boiling (MEB) system is the oldest method used to desalinate seawater in large quantities (Darwish *et al.*, 2006). MEB has the advantage of using a low-temperature heat source (steam or hot water) when it operates at a low top brine temperature (*TBT*), and this can result in much lower equivalent work or available consumed energy than MSF units. This system consumes about half the pumping energy of MSF (Darwish *et al.*, 2006). MEB can work with a low-grade energy source (up to 70°C) and, as described by Darwish *et al.* (2006), vapor is generated from hot water by flashing as it enters a flash chamber upstream of the first effect. They also show that hot water produced by phosphoric acid fuel cells (PAFC) at 60–65°C can be used as a heat source to operate an MEB system.

### 7.3.6    *Integration with mechanical vapor compression desalination unit*

In the mechanical vapor compression (MVC) process, the heat for evaporating seawater is generated through vapor compression (Ghalavand *et al.*, 2015). Two methods are used to condense the water vapor and produce the amount of heat necessary to evaporate the incoming seawater (e.g. mechanical compressor and a steam jet). In one method, illustrated in Figure 7.8, seawater held at 100°C is evaporated and the vapor is passed through a compressor, which leads to an increase in vapor dew point, so vapor can be condensed by seawater indirectly.

MVC can work with a high-grade energy source (up to 100°C). The power consumption in the MVC desalination method is high because of compressor usage (Ghalavand *et al.*, 2015), but

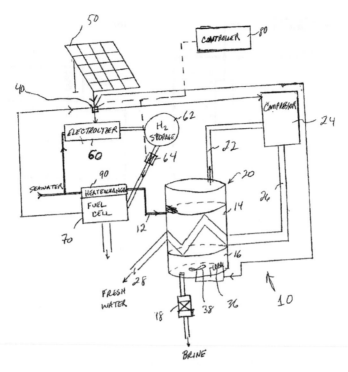

Figure 7.8. Integration of mechanical vapor compression (MVC) process with phosphoric acid fuel cell (PAFC) (US Patent 2003/0132097 A1, Kenet and Belmar, 2003) [Component details are as follows: 10, vapor compression desalinator; 12, water input line; 14, heat exchange section; 16, brine section; 18, valve; 20, evaporator; 22, vapor line; 24, compressor; 36, heater; 38, stirring device; 40, distributor; 50, renewable energy source; 60, electrolyzer; 62, hydrogen storage tank; 64, valve; 70, fuel cell; 80, controller; 90, heat exchanger]. (*Note:* this patented art is original).

it can become economically feasible when energized by fuel cells such as PAFCs, as proposed by Kenet and Belmar (2003), which operate at higher temperatures than PEMFCs. In MVC, the desalinated water vapor ideally passes through at least one compressor, which heats the water vapor, and then returns through the evaporator to heat brine in the evaporator. The evaporator of the desalinator may operate at approximately 40–45°C, and the input brine is typically at 18–25°C. Although PAFCs have been found to be less desirable than PEMFCs for many applications due to their acidic content and high temperatures (about 80–200°C), Kenet and Belmar (2003) suggested that PAFCs are preferred here because the fuel cell is stationary and the greater amount of waste heat is actually desirable for heating the desalinator. However, Kenet and Belmar (2003) also suggested that PEMFCs, which also operate at relatively elevated temperatures (about 80°C), may be used as well.

### 7.3.7 *Other examples of fuel cell integration with desalination units*

There are only a few examples of the integration of other types of fuel cells, such as microbial fuel cells (MFCs), with desalination units. As summarized by Gude *et al.* (2013), an MFC operates in a galvanic mode where it employs microbial catalysts to extract oxidation current from waste organic matter in the anodic half-cell and uses chemical catalysts in the cathodic half-cell to consume electrons in the presence of protons and a terminal electron acceptor. In MFCs, the anode is normally designed for treatment of municipal waste streams and high-strength organic

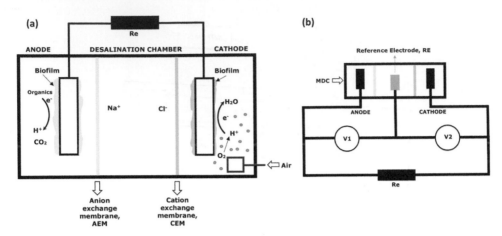

Figure 7.9.    Integration of microbial desalination cell (MDC): (a) biocathode MDC; (b) measuring circuit (adapted and reproduced with permission from Wen *et al.*, 2012).

wastes (e.g. from cattle farms, breweries, landfills, chocolate factories and food processors) and the reducing conditions in the cathodic half-cell provide a legitimate route for treating oxidized contaminants (e.g. nitrates, chromium) in water bodies.

Cao *et al.* (2009) modified an MFC to develop a new integrated water desalination method. The authors performed experiments with different concentrations of salt (up to $35\,g\,L^{-1}$) and found that a single desalination cycle successfully removed approximately 90% of the salts from water. Chen *et al.* (2011) extended this work by using a stacked MFC-based desalination unit (i.e. a microbial desalination cell or MDC), where they found that desalination cell numbers and external resistance had a significant effect on the total desalination rate. It is important to note that the MDC is a variant of an MFC in which an additional middle chamber for sustainable energy production (from organic wastes) and water desalination has been included (Gude *et al.*, 2013). MDCs can be designed for treatment of organic waste and simultaneous desalination of saltwater. As shown in Figure 7.9, Wen *et al.* (2012) used a biocathode in an MDC to improve the cell performance in desalination and wastewater treatment. Meng *et al.* (2014) found that using dewatered sludge as the anodic substrate improves the stability of the MDC biocathode for desalination purposes.

Mohanakrishna *et al.* (2010) investigated the use of an open-air cathode MFC for desalination and power generation purposes, while Wang *et al.* (2012) investigated the performance of an integrated MFC-membrane bioreactor system for wastewater treatment and concluded that the low cost and efficiency of the integrated system demonstrated the prospect of using this technology for wastewater treatment in future. Ghasemi *et al.* (2013) performed an economic comparison study between the use of sulfonated poly(etheretherketone) (SPEEK) and Nafion 117 membranes in MFC for wastewater treatment and suggested that the higher power density and lower cost of the SPEEK membrane demonstrated the potential for MFC implementation at an industrial level.

Another unusual but interesting example of fuel cell integration involves a novel cogeneration unit incorporated into a water purification scheme (Huicochea *et al.*, 2011). This system consists of a PEMFC and an absorption heat transformer. Huicochea *et al.* (2011) found that the cogeneration system offered an efficiency improvement of approximately 12.4% compared to a standalone fuel cell. Similarly, Klaysom *et al.* (2010, 2011) reported a new type of composite ion-exchange membrane for water purification application using electrodialysis desalination, where a reduction in power consumption of almost 50% was observed compared to pristine, unaltered membrane.

## 7.4   ENERGY, ENVIRONMENTAL AND ECONOMIC FACTORS

It is recognized that carbon dioxide emissions are at an all-time high, with global estimates of around 36 billion metric tons in 2013 (Olivier *et al.*, 2013). Desalination has been proved to provide a reliable water supply in many countries around the world, with a total global desalination capacity of ~60 million $m^3 day^{-1}$ in 2013 (Ziolkowska and Reuben, 2016). Hydrocarbon-based energy consumption during desalination makes a significant contribution to total $CO_2$ emissions and it is, therefore, very important to explore in more detail the energy-saving, environmental and economic effects of implementing any novel concept of integrated fuel cell/desalination systems. A strategy based on such systems can substantially reduce the costs of seawater desalination, particularly when integrating the latest, environmentally friendly fuel cell technologies (e.g. Table 7.1) with desalination systems (Table 7.2) in dual-purpose facilities that simultaneous produce electricity and water. As summarized by Almarzooqi *et al.* (2014), desalination costs depend on many factors, from type of technology, energy cost, feed water salinity and capacity, to many other site-specific factors. Due to the large number of variables affecting desalination costs, it is often difficult to derive a completely objective comparison between the different technologies.

As can be seen through the sampling of literature on the integration of fuel cells with desalination units, integration methods are well-defined and, to an extent, established for some technologies, while others remain largely unexplored. However, for commercial or economic success, the integration methodologies must not only be operational but must also be cost-effective, highlighting a potential contradiction between the need for technological sophistication and ease of potable water production. The rationale for developing integrated energy-water systems is the reduction of capital costs, energy consumption and the cost of desalinating seawater, with 50–60% of an RO system's operating cost being due to energy consumption (Singh, 2008). As discussed by Khawaji *et al.* (2008), from an economic point of view a cogeneration scheme is essential in conjunction with power generation. The industrial goal of global desalination is to produce desalinated water at 50 US cents per $m^3$ of water and power at 2 US cents per kWh. Thus, the estimated water production cost for a seawater RO plant project with a capacity of 94,600 $m^3 day^{-1}$ in Tampa, USA was reported to be at 55 US cents per $m^3$ (Wilf and Klinko, 2001).

As shown above, many improvements have been made in fuel cell efficiencies in an effort to reduce harmful emissions and decrease the cost of desalination operation. Further work is necessary to investigate the economic benefits and performance improvements and to quantify these for specific markets and applications (Khawaji *et al.*, 2008).

## 7.5   CONCLUDING REMARKS

As shown in this chapter, there are encouraging developments in the methodologies for integration of fuel cells with desalination units. In recent years, considerable progress has been made in the development of knowledge about the interrelationship between applied and advanced materials (e.g. nanostructures) in fuel cells and the resulting desalinated water quality. In addition, many different material manufacturing methods (including large-scale production) have been established and used to make fuel cell components and also entire cells. However, as seen in this chapter, only a few have proven their potential for reliable and efficient production of desalinated water. There is great promise, though, for a technical breakthrough in mass production where continuous operation with high throughput and yield will produce the desired water quality. Likewise, work related to mass production of integrated fuel cell/desalination systems is also lacking, and considerable investment is necessary for a serious market entrance which would deliver a breakthrough success and give rise to the cost-effective solutions required. Long term, fundamental multidisciplinary research leading to technology development programs is needed in order to make the integration of seawater desalination techniques with fuel cell systems affordable worldwide.

ACKNOWLEDGEMENTS

The authors acknowledge the support of the project entitled "Advance anode materials for direct hydrocarbon proton-conducting solid oxide fuel cell (PC-SOFC) in auxiliary power unit", funded by Saudi Aramco (Contract number 6000074197).

REFERENCES

Al-Hallaj, S. & Kiszynski, K. (2011) Hybrid hydrogen systems. In: *Green Energy and Technology*. Springer, London. pp. 109–129.

Al-Hallaj, S., Alasfour, F., Parekh, S., Amiruddin, S., Selman, J.R. & Ghezel-Ayagh, H. (2004) Conceptual design of a novel hybrid fuel cell/desalination system. *Desalination*, 164, 19–31.

Alkhudhiri, A., Darwish, N. & Hilal, N. (2012) Membrane distillation: a comprehensive review. *Desalination*, 287, 2–18.

Almarzooqi, F.A., Al-Ghaferi, A.A., Saadat, I. & Hilal, N. (2014) Application of capacitive deionisation in water desalination: a review. *Desalination*, 342, 3–15.

Al-Shammiri, M. & Safar, M. (1999) Multi-effect distillation plant: state-of-the-art. *Desalination*, 126, 45–49.

Al-Shayji, K.A. (1998) *Modeling, Simulation, and Optimization of Large-scale Commercial Desalination Plants*. PhD Thesis, Virginia Polytechnic Institute and State University, Blacksburg, VA.

Bataineh, K.M. (2016) Multi-effect desalination plant combined with thermal compressor driven by steam generated by solar energy. *Desalination*, 385, 39–52.

Cao, X., Huang, X., Liang, P., Xiao, K., Zhou, Y., Zhang, X. & Logan, B.E. (2009) A new method for water desalination using microbial desalination cells. *Environmental Science and Technology*, 43(18), 7148–7152.

Chen, X., Xia, X., Liang, P., Cao, X., Sun, H. & Huang, X. (2011) Stacked microbial desalination cells to enhance water desalination efficiency. *Environmental Science and Technology*, 45(6), 2465–2470.

Chittu, K.G. & Jeyaprabha, S.B. (2013) Design and simulation of fuel cell energy system to power RO desalination plant. *International Journal of Inventive Engineering and Sciences (IJIES)*, 1(4), 9–14.

Darwish, M.A., Al-Juwayhel, F. & Abdulraheim, H.K. (2006) Multi-effect boiling systems from an energy viewpoint. *Desalination*, 194, 22–39.

Faisal, N.H., Ahmed, R., Katikaneni, S.P., Souentie, S. & Goosen, M.F.A. (2015) Development of plasma sprayed molybdenum carbide-based anode layers with various metal oxide precursors for SOFC. *Journal of Thermal Spray Technology*, 24(8), 1415–1428.

Ghaffour, N., Bundschuh, J., Mahmoudi, H. & Goosen, M.F.A. (2015) Renewable energy-driven desalination technologies: a comprehensive review on challenges and potential applications of integrated systems. *Desalination*, 356, 94–114.

Ghalavand, Y., Hatamipour, M.S. & Rahimi, A. (2015) A review on energy consumption of desalination processes. *Desalination and Water Treatment*, 54, 1526–1541.

Ghasemi, M., Daud, W.R.W., Ismail, A.F., Jafari, Y., Ismail, M., Mayahi, A. & Othman, J. (2013) Simultaneous wastewater treatment and electricity generation by microbial fuel cell: performance comparison and cost investigation of using Nafion 117 and SPEEK as separators. *Desalination*, 325, 1–6.

Goosen, M.F., Mahmoudi, H. & Ghaffour, N. (2014) Today's and future challenges in applications of renewable energy technologies for desalination. *Critical Reviews in Environmental Science and Technology*, 44(9), 929–999.

Gryta, M. (2012) Effectiveness of water desalination by membrane distillation process. *Membranes*, 2, 415–429.

Gude, V.G., Kokabian, B. & Gadhamshetty, V. (2013) Beneficial bioelectrochemical systems for energy, water, and biomass production. *Journal of Microbial and Biochemical Technology*, S6(005), 1–14.

Huicochea, A., Romero, R.J., Rivera, W., Gutierrez-Urueta, G., Siqueiros, J. & Pilatowsky, I. (2013) A novel cogeneration system: a proton exchange membrane fuel cell coupled to a heat transformer. *Applied Thermal Engineering*, 50(2), 1530–1535.

Jones, E.A. (2013) *Modeling of a Polymer Electrolyte Membrane Fuel Cell – Membrane Distillation Desalination Cogeneration System*. MSc Thesis, Duke University, Durham, NC.

Kalogirou, S.A. (2005) Seawater desalination using renewable energy sources. *Progress in Energy and Combustion Science*, 31(3), 242–281.

Kenet, B. & Belmar, P.J.S. (2003) Fuel-cell powered desalination device. US Patent 2003/0132097 A1, 17 July 2003.

Khawaji, A.D., Kutubkhanah, I.K. & Wie, J.M. (2008) Advances in seawater desalination technologies. *Desalination*, 221, 47–69.

Klaysom, C., Marschall, R., Ladewig, B.P., Lu, G.Q.M. & Wang, L. (2010) Synthesis of composite ion-exchange membranes and their electrochemical properties for desalination applications. *Journal of Materials Chemistry*, 20, 4669–4674.

Klaysom, C., Moon, S.H., Ladewig, B.P., Lu, G.Q.M. & Wang, L. (2011) The influence of inorganic filler particle size on composite ion-exchange membranes for desalination. *The Journal of Physical Chemistry C*, 115(31), 15124–15132.

Kwak, S.Y. & Ihm, D.W. (1999) Use of atomic force microscopy and solid-state NMR spectroscopy to characterize structure-property-performance correlation in high-flux reverse osmosis (RO) membranes. *Journal of Membrane Science*, 158, 143–153.

Lee, K.P., Arnot, T.C. & Mattia, D. (2011) A review of reverse osmosis membrane materials for desalination – development to date and future potential. *Journal of Membrane Science*, 370, 1–22.

Lunghi, R., Ubertini, S. & Desideri, U. (2001) Highly efficient electricity generation through a hybrid molten carbonate fuel cell-closed loop gas turbine plant. *Energy Conservation Management*, 42, 1657–1672.

Maddocks, A., Young, R.S. & Reig, P. (2015) Ranking the world's most water-stressed countries in 2040. World Resources Institute, Washington, DC. Available from: http://www.wri.org/blog/2015/08/ranking-world%E2%80%99s-most-water-stressed-countries-2040 [accessed November 2016].

Meng, F., Jiang, J., Zhao, Q., Wang, K., Zhang, G., Fan, Q., Wei, L., Ding, J. & Zheng, Z. (2014) Bioelectro-chemical desalination and electricity generation in microbial desalination cell with dewatered sludge as fuel. *Bioresource Technology*, 157, 120–126.

Meratizaman, M., Monadizadeh, S. & Amidpour, M. (2014) Introduction of an efficient small-scale freshwater-power generation cycle (SOFC-GT-MED), simulation, parametric study and economic assessment. *Desalination*, 351, 43–58.

Meratizaman, M., Monadizadeh, S. & Amidpour, M. (2016) Simulation and economic evaluation of small-scale SOFC-GT-MED. *Desalination and Water Treatment*, 57, 4810–4831.

Mohanakrishna, G., Mohan, S.V. & Sarma, P.N. (2010) Bio-electrochemical treatment of distillery wastewater in microbial fuel cell facilitating decolorization and desalination along with power generation. *Journal of Hazardous Materials*, 177(1–3), 487–494.

Olivier, J.G.J., Janssens-Maenhout, G., Muntean, M. & Peters, J.A.H.W. (2013) Trends in global $CO_2$ emissions, 2013 report. PBL Netherlands Environmental Assessment Agency, The Hague, The Netherlands. Available from: http://edgar.jrc.ec.europa.eu/news_docs/pbl-2013-trends-in-global-co2-emissions-2013-report-1148.pdf [accessed November 2016].

Shao, Z., Haile, S.M., Ahn, J., Ronney, P.D., Zhan, Z. & Barnett, S.A. (2005) A thermally self-sustained micro solid-oxide fuel-cell stack with high power density. *Nature*, 435, 795–798.

Singh, R. (2008) Sustainable fuel cell integrated membrane desalination systems. *Desalination*, 227, 14–33.

Tamiotti, L. (2009) Trade and climate change: a report by the United Nations Environment Programme and the World Trade Organization. UNEP/Earthprint. p. 68.

Tarifa, E.E. & Scenna, N.J. (2001) A dynamic simulator for MSF plants. *Desalination*, 138, 349–364.

Wali, F. (2014) The future of desalination research in the Middle East. *Nature Middle East*, 26 November 2014. doi:10.1038/nmiddleeast.2014.273.

Wang, B., Liu, Y. & Liu, S. (2011) Self-sustained desalination in combination with wastewater treatment – hybrid microbial desalination system. 2011 Ralph Desch Technical Writing Award for Colorado Science and Engineering Fair, Colorado State University, Fort Collins, CO.

Wang, Y.P., Liu, X.W., Li, W.W., Li, F., Wang, Y.K., Sheng, G.P., Zeng, R.J. & Yu, H.Q. (2012) A microbial fuel cell-membrane bioreactor integrated system for cost-effective wastewater treatment. *Applied Energy*, 98, 230–235.

WateReuse Association (2011) Seawater desalination power consumption. WateReuse Association Desalination Committee White Paper, WateReuse Association, Alexandria, VA.

Wen, Q., Zhang, H., Chen, Z., Li, Y., Nan, J. & Feng, Y. (2012) Using bacterial catalyst in the cathode of microbial desalination cell to improve wastewater treatment and desalination. *Bioresource Technology*, 125, 108–113.

Wilf, M. & Klinko, K. (2001) Search for the optimal SWRO design. *Journal of Water Reuse and Desalination*, 11(3), 15–20.

Wu, L.Y., Xiao, S.N. & Gao, C.J. (2012) Simulation of multi-stage flash (MSF) desalination process. *Advances in Materials Physics and Chemistry*, 2, 200–205.

Ziolkowska, J.R. & Reuben, R. (2016) Impact of socioeconomic growth on desalination in the US. *Journal of Environmental Management*, 167, 15–22.

# CHAPTER 8

## Wind turbine electricity generation for desalination: design, application and commercialization

Zhao Yong, Zheng Shuai & Chua Leok Poh

### 8.1 INTRODUCTION

The wider use of desalination technologies is still limited, primarily because of high energy requirements that are currently met with expensive fossil fuels (Ghaffour et al., 2015). The use of alternative energy sources, such as wind and solar, are essential to meet the growing demands for water desalination. However, the expansion of these efforts towards large-scale plants is hampered by techno-economic challenges. For example, many plants are connected to the electrical grid to assure a continuous energy supply for stable operation. The successful application of renewable technology requires an understanding of sustainable development. In particular, there is a need to find a balance between three sets of goals: social, economic and environmental.

The Kwinana desalination plant, located south of Perth in Western Australia, is one example where grid-connected wind power and a reverse osmosis desalination plant have been successfully combined (Goosen et al., 2014). The plant produces well over 100 megaliters of drinking water per day. Research is also underway on micro/nano wind-gathering devices of less than an inch that can be mounted on rollable or stacked sheets for easy installation in residential homes and business (Fein and Merritt, 2011). However, it remains to be seen whether such devices can be deployed in a cost-effective way. In order to aid commercialization, different types of governmental policy instruments (e.g. tax breaks, low-interest loans) can be effective for different renewable energy sources (Frondel et al., 2010).

In 2008, the estimated share of wind power in Germany's electricity production amounted to 6.3%, followed by biomass-based electricity generation and water power, whose shares were around 3.6 and 3.1%, respectively (Frondel et al., 2010). In contrast, the amount of electricity produced through solar PV was negligible: its share in 2008 was just 0.6%. One worrisome observation, noted by Goosen et al. (2014), is that renewable energy sources have consistently accounted for only 13% of total energy use over the past 40 years. Nevertheless, large-scale electricity generation using wind turbines shows great promise in solving the energy demands of desalination plants worldwide. The technology is easy to scale up, it is environmentally friendly and it can be easily integrated with other renewable energy systems such as solar and geothermal. Furthermore, the principal motivation for using wind turbines to generate electricity is low $CO_2$ emissions, which can help in slowing down the rate of global warming. These merits make wind energy a suitable renewable energy to support the sustainable development of all countries, especially those that lack natural fuel and energy resources, such as island countries like Singapore (GWEC, 2015).

A wind turbine which converts kinetic energy absorbed from the wind into mechanical energy can be classified into two categories: horizontal-axis wind turbines (HAWTs) and vertical-axis wind turbines (VAWTs). Currently, the majority of commercial wind turbines in operation are HAWTs, thanks to their high efficiency, stable performance and relatively low cost (Jamieson, 2011). This chapter will focus on HAWTs.

The wind turbine's blade is the most important component used to extract energy from wind. The blade design and operation should thus be targeted for high power efficiency and long lifespan.

Currently, wind turbine blades are designed according to the so-called improved blade element momentum (BEM) method, after taking the tip and hub loss coefficients into consideration. The BEM method is derived from the laws of mass conservation, momentum conservation and energy conservation (Hansen, 2008). The method assumes that the wind turbine blade is operating under steady-state conditions and the flow field on each blade element does not interact with other elements. In the real-world situation, there are three main loads acting on a wind turbine blade: gravitational load, inertial load and aerodynamic load. All three loads vary over time. The change in load over time affects the deformation of each wind turbine blade, which, in turn, will affect the flow field and the wake around the blade, thus resulting in a continuous variation of the load. This type of two-way coupling between blade and flow is known as fluid-structure interaction (FSI) (Wang, 2008). The FSI of wind turbine blades will affect the power-generating efficiency, stability and lifespan of a wind turbine.

The current chapter will focus on a detailed analysis of the behavior of the FSIs of wind turbines. In addition to the design theory, the application and commercialization of wind turbines in relation to water desalination will also be discussed.

## 8.2   INSTALLED WIND CAPACITIES AND THE IMPORTANCE OF TURBINE CONFIGURATION AND ROTOR DESIGN

### 8.2.1   *Global installed wind capacities and commercialization of the renewable energy industry*

The modern wind turbine first appeared in 1920 when Betz employed actuator disc theory to show that a maximum of 59.3% of kinetic wind energy could be converted into mechanical energy by a free-stream wind turbine (Gasch and Twele, 2012). After the Betz limit theory was proposed, the development of wind turbines accelerated and many promising approaches to modern wind turbine design have since emerged. According to the Global Wind Report 2014, the newly installed wind energy capacity in 2014 was 51 GW, which makes the cumulative global wind energy capacity 369.6 GW (GWEC, 2015). Figure 8.1a and Figure 8.1b show, respectively, the newly installed and cumulative global wind energy capacities between 1997 and 2014.

The majority of newly installed wind energy capacity is located in developing countries, such as China, Brazil and India (Fig. 8.2). These data demonstrate that the wind energy industry has a huge market with a promising future. It can be argued, therefore, that it is well worth investing in wind-turbine-related studies in order to improve device performance. Even a small improvement in turbine performance could bring significant economic benefits, due to the huge number of new wind energy installations. Thus, many scientists and institutions have expended tremendous time

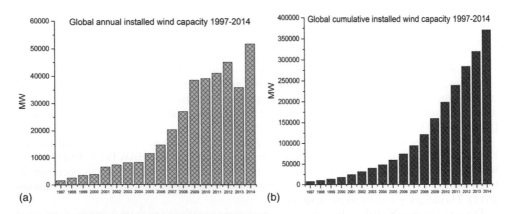

Figure 8.1.    Global annual and cumulative installed wind capacities between 1997 and 2014 (GWEC, 2015).

and resources on this field (GWEC, 2015). Furthermore, due to the complexity of wind turbine design problems and resource constraints, there are still a lot of critical issues that have not been fully solved. FSIs are one of these important issues and will be described in this chapter, because they are closely related to wind turbine power efficiency, fatigue failure and stability of operation. With the advance of the global wind energy industry, there will be thousands of new wind turbines designed and installed all over the world. It is anticipated that research work in the FSI field could be valuable in improving wind turbine design and should enhance turbine efficiency and lifespan. To pursue our investigation, some basic concepts of wind turbine and aerodynamics will be necessary. In the following sections, wind turbine configuration, the aerodynamics of HAWTs and wind turbine rotor design will all be reviewed.

### 8.2.2  *From sailing ships and windmills to electricity generation*

Wind energy has been used to drive the sailing ship since the 7th century. The single-mast sailing ship first appeared at that time and the multiple-mast sailing ship was developed in the 15th century (see Fig. 8.3a and Fig. 8.3b, respectively). The use of wind energy as the thrust power for ships was a significant development in human history. It solved the issue of slow-moving ships as well as removing the historical reliance on manpower to swing a paddle. Long-distance shipping became possible after the sailing ship appeared and this enabled significant increases in trade and commerce. With the invention of the steam engine in the 18th century, the sailing ship gradually disappeared from ocean transportation.

The windmill, another early application of wind technology, was invented by the Persians in the 6th century and was employed for irrigation and grain milling. Two more modern types of windmill structure are shown in Figure 8.3c and Figure 8.3d. The energy generated by windmills played an important role in Europe before and during the period of the Industrial Revolution (Manwell *et al.*, 2002). In Holland, for example, there were more than 12,000 windmills working in the 18th century, being used not only to grind grain and press oil from seed, but also to drain water from wetlands, as more than two-thirds of the country was below sea level. Although the windmill became less important with the development of the combustion engine in the 20th century, the manpower-free and sustainable nature of wind energy means that the windmill is still very popular in Holland today.

Wind energy can also be used to provide home heating energy requirements. The wind turbine can generate electricity which can be converted into heat. Alternatively, the wind turbine can drive

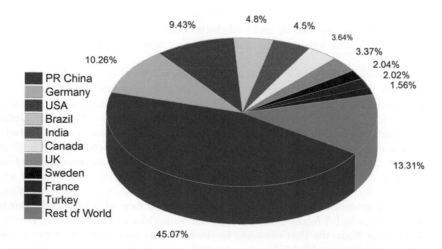

Figure 8.2.    Share by country of newly installed wind capacity in 2014 (GWEC, 2015).

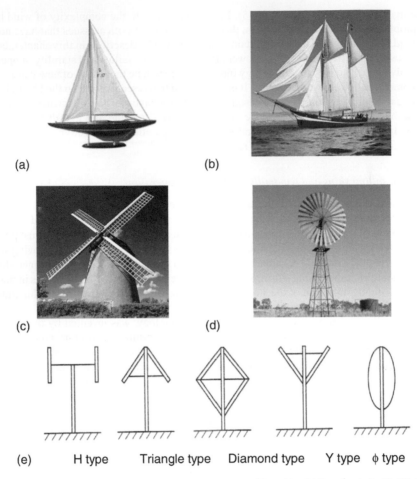

(a)                                    (b)

(c)                                    (d)

(e)        H type      Triangle type    Diamond type    Y type    φ type

Figure 8.3.   (a) Single-mast sailing boat; (b) multiple-mast sailing ship; (c) Dutch windmill; (d) American windmill; (e) different types of Darrieus vertical-axis wind turbines (Manwell *et al.*, 2002).

a compressor engine to condense air under adiabatic conditions; heat is then released. A third approach is to employ the wind turbine to drive blenders to mix liquids inside a container to generate heat.

Research into wind turbines for electricity generation started at the beginning of the 19th century. Charles F. Brush was the pioneer of wind electricity research and installed the first automatic wind turbine to generate electricity in Ohio (USA) (Manwell *et al.*, 2002). This turbine was rated at 12 kW, with a horizontal axis and 144 wooden blades. Poul la Cour from Denmark made a huge contribution to the development of modern high-speed horizontal-axis wind turbines by discovering that a turbine with fewer blades and a high rotating speed was more efficient. In addition, the vertical-axis wind turbine appeared in the 1930s thanks to the invention of a Finnish engineer, Sigurd Johannes Savonius. However, this kind of turbine, which consisted of several S-shaped blades, had the disadvantage of lower power efficiency than HAWTs, resulting in reduced competitiveness in the global energy market. Another form of VAWT is the Darrieus-type, which is driven by lift forces generated from the blade (Manwell *et al.*, 2002). Depending on the shape of the blade, the Darrieus could be classified into several subcategories as shown in Figure 8.3e. The Darrieus VAWT has the advantage of high speed, simple structure and low build-cost; disadvantages include small starting torque and relatively low power efficiency.

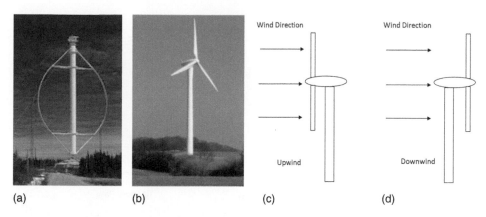

Figure 8.4.   (a) Vertical-axis wind turbine; (b) horizontal-axis wind turbine; (c) upwind rotor configuration; (d) downwind rotor configuration (Hansen and Butterfield, 1993).

Neither Savonius nor Darrieus VAWTs could satisfy the requirement for high power efficiency and gradually disappeared from the global market. From the 1980s, the modern high-speed three-blade HAWT became the mainstay of the market and displaced all other forms of wind turbine. The capacity of the wind turbine has steadily increased and those with a capacity of ~1 MW are currently the most common in the market (Manwell *et al.*, 2002). The development of the wind turbine demonstrates the trend towards high-speed single-machine capability. Wind farms with hundreds of installed wind turbines have appeared, many of them offshore (Goosen *et al.*, 2014).

### 8.2.3   *Influence of wind turbine configuration and control strategy on efficiency*

According to recent statistics (GWEC, 2015), most of the units which have been installed worldwide are HAWTs. The HAWT was therefore chosen for detailed study in this chapter. As we have already seen, turbines can be classified into two categories according to the orientation of the rotor. Thus, a HAWT (Fig. 8.4b) has to be installed facing the wind direction while a VAWT (Fig. 8.4a) is independent of wind path. When the turbine is working under a constant wind speed, the lift force on a HAWT rotor is almost constant whereas the lift force on a VAWT rotor varies periodically, which will result in power output fluctuation for the latter. Likewise, the wind shear effect will also exert significant influence on a VAWT because its blade is located near the ground, which leads to fatigue problems, whereas HAWTs are little affected by this issue. All of these merits make HAWTs more efficient and economical than VAWTs (Vries, 1983). In addition, there are two kinds of HAWT, distinguished by their rotor configurations, that is, whether they are upwind or downwind of the supporting tower, as shown in Figure 8.4c and Figure 8.4d, respectively. The downwind configuration allows the blades to deflect from the tower when the thrust loading increases, but for upwind turbines the yaw (i.e. oscillation about a vertical axis) is supposed to be controlled. The downwind wind turbine will generate a lot of noise when the wind passes by the tower and generates addition moments which will cause yaw loads when the rotor changes its orientation relative to wind direction.

Another important parameter that needs to be considered is the tip speed ratio, $\lambda$, which refers to the relation of the circumferential velocity of the blade tip to the wind speed:

$$\lambda = \omega R / V_1 \tag{8.1}$$

where $\omega$ is the rotor angular velocity, $R$ is the radius of the rotor and $V_1$ is the incoming wind speed. A typical performance curve for a modern, high-speed wind turbine is shown in Figure 8.5a. This curve shows the power coefficient, $C_p$, versus tip speed ratio, $\lambda$. This figure demonstrates that

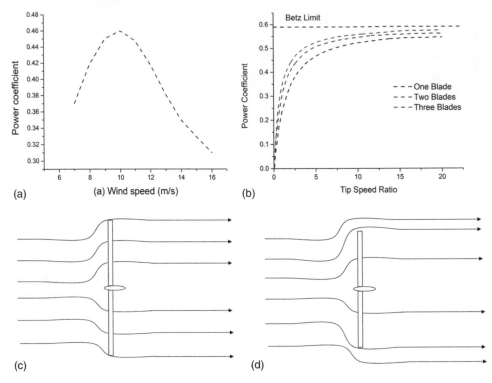

Figure 8.5.    (a) Power coefficient versus tip speed ratio curve; (b) maximum achievable power coefficient as a function of the number of blades; (c) wind passage through low-solidity rotor; (d) wind passage through high-solidity rotor.

the maximum achievable power coefficient will appear when the tip speed ratio is around 6. Furthermore, most modern high-speed HAWTs have three blades. To clearly explain the reason for choosing three blades, the concept of rotor solidity will be introduced. Rotor solidity is defined as the ratio of the area projected by the wind turbine blades in the wind speed direction to the entire area through which the wind turbine rotor sweeps. Put simply, the more same-sized blades on a rotor, or the wider those blades, the higher the rotor solidity. Figure 8.5c and Figure 8.5d show the differences when the incoming wind passes through low-solidity and high-solidity rotors. In Figure 8.5c it can be seen that when the wind encounters a low-solidity rotor, most of the wind will pass through the rotor and generate torque to push the wind turbine blade to rotate. The wind speed will dramatically decrease after passing through the rotor due to energy losses and the volume of airflow will expand as a result of the drop in wind speed. Figure 8.5d illustrates the scenario of wind passing a high-solidity rotor. Due to the relatively high number of blades, which increase the drag when wind passes through, most parts of the airflow will bypass the rotor and will not generate any torque on the wind turbine blade. These figures demonstrate that the number of wind turbine blades needs to be relatively low. The reason for choosing three blades is illustrated in Figure 8.5b. A three-blade rotor has a higher maximum power coefficient than a two-blade rotor. Although the maximum achievable power coefficient will go up as the number of blades increases, the size of this gain progressively diminishes under the same tip speed ratio (Wilson *et al.*, 1976). Furthermore, the three-blade rotor has a simple structure and yields the lowest system cost (Hansen and Butterfield, 1993).

Based on rotating speeds, wind turbines can be classified into two categories: constant speed and variable speed. The wind turbine can also be divided into two groups according to its power output control strategy. Thus, there is a stall-control wind turbine, which means that when the

Table 8.1.   Four types of wind turbine control strategy.

|  | Stall control | Variable pitch angle control |
|---|---|---|
| Constant speed | Constant speed stall control | Constant speed variable pitch angle |
| Variable speed | Variable-speed stall control | Variable-speed variable pitch angle |

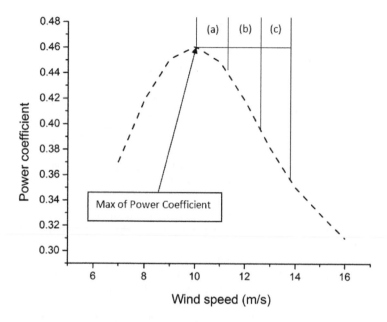

Figure 8.6.   Different control strategies under different power coefficients.

wind speed goes above the rated value, the wind turbine will make use of certain blade features to automatically switch into a stall configuration to limit the output power. The second group involves variable pitch control, in which the wind turbine changes the pitch angle of its blades to control power output when the wind speed goes above the rated value. The speed control and the power output control are independent from one another and in combination give four types of overall control strategy, as shown in Table 8.1.

For large, modern, high-speed HAWTs, the most common control strategy is a comprehensive control method. As shown in Figure 8.6, the working range of a wind turbine can be divided into three zones. In zone (a) the wind turbine rotation speed will increase as the wind speed increases to ensure that relatively high power coefficient is obtained. As the wind speed increases, the wind turbine will enter zone (b) in which the rotating speed will keep increasing until the maximum value of rotating speed is achieved. After that, as long as the generating power is less than the allowed maximum output power, then the rotating speed will remain unchanged. In zone (b) the power coefficient keeps decreasing but the output power is still increasing. When the wind turbine functions in zone (c), the output power will be limited to a constant by adjusting the pitch angle to maintain stable power generation.

### 8.2.4   *Wind turbine rotor design and momentum theory*

The HAWT has to be oriented in such a way that its axis is parallel to the wind flow, which is the stream-wise direction (i.e. $x$ axis). The wind turbine blades are also long and slim structures, hence the span wise velocity component is much lower than the stream-wise component. It is

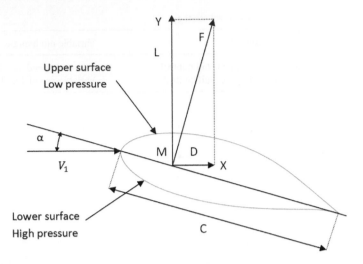

Figure 8.7.    Schematic of lift, drag, angle of attack, chord and aerodynamic moment of aerofoil. Note that the aerofoil is rotating in an anticlockwise direction around axis $Y$.

reasonable, therefore, to assume that flow within a small length interval along the blade in the radial direction is two-dimensional. For example, Figure 8.7 shows a two-dimensional flow over an aerofoil. The force, $F$, acting on this element of aerofoil, can be resolved into two components, one being the lift force, $L$, which is perpendicular to the wind velocity, $V_1$, and the other the drag force, $D$, which is parallel to $V_1$. In the case of a HAWT, the lift force is the main component in rotating the wind turbine blade and thus generating electricity, and the drag force does not contribute to the output. Therefore, the main objective is to maximize the ratio of lift force to drag force. The chord, $c$, in Figure 8.7 is the length of a straight line from the leading edge to the trailing edge of an aerofoil, and angle of attack, $\alpha$, is the angle between the chord and the direction of incoming wind flow. The aerodynamic moment, $M$, is the product of resultant force, $F$, and the length to the aerodynamic centre, which is usually located on the chord at $c/4$ from the leading edge. Lift and drag coefficients, $C_l$ and $C_d$, are defined as:

$$C_l = 2L/\rho V_1^2 c \tag{8.2}$$

$$C_d = 2D/\rho V_1^2 c \tag{8.3}$$

where $\rho$ is the density of air.

A streamline is the airflow around the shape of an aerofoil (White, 1991). It is known that when air passes over a curved surface, a pressure gradient will result. This pressure gradient causes a centripetal force in the radial direction. The flow over the upper surface has a tendency to leave the surface which will result in lowered pressure, and the flow over the lower, flatter surface pushes on the surface which will lead to heightened pressure. The difference in pressure above and below the aerofoil results in a lifting force (White, 1991). This lift force is the driving force for wind turbine rotation and should be enhanced as much as possible. On the other hand, a drag force results from friction on the aerofoil surface, the pressure differential between the leading edge and the trailing edge, and boundary-layer separation from the upper surface of the aerofoil, which generates a lot of viscous vortex bubbles. These vortex bubbles will consume some of the total wind energy and do not contribute to electricity power generation. The drag force not only reduces turbine lifespan but also induces flow instability which is harmful to wind turbine operation and should be suppressed (White, 1991).

The coefficients $C_l$ and $C_d$ are functions of angle of attack, $\alpha$, Mach number, $Ma$, which represents the ratio between $V_1$ and the speed of sound and the Reynolds number, $Re$. The

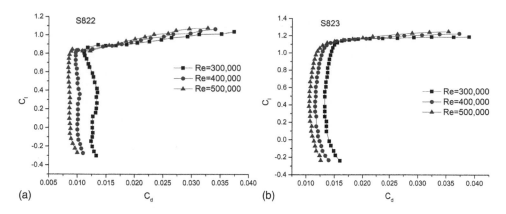

Figure 8.8.    Polar curve of aerofoils (a) S822 and (b) S823 (adapted from Selig and McGranahan, 2004).

incoming wind speed is $10\,\mathrm{m\,s^{-1}}$ and the Mach number is around 0.03. According to Houghton and Carpenter (2003), the effect of Mach number can be ignored. Note that $Re$ is defined as:

$$Re = (cV_1)/v \qquad (8.4)$$

where $v$ is the kinematic viscosity.

The lift coefficient, $C_1$, increases linearly with angle of attack, $\alpha$, to reach a maximum value. It can maintain the maximum for some $\alpha$ and then decreases dramatically, which means that the aerofoil has entered a stall regime. These depend on the Reynolds number. For the drag coefficient, $C_d$, this keeps increasing for all the Reynolds numbers, even after the aerofoil steps into the stall condition. Figure 8.8a and Figure 8.8b show the curves of lift coefficient ($C_1$) versus drag coefficient ($C_d$) for aerofoils S822 and S823, which will be used as an example for this chapter (Selig and McGranahan, 2004). It can be observed that when selecting the normal working angle of attack for each element of the blade, not only $C_1$ should be considered, but also $C_d$. It is necessary to select the angle of attack at the point where the $C_1/C_d$ ratio is at a maximum to ensure that each element of the blade will deliver the best performance.

For the aerofoils S822 and S823, the drag coefficient, $C_d$, increases more slowly than the lift coefficient, $C_1$, from the point of minimum $C_1/C_d$ ratio to the point where $C_1$ reaches a maximum value. According to Burton *et al.* (2011), the drag effect due to skin friction can be ignored as long as the flow remains attached to the blade because the losses caused by drag are only critical when the wind turbine is operating at a high tip speed ratio. The variation of maximum power coefficient versus tip speed ratio is shown in Figure 8.9a. The wind turbine power coefficient, $C_p$, is the ratio of the power generated by a wind turbine to the total wind power. Since the lift-to-drag ratios and design tip speed ratio for both S822 and S823 are relatively low, the angle of attack will be selected near the point where $C_1$ achieves its maximum value.

The stall status of the aerofoil is not only related to angle of attack, $\alpha$, but also depends on the Reynolds number. Schlichting (1968) showed that Reynolds number dependency is related to the point on an aerofoil where the boundary layer transits from laminar flow to turbulent flow, as shown in Figure 8.9b. As the stall phenomenon is closely related to boundary-layer separation, it implies that the main target in blade design is to reduce drag, as well as trying to make the flow attach to the blade surface for as long as possible. In other words, blade design has to ensure that the aerofoil is working below the angle of attack at which the $C_1/C_d$ ratio is at a maximum (Bertin and Smith, 1989).

To ensure that the aerofoil elements which constitute the wind turbine blade are not stepping into a stall situation, we consider a small element $(r, r+dr)$ from a rotating wind turbine rotor, as shown in Figure 8.9c, in order to perform flow analysis. The velocity triangle is shown in

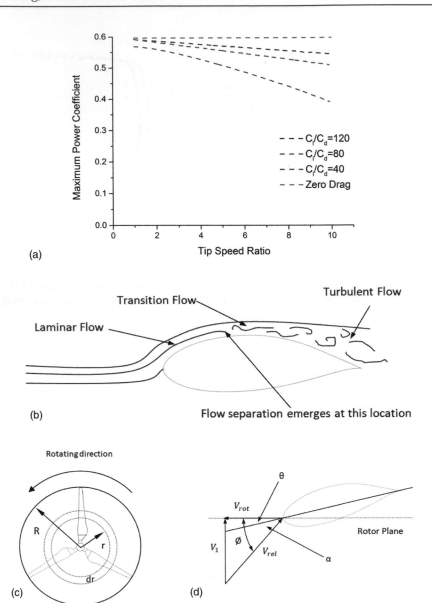

Figure 8.9.    (a) Variation of maximum $C_p$ with design $\lambda$ for different lift-to-drag ratios; (b) transition of flow from laminar to turbulent along upper surface of aerofoil; (c) rotor of wind turbine showing annular ring $(r, r+dr)$; (d) velocity analysis of annular ring section $(r, r+dr)$.

Figure 8.9d. Note that in this figure, $V_1$ is wind velocity (which flows perpendicularly into the paper in Figure 8.9c), $V_{rot}$ is the blade rotating velocity, $V_{rel}$ is the relative wind velocity, which is the vector difference between $V_1$ and $V_{rot}$, $\alpha$ is angle of attack, $\theta$ twist angle and $\emptyset$ relative wind velocity angle. The relationship between these three angles is:

$$\alpha + \theta = \emptyset \tag{8.5}$$

To simplify the illustration, it is assumed that there are no tip and root losses. If the wind speed and angular velocity of the rotor is fixed, then it can be expected that the relative wind direction angle Ø of each element of the blade will increase from tip to root. To maintain a constant angle of attack, the twist angle $\theta$ has to increase along the blade in a radial direction from tip to root based on the relationship of Equation (8.5). This is the main reason why most HAWT blades are twisted. Tong (2010) demonstrated that the tip and body portions of a turbine blade contribute the most to power generation. In wind turbine blade design, it is paramount to ensure that the tip and body portions work at a maximum angle of attack. However, as the twist angle has to increase from the tip to the root of the blade, it is unavoidable that the performance of elements near the root has to be sacrificed because when the tip portion is working at the rated wind speed the root portion steps into the stall condition earlier than the tip. Although, theoretically, the angle of attack at the blade tip should be selected as the maximum, in practice the designed angle of attack will be afforded a small buffer from the maximum value, that is, smaller than the maximum angle of attack. This is to avoid the wind turbine blade operating under stall conditions when encountering a gust, which may result in an unfavorable angle of attack (Gasch and Twele, 2012).

There are a lot of wind turbine blade design methods and among these the Wilson method is the most widely employed and has incorporated the Betz theory, wake theory, blade element theory and momentum theory (Wilson and Lissaman, 1974). The theory established by Betz in 1919 was the first to describe the amount of energy that a turbine can extract from the wind. The important assumptions of this theory are that: the ideal rotor has an infinite number of blades; the flow can be simplified as an annular stream tube, as shown in Figure 8.10a; the flow is stationary, incompressible and frictionless; there is no rotation in the wake; the force applied on the rotor is distributed evenly.

Let us now consider an annular stream tube containing a wind turbine, as shown in Figure 8.10a. Note that $V$ is the wind speed after the rotor, $V_2$ is the wind speed far downstream, $S_1$ is the area of actuator disc before the rotor, $S$ is the area of the rotor and $S_2$ is the area of actuator disc after the rotor.

From the mass conservation equation it can be shown that:

$$S_1 V_1 = S_2 V_2 = SV \tag{8.6}$$

The force, $F$, acting on a wind turbine blade can be derived as:

$$F = \rho SV(V_1 - V_2) \tag{8.7}$$

Then the power of the wind turbine rotor, $P$, can be expressed as:

$$P = FV = \rho SV^2(V_1 - V_2) \tag{8.8}$$

The axial momentum equation is applied within the circular control volume and combined with the mass conservation law. It can be expressed as follows: firstly, $V$ can be taken as an average of $V_1$ and $V_2$:

$$V = (1/2)(V_1 + V_2) \tag{8.9}$$

Substituting Equation (8.9) into (8.8), which is the power equation, we have:

$$P = FV = (1/4)\rho S(V_1 + V_2)(V_1^2 - V_2^2) \tag{8.10}$$

By taking the derivative of Equation (8.10) with respect to $V_2$ (because $V_1$, $\rho$ and $S$ are constants, hence power is only a function of $V_2$):

$$dP/(dV_2) = (1/4)\rho S(V_1^2 - 2V_1 V_2 - 3V_2^2) \tag{8.11}$$

To obtain the maximum power, set Equation (8.11) to zero and we have:

$$V_2 = (1/3)V_1 \tag{8.12}$$

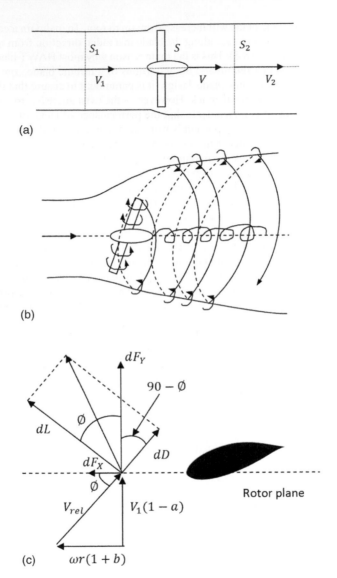

(a)

(b)

(c)    $\omega r(1 + b)$

Figure 8.10.    (a) Energy-extracting actuator and annular stream tube; (b) the vortex behind wind turbine; (c) local force and velocity on a wind turbine blade element at radius $r$.

Substituting Equation (8.12) into (8.10), the maximum power of an ideal wind turbine rotor is as follows:

$$P_{\max} = (8/27)\rho S V_1^3 \tag{8.13}$$

Since the power coefficient is defined as follows:

$$C_p = P/(0.5\rho S V_1^3) \tag{8.14}$$

then the maximum power coefficient $C_{p(\max)} = P_{\max}/(0.5\rho S V_1^3) = 16/27 \approx 0.593$ which is the Betz limit. This conclusion demonstrates that in ideal circumstances the wind turbine power coefficient cannot be higher than 0.593.

The rotor wake theory takes the flow rotation induced by the rotor into account and assumes that: the effect of aerofoil drag and blade tip loss were ignored; the periodic effect of the flow due to the limited number of blades was ignored; the blade elements in a radial direction are independent of one another. Likewise, the induced wind speed can be treated as the combination of three types of vortex, shown in Figure 8.10b: a center vortex, which is located behind the rotating shaft; an attached vortex, which is located at the blade's main body; the helical tip vortex, which is located at the blade tip. Due to the presence of these vortices, the axial and tangential directions of the wind velocity will change. The induction factors of axial and rotational direction are denoted as $a$ and $b$, respectively, and will be introduced to represent the effect of the rotor on the flow. From the vortex theory of Lugt (1996), it can be shown that at the rotating plane, the axial wind velocity is:

$$V = (1-a)V_1 \tag{8.15}$$

where $a$ is introduced into the equation to represent the obstruction effect of the wind turbine rotor on the incoming wind speed ($V_1$).

In terms of the rotational direction induction factor due to the vortex motion, there will be a tangential rotating angular velocity downstream and zero angular velocity upstream. When the blade rotates at constant angular velocity $\omega$, then – based on rotor wake theory (Burton *et al.*, 2011) – the flow that passes through the turbine rotor will be induced to rotate at an angular velocity $\Omega$ in the same direction as the rotor, where $\Omega$ represents the induced angular velocity. The expression of induced tangential velocity is:

$$V_t = b\omega r \tag{8.16}$$

Based on the Betz theory (Betz, 1966), the following equation of flow angular velocity at the location of the rotor plane can be derived:

$$\omega + \Omega/2 = (1+b)\omega \tag{8.17}$$

Then the rotational induced factor will be:

$$b = \Omega/(2\omega) \tag{8.18}$$

Taking the induced tangential velocity into account, the tangential velocity at the location of radius $r$ is:

$$U_T = V_t + \omega r = b\omega r + \omega r = (1+b)\omega r \tag{8.19}$$

In blade element theory, the blade is divided into multiple small blade elements in the radial direction. Each blade element is then treated as a two-dimensional aerofoil and the assumption is made that there is no interaction between the elements. The integration is carried out for the force and torque acting on each element along the radial direction as shown in Figure 8.10c, and then the total force and torque on the whole blade can be obtained (Manwell *et al.*, 2002). The equations for the forces on the blade element are:

$$dL = (1/2)\rho V_{rel}^2 c C_l dr \tag{8.20}$$

$$dD = (1/2)\rho V_{rel}^2 c C_d dr \tag{8.21}$$

$$V_{rel} = V_1(1-\alpha)/\sin \emptyset \tag{8.22}$$

$$dF_X = (1/2)\rho V_{rel}^2 c dr (C_l \cos \emptyset + C_d \sin \emptyset) \tag{8.23}$$

$$dF_Y = (1/2)\rho V_{rel}^2 c dr (C_l \sin \emptyset - C_d \cos \emptyset) \tag{8.24}$$

The expressions for thrust and torque on the blade element at the location of radius $r$ are:

$$dT = (1/2)\rho V_{rel}^2 Bcdr(C_1 \cos \emptyset + C_d \sin \emptyset) \tag{8.25}$$

$$dM = (1/2)\rho V_{rel}^2 Bcrdr(C_1 \sin \emptyset - C_d \cos \emptyset) \tag{8.26}$$

where $T$ is the thrust on the blade, $M$ is the torque on the blade and $B$ the number of blades in the whole rotor.

Wind turbine momentum was first suggested by William Rankine to describe the relation between the force acting on a wind turbine and the incoming wind speed. As shown in Figure 8.9c, the thrust and moment applied on the annular ring zone $(r, r+dr)$ of the wind turbine blade (Spera, 1994) is:

$$dT = 4\pi \rho r V_1^2 (1 - a)adr \tag{8.27}$$

$$dM = 4\pi \rho r^3 V_1 \omega (1 - a)bdr \tag{8.28}$$

By connecting Equations (8.25) and (8.27), the following can be obtained:

$$a/(1 - a) = Bc(C_1 \cos \emptyset + C_d \sin \emptyset)/(8\pi r(\sin \emptyset)^2) \tag{8.29}$$

Similarly for equating Equations (8.26) and (8.28), we end up with:

$$b/(1 + b) = \omega Bc(C_1 \sin \emptyset - C_d \cos \emptyset)/(8\pi r(\sin \emptyset \cos \emptyset)) \tag{8.30}$$

From the velocity triangle shown in Figure 8.10c, it can be shown that:

$$\tan \emptyset = (1 - a)/((1 + b)\lambda_r) \tag{8.31}$$

where $\lambda_r$ is called the local tip speed ratio and is defined as:

$$\lambda_r = \omega r/V_1 \tag{8.32}$$

Applying Bernoulli's equation in front of and behind the rotor in Figure 8.10a gives the pressure drop (Spera, 1994) as:

$$P' = \rho r^2 (\omega + \Omega/2)\Omega \tag{8.33}$$

With Equation (8.33) derived, $dT$ can also be written as:

$$dT = P'dS = 2\pi \rho r^3 (\omega + \Omega/2)\Omega dr \tag{8.34}$$

From Equations (8.17), (8.18), (8.27), (8.32) and (8.34), the following expression can be obtained:

$$b(1 + b)\lambda_r^2 = a(1 - a) \tag{8.35}$$

From Equation (8.28), the power, $dP_{ele}$, of a blade element $dr$ at location $r$ can be derived:

$$dP_{ele} = \omega dM = 4\pi \rho r^3 V_1 \omega^2 (1 - a)bdr \tag{8.36}$$

Therefore, the expression of the power coefficient of the blade element is:

$$dC_p = (8b(1 - a)\lambda_r^3/\lambda^2)d\lambda_r \tag{8.37}$$

The power coefficient derived in Equation (8.37) is based on the assumption of an infinite number of blades. The vortex distribution in the wake of a rotor with a finite number of blades is different from that of a rotor with an infinite number. Therefore, the losses at the locations of blade tip and hub should not be ignored. Prandtl derived a total loss factor $F_T$ to include such effects (Hansen, 2008):

$$F_T = F_{tip}F_{hub} \tag{8.38}$$

$$F_{tip} = (2\cos^{-1} e^{-(B(R-r)/(2R\sin\phi))})/\pi \tag{8.39}$$

$$F_{hub} = (2\cos^{-1} e^{-(B(r-r_{hub})/(2r_{hub}\sin\phi))})/\pi \tag{8.40}$$

where $F_{tip}$ is the tip loss factor, $F_{hub}$ is the hub loss factor and $r_{hub}$ is the radius of the hub.

By considering the total correction factor and according to Glauert (1935), Equations (8.4), (8.35) and (8.37) will be transformed to:

$$Re = V_1(1-a)c/(\upsilon\sin\phi) \tag{8.41}$$

$$b(1+b)\lambda_r^2 = a(1-aF_T) \tag{8.42}$$

$$dC_p = (8F_T b(1-a)\lambda_r^3/\lambda^2)d\lambda_r \tag{8.43}$$

Finally, it is important to remember that the core mission in the design process of a wind turbine blade is to obtain the maximum value of $dC_p$ in equation (8.43) under the constraints of Equations (8.31), (8.38), (8.39), (8.40) and (8.42) and based on the given conditions, and then to use the iterative method to calculate chord and twist angle (Ingram, 2005). Based on the BEM method, the wind turbine blade can be designed with a relatively high power coefficient which ensures that wind energy plays an important role in the renewable energy industry. An outline of the design process is shown in Figure 8.11.

## 8.3 FLUID-STRUCTURE INTERACTIONS AND IMPROVING POWER EFFICIENCY, LIFESPAN AND STABILITY DURING WIND TURBINE OPERATION

Dabiri *et al.* (2015) argued that there remains a large gap between the immense available global wind energy resource and the actual infiltration of current wind energy technologies as a means for electricity generation worldwide. They reported on technology development that is focused on large arrays of small wind turbines that can harvest wind energy at low altitudes, installed in many places that larger systems cannot be, especially urban areas. Dabiri *et al.* (2015) went on to note that these features (i.e. larger arrays of small turbines) can be leveraged to achieve social acceptance and rapid adoption of this new technology. The authors claimed that favorable economics would stem from an order-of-magnitude reduction in the number of components in a new generation of simple, mass-manufacturable vertical-axis wind turbines. However, it was noted that this vision can only be attained by overcoming some substantial scientific challenges that have limited such growth in previous decades. One of these challenges is fluid-structure interaction, which is a very important phenomenon closely related to power efficiency, lifespan and stability in wind turbine operation. In order to have a better understanding of how flow causes blade deformation, a literature review of the fluid-structure interactions of wind turbines is presented in this section.

Fluid-structure interaction problems can be described by a set of highly non-linear equations. When solving the fluid-structure interaction problem, the instantaneous solutions of the coupling equations should be obtained. These solutions should fulfill the boundary conditions at the interface of the fluid and the structure (Bathe *et al.*, 1999). One of the challenges is that the fluid and structure are described in different ways, as the fluid is treated with the Eulerian method and the structure is described with the Lagrangian method. In the Navier-Stokes equations, which are written in Eulerian frame, the field parameters such as velocity and pressure are related to fixed

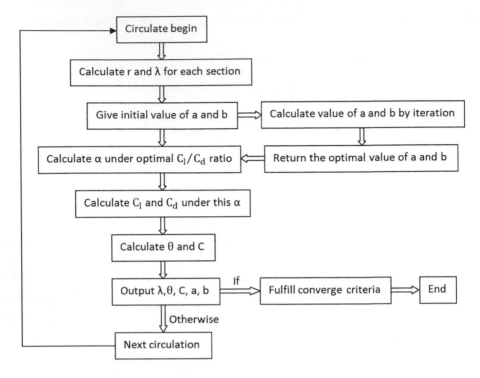

Figure 8.11.    Flow chart of wind turbine blade design process.

locations. In fluid-structure interaction, the flow domain is changing with time and, therefore, the Navier-Stokes equations cannot be directly applied. To adapt to a time-dependent flow domain, the most common approach is to combine the Eulerian method with the Lagrangian method. The arbitrary Lagrangian-Eulerian (ALE) method is widely adopted to solve for moving boundary conditions (Donea, 1983).

The two main methods used to solve the interaction problems are the direct approach and the iterative one. The direct approach uses and solves the fluid and structure equations together, simultaneously. In the iterative approach the equations of the fluid and structure are solved separately and the solution parameter is passed iteratively from one domain to the other until a convergence of computation is obtained. Normally, the direct approach will be applied to strongly coupled domains and the iterative one will be employed when the interaction between fluid and structure is weak (Rugonyi and Bathe, 2001).

The equations of motion for a compressible Newtonian fluid using the ALE method (Rugonyi and Bathe, 2001) are as follows.

Momentum equation:

$$\rho(D\mathbf{v}/Dt) + \rho[(\mathbf{v} - \hat{\mathbf{v}}) \cdot \nabla]\mathbf{v} = \nabla \cdot \tau + \mathbf{f^B} \qquad (8.44)$$

Mass conservation equation:

$$(D\mathbf{v}/Dt) + \rho(\upsilon - \hat{\upsilon}) \cdot \nabla + \nabla \cdot \mathbf{v} = 0 \qquad (8.45)$$

Energy equation:

$$\rho(De/Dt) + \rho(\mathbf{v} - \hat{\mathbf{v}}) \cdot \nabla e = \tau \mathbf{f^{st}} - \nabla \cdot \mathbf{q} + q^B \qquad (8.46)$$

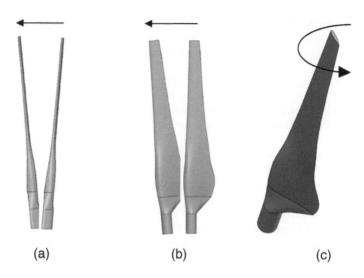

Figure 8.12.   Three main types of blade deformation: (a) flap; (b) swing; (c) twist.

where $D/Dt$ is the total time derivative in the ALE frame; $\mathbf{v}$ is the velocity vector; $\hat{\mathbf{v}}$ is the velocity vector in ALE frame; $\boldsymbol{\tau}$ is the fluid stress tensor; $\mathbf{f^B}$ is the body force vector; $\mathbf{f^{st}}$ is the velocity strain vector where $2f^{st} = \nabla \mathbf{v} + (\nabla \mathbf{v})^T$; $\mathbf{q}$ is the heat flux tensor; $q^B$ is heat per unit volume; $(\nabla \cdot)$ is divergence operator, $\nabla$ gradient operator and $(\cdot)$ dot product. If $\hat{\mathbf{v}} = 0$, then it means that the frame is not moving and all the equations take Eulerian form. If $\mathbf{v} = \hat{\mathbf{v}}$, then the fluid is moving together with the frame and the Lagrangian formulation will be recovered.

The governing equation of the motion of the structure (Rugonyi and Bathe, 2001) is defined as:

$$\rho D^2 \mathbf{u}/Dt^2 = \nabla \cdot \boldsymbol{\tau} + \mathbf{f^B} \tag{8.47}$$

where $\mathbf{u}$ is the vector of the structure displacement. Furthermore, to solve the fluid-structure interaction problem, Equations (8.44) to (8.47) are calculated together based on the determined boundary conditions and solution approach.

The blade is the component of a wind turbine that sustains the most complicated forces. Aerodynamic loading, inertial loading and gravitational loading are the three most crucial loads acting on a wind turbine blade. The combined effects of these three loads will cause blade deformations, of which there are three main types (Fig. 8.12). The interaction between these three types of deformations and aerodynamic loading will lead to aero-elastic problems (Rasmussen *et al.*, 2003). When a blade is deformed, the bending will cause a change in its angle of attack, which, in turn, will result in lift force variation, and subsequently will alter the aerodynamic torque. This will result in feedback of the bending into bending-twist coupling vibration. Then there is self-excited oscillation which can cause severe damage to the wind turbine blade. This unstable aero-elastic vibration is called the fluttering phenomenon.

When the excitation frequency of the blade is approaching its own natural frequency, resonance will occur. To avoid this situation, within its main operating range the designed blade's natural frequency should have sufficient difference from the excitation frequency. According to Tong (2010), the behavior of the wind turbine blade in fluid-structure interaction is similar to that of a cantilever beam. Based on the mechanical vibration theory, the blade can be treated as a distributed-parameter system. Modal analysis can be used to obtain the natural frequency and the mode of the blade. Therefore, the displacement function of the blade $y(x, t)$ is expressed as a product of a space function, $\gamma(x)$, and a time function, $q(t)$:

$$y(x, t) = \gamma(x)q(t) \tag{8.48}$$

where $\gamma(x)$ is the mode of the elastic object, which is only related to its particle position, and $q(t)$ is the pattern of the vibration and only depends on time. The natural frequency and vibration mode can then be obtained by solving the following equation (Meirovitch, 2001):

$$d^2(EI(d^2\gamma/dx^2))/dx^2 = m\omega_n^2\gamma \qquad (8.49)$$

where $E$ is Young's modulus, $I$ is moment of inertia, $x$ is distance from origin, $\omega_n$ is natural frequency, and $m$ is mass. Each order of the natural frequency and normalized mode can then be obtained.

To sum up, the wind turbine blade fluid-structure interaction involves a strongly coupled distributed-parameter system. The analytical solutions of the equations, including the governing equations for the fluid and solid domains as well as the vibration equations, can only be obtained under special simple cases. In the present work, the aero-elastic phenomenon is very complicated; its numerical simulation requires efficient use of computing resources in order to obtain detailed blade deflections and information for the whole flow field.

Load predictions for wind turbines are frequently based on engineering design methods that use overly simplified physical models. The wind engineering community has been working on aero-elastic stability problems for quite some time (Bjorck, 1995). However, most of these works focused on flap-wise vibrations only, for which a variety of engineering-type dynamic stall models have been developed. Thus, Simms *et al.* (2001) introduced a quasi-steady linearized flap/lead-lag model to study edgewise vibrations in terms of blade and wind turbine parameters (Petersen *et al.*, 1998). Nevertheless, all of the engineering design tools currently employed are incapable of taking into account dynamic stall phenomena as well as obtaining detailed information. For instance, two comparative studies of unsteady flow and fluid-structure interaction solvers for wind turbine applications, performed by Chaviaropoulos *et al.* (2003) with researchers from many countries, found major uncertainty and efficiency problems with these sophisticated simulation tools. Simms *et al.* (2001) compared the wind tunnel experimental results of the National Renewable Energy Laboratory (NREL) with 19 simulation packages, two of which were unsteady Navier-Stokes solvers, and were surprised by the wide variation between the experimental results and the various code predictions. One three-dimensional Navier-Stokes solver did the best job of predicting aerodynamic forces under the simplest upwind zero-yaw conditions. However, the investigators indicated that success with this computational fluid dynamics (CFD) code tended to be uncertain, varying significantly and very much dependent on the type of aerofoil used. Simms *et al.* (2001) reported that they had very good results in this effort with the S809 aerofoil (Tangler and Somers, 1996) used by NREL, but had previously had poor results with some other aerofoils. Another problem was that very high computational power was required for even the simplest cases. However, Simms *et al.* (2001) suggested that the success with this CFD code did show the potential benefit of pursuing CFD and that further work in this area was warranted.

## 8.4   THE ECONOMICS OF WIND TECHNOLOGY AND DESALINATION: FUTURE RESEARCH DIRECTIONS

A major long-term goal is to extensively deploy environmentally friendly and optimized wind turbine technology on a large scale to generate electricity to run water desalination plants. Abusharkh *et al.* (2015) presented a review of the use of wind energy in desalination, with a focus on sustainable commercial processes. In terms of future research directions, they argued that although wind energy is considered to be the most environmentally friendly renewable energy, problems might still arise from the application of wind energy to desalination because of the intermittency of wind speed. Such fluctuations would lead to variation in the renewable energy generated and an inability to accurately predict the specific cost of potable water production. However, the hybridization of wind energy with other energy sources to stabilize the power output is a possible solution. Abusharkh *et al.* (2015) went on to report that the design of a combined renewable energy farm and desalination plant is also another important factor that determines the efficiency of wind power as an energy source for desalination. For example, a significant reduction in capital costs can be achieved by using lighter but stronger wind blades. In addition, the ability to reuse or recycle the residual energy in all outputs from desalination plants would go a long way to minimizing the energy and cost requirements of these plants. Future research needs to be focused on designs for wind-powered desalination plants that can adequately avoid energy loopholes or unnecessary energy sinks and ensure optimum efficiency during the consumption of the energy generated.

The review by Abusharkh *et al.* (2015) was able to show that wind and geothermal energy can be integrated with other energy sources to improve the economics and reduce the environmental impacts of desalination processes. The coupling of wind energy with thermal and photovoltaic energy systems, and the use of wind energy to supply the gravitational energy potential required for desalination, yielded impressive results. The major benefit of coupled systems is that a year-round power supply can be achieved, and pure water production can be accurately monitored on a continuous basis. Abusharkh *et al.* (2015) recommended that future investigations be directed towards the design of desalination systems with such integrated energy sources, and a similar conclusion was reached by Ghaffour *et al.* (2015) and Goosen *et al.* (2014).

In a related study dealing with the opportunities and challenges facing the commercialization of wind energy technology, Dabiri *et al.* (2015) maintained that favorable economics would stem from an order-of-magnitude reduction in the number of components in a new generation of simple, mass-manufacturable VAWTs. This would be based on large arrays of small, simple VAWTs that could, in many cases, supplant conventional HAWTs or simply be used in locations and applications that are incompatible with HAWTs. The authors noted that there remains significant additional research to be completed to achieve the full, global potential of wind energy.

## 8.5 CONCLUDING REMARKS

The search for inexpensive, reliable and clean renewable energy sources for desalination of sea-water and brackish water is a major challenge facing mankind. Wind turbines show great promise for providing electricity on a large scale, not only for desalination but also for applications such as heating and cooling in urban areas. Nowadays, wind turbine technology is developing quickly and commercialized wind farms have sprung up all over the world. More and more companies are manufacturing wind turbines and others are expanding wind farms from land to offshore locations. Turbine control strategy is also evolving, from the use of a passive fixed-speed stall-control method to adoption of a positive variable-speed pitch angle control method.

In future, the trend will be for wind farms located away from land and in the sea. The potential of offshore wind resource is huge and the total resource available is more than three times that on land. Offshore wind resources have many advantages, such as higher and more stable wind speeds, higher overall annual availability, no use of land resources and less impact on the environment. However, offshore wind turbines have challenges in terms of safety, reliability, maintenance and operational cost control that require detailed analysis during the process of wind turbine design, production and installation.

The merits of the HAWT, such as high power efficiency, compact structure, long lifespan and relatively low costs, make it a popular choice. HAWTs account for more than 95% of the global wind turbine market. Over the same time period, VAWTs have encountered technological problems, such as low power efficiency, short lifespans and high installation costs, all of which have made it less favored in the wind turbine market.

As demonstrated in this chapter, the wind turbine blade faces very complicated aerodynamic conditions due to flow separation, wake effects and fluid-structure interactions (FSI). As the wind turbine reaches the megawatt (MW) level of electricity generation, blade lengths will extend to around one hundred meters. This will dramatically increase blade flexibility and thus make the wind turbine more vulnerable, which may cause catastrophic results, such as reduction in wind turbine blade lifespans and even blade fracture. Currently, the most widely employed wind turbine design technique, the BEM method, does not take the FSI effect into account and therefore could not precisely predict wind turbine performance. Wind turbine aerodynamic optimization will need to include these effects as blade lengths grow larger.

In closing, the optimization and scaling up of electricity generation using wind turbine technology for applications in desalination is crucial. The emphasis and focus needs to be on sustainable commercial processes. In terms of future research directions, wind energy may be considered the most environmentally friendly renewable energy, but problems might still arise because of its intermittency. Such fluctuations would, for example, lead to an inability to accurately predict the specific cost of potable water production. However, the hybridization of wind energy with other energy sources to stabilize power output is one possible solution, and large-scale electricity generation using wind turbines has great promise in solving the energy demands of desalination plants worldwide. The technology is easy to scale up, it is environmentally friendly and it can be readily integrated with other renewable energy systems such as solar and geothermal.

## NOMENCLATURE

| | |
|---|---|
| $a$ | axial induction factor |
| $b$ | rotational induction factor |
| $c$ | chord length [m] |
| $C_1$ | lift coefficient |
| $C_d$ | drag coefficient |
| $C_p$ | power coefficient |
| $D$ | drag force [N] |
| $E$ | Young's modulus [Pa] |
| $\mathbf{f_B}$ | body force vector [N] |
| $\mathbf{f_{st}}$ | velocity strain vector |
| $F$ | force on blade [N] |
| $F_T$ | total loss factor |
| $F_{tip}$ | tip loss factor |
| $F_{hub}$ | hub loss factor |
| $F_X$ | force in $x$ direction [N] |
| $F_Y$ | force in $y$ direction [N] |
| $I$ | moment of inertia [kg m$^2$] |
| $L$ | lift force [N] |
| $m$ | mass [kg] |
| $M$ | moment [N·m] |
| $P$ | power output of wind turbine [W] |
| $P_{ele}$ | power of the blade element [W] |
| $P_{max}$ | maximum power for an ideal wind turbine [W] |
| $P'$ | pressure drop [Pa] |
| $\mathbf{q}$ | heat flux tensor [W m$^{-2}$] |
| $q^B$ | heat per unit volume [W m$^{-3}$] |
| $q(t)$ | pattern of vibration |
| $r_{hub}$ | radius of hub [m] |
| $R$ | radius of rotor [m] |
| $S$ | area of rotor [m$^2$] |
| $S_1$ | area of actuator disc before rotor [m$^2$] |
| $S_2$ | area of actuator disc after rotor [m$^2$] |
| $T$ | thrust on wind turbine [N] |
| $U_T$ | tangential velocity [m s$^{-1}$] |
| $\mathbf{v}$ | velocity vector [m s$^{-1}$] |

$\hat{\mathbf{v}}$    fluid velocity in ALE frame [m s$^{-1}$]
$V$    wind speed after rotor [m s$^{-1}$]
$V_1$    incoming wind speed [m s$^{-1}$]
$V_2$    wind speed far downstream [m s$^{-1}$]
$V_t$    induced tangential velocity [m s$^{-1}$]
$x$    displacement [m]

*Greek symbols*

$\alpha$    angle of attack [rad]
$\gamma(x)$    vibration mode
$\theta$    twist angle [rad]
$\lambda$    tip speed ratio
$\lambda_r$    local tip speed ratio
$\tau$    fluid stress tensor [Pa]
$\varnothing$    wind direction angle [rad]
$\omega$    rotor angular velocity [rad s$^{-1}$]
$\omega_n$    natural frequency [Hz]
$\Omega$    induced angular velocity [rad s$^{-1}$]

Others

$D/Dt$    time derivative in ALE frame
$(\nabla\cdot)$    divergence operator
$\nabla$    gradient operator
$(\cdot)$    dot product

## ACKNOWLEDGEMENTS

Zheng Shuai and Chua Leok Poh would like to express their sincere thanks for the financial support provided by the Ministry of Education (Singapore) Tier 1 Research Grant of Project reference no. RG 121/14(EP5).

## REFERENCES

Abusharkh, A.G., Giwa, A. & Hasan, S.W. (2015) Wind and geothermal energy in desalination: a short review on progress and sustainable commercial processes. *Industrial Engineering and Management*, 4, 175.

Bathe, K.J., Zhang, H. & Ji, S. (1999) Finite element analysis of fluid flows fully coupled with structural interactions. *Computers & Structures*, 72, 1–16.

Bertin, J.J. & Smith, M.L. (1989) *Aerodynamics for Engineers*. Prentice Hall, Englewood Cliffs, NJ.

Betz, A. (1966) *Introduction to the Theory of Flow Machines*. Pergamon, Oxford.

Bjorck, A. (1995) Dynamic stall and three-dimensional effects (DNS3D). EC DXGII Joule II Project No. JOU2-CT93-0345 Report. The Aeronautical Research Institute of Sweden, Stockholm. pp. 231–238.

Burton, T., Jenkins, N., Sharpe, D. & Bossanyi, E. (2011) *Wind Energy Handbook*, 2nd edn. John Wiley & Sons, Chichester, UK.

Chaviaropoulos, P.K., Nikolaou, I.G., Aggelis, K.A., Soerensen, N.N., Johansen, J., Hansen, M.O.L., Gaunaa, M., Hambraus, T., Von Geyr, H., Hirsch, C., Shun, K., Voutsinas, S.G., Tzabiras, G., Perivolaris, Y. & Dyrmose, S.Z. (2003) Viscous and aeroelastic effects on wind turbine blades. The VISCEL project, Part I: 3D Navier-Stokes rotor simulations. *Wind Energy*, 6, 365–385.

Dabiri, J.O., Greer, J.R., Koseff, J.R., Moin, P. & Peng, J. (2015) A new approach to wind energy: opportunities and challenges. *AIP Conference Proceedings, 8–9 March 2014, Berkeley, CA*. 1652, 51.

Donea, J. (1983) Arbitrary Lagrangian Eulerian methods. In: Belytschko, T. & Hughes, T.J.R. (eds) *Computational Methods for Transient Analysis*, Volume 1. North Holland, Amsterdam. pp. 58–62.

Fein, G.S. & Merritt, E.T. (2011) Method for creating micro/nano wind energy gathering devices. Patent US 7950143 B2, 31 May 2011.

Frondel, M., Ritter, N., Schmidt, C.M. & Vance, C. (2010) Economic impacts from the promotion of renewable energy technologies – the German experience. *Energy Policy*, 38(8), 4048–4056.

Gasch, R. & Twele, J. (2012) *Wind Power Plants – Fundamentals, Design, Construction and Operation*, 2nd edn. Springer, Heidelberg, Germany.

Ghaffour, N., Bundschuh, J., Mahmoudi, H. & Goosen, M.F. (2015) Renewable energy-driven desalination technologies: a comprehensive review on challenges and potential applications of integrated systems. *Desalination*, 356, 94–114.

Glauert, H. (1935) Airplane propellers. *Aerodynamic Theory*, 4, 169–360.

Goosen, M.F., Mahmoudi, H. & Ghaffour, N. (2014) Today's and future challenges in applications of renewable energy technologies for desalination. *Critical Reviews in Environmental Science and Technology*, 44(9), 929–999.

GWEC (2015) Global wind report – annual market update 2014. Global Wind Energy Council, Brussels.

Hansen, A. & Butterfield, C. (1993) Aerodynamics of horizontal-axis wind turbines. *Annual Review of Fluid Mechanics*, 25, 115–149.

Hansen, M.O.L. (2008) *Aerodynamics of Wind Turbines*, 2nd edn. Earthscan, London.

Houghton, E.L. & Carpenter, P.W. (2003) *Aerodynamics for Engineering Students*. Butterworth-Heinemann, Oxford.

Ingram, G. (2005) *Wind turbine blade analysis using the blade element momentum method*. Course note, School of Engineering, Durham University, UK. Available from: https://community.dur.ac.uk/g.l.ingram/download/wind_turbine_design.pdf [accessed November 2016].

Jamieson, P. (2011) *Innovation in Wind Turbine Design*. John Wiley & Sons, Chichester, UK.

Lugt, H.J. (1996) *Introduction to Vortex Theory*. Vortex Flow Press, Potomac, MD.

Manwell, J.F., McGowan, J.G. & Rogers, A.L. (2002) *Wind Energy Explained: Theory, Design and Application*. John Wiley & Sons, Chichester, UK.

Meirovitch, L. (2001) *Fundamentals of Vibrations*, 2nd edn. McGraw-Hill, New York.

Petersen, J.T., Madsen, H.A., Bjorck, A., Enevoldsen, P., Oye, S., Ganander, H. & Winkelaar, D. (1998) Prediction of dynamic loads and induced vibrations in stall. Risø-R-1045(EN), Risø National Laboratory, Roskilde, Denmark.

Rasmussen, F., Hansen, M.H., Thomsen, K., Larsen, T.J., Bertagnolio, F., Johansen, J., Madsen, H.A., Bak, C. & Hansen, A.M. (2003) Present status of aeroelasticity of wind turbines. *Wind Energy*, 6, 213–228.

Rugonyi, S. & Bathe, K. (2001) On finite element analysis of fluid flows fully coupled with structural interactions. *Computer Modeling in Engineering and Sciences*, 2, 195–212.

Schlichting, H. (1968) *Boundary-Layer Theory*. McGraw-Hill, New York.

Selig, M.S. & McGranahan, B.D. (2004) Wind tunnel aerodynamic tests of six aerofoils for use on small wind turbines. *Journal of Solar Energy Engineering*, 126, 986–1001.

Simms, D.A., Schreck, S., Hand, M. & Fingersh, L. (2001) NREL unsteady aerodynamics experiment in the NASA-Ames wind tunnel: a comparison of predictions to measurements. National Renewable Energy Laboratory, Golden, CO.

Spera, D.A. (1994) *Wind Turbine Technology*. CRC Press, Boca Raton, FL.

Tangler, J.L. & Somers, D.M. (1996) NREL aerofoil families for HAWTs. National Renewable Energy Laboratory, Golden, CO.

Tong, W. (2010) *Wind Power Generation and Wind Turbine Design*. WIT Press, Southampton, UK.

Vries, O. (1983) On the theory of the horizontal-axis wind turbine. *Annual Review of Fluid Mechanics*, 15, 77–96.

Wang, X.D. (2008) *Fundamentals of Fluid-Solid Interactions: Analytical and Computational Approaches*, 1st edn. Elsevier, Amsterdam.

White, F.M. (1991) *Viscous Fluid Flow*. McGraw-Hill, New York.

Wilson, R.E. & Lissaman, P.B.S. (1974) Applied aerodynamics of wind power machines. Review paper, Oregon State University, Corvallis, OR. Available from: http://ir.library.oregonstate.edu/xmlui/bitstream/handle/1957/8140/WilsonLissaman_AppAeroOfWindPwrMach_1974.pdf [accessed November 2016].

Wilson, R.E., Lissaman, P.B.S. & Walker, S.N. (1976) Aerodynamic performance of wind turbines. Energy Research and Development Administration, Washington, DC.

# CHAPTER 9

## A critical review of fuel cell commercialization and its application in desalination

Yousef Alyousef, Mattheus Goosen & Youssef Elakwah

## 9.1 INTRODUCTION

The largest markets for fuel cells for energy generation are in stationary power, portable power, auxiliary power units, and material-handling equipment (DOE, 2012, 2013). Fuel cells convert chemical energy into electrical current and heat without combustion. According to Staffell (2015), although prototype hydrogen cars have commanded greater media attention, combined heat and power (CHP) is the largest and most established market for fuel cells. In 2009, companies began selling thousands of units per year, marking the switch to mass manufacture. As reported by Staffell (2015), nearly 60,000 systems were sold in Japan during a four-year period from 2009 to 2013. He claimed that, by 2015, residential fuel cells would be sold in Japan without government subsidies, which would make fuel cell CHP commercially viable, at least in Japan. Although hydrogen and fuel cells are not strictly renewable energy resources, their components are abundant and are very low in pollution when operated. Hydrogen, for example, can be burned as a fuel, normally in a vehicle or in a fuel cell, with only water as the combustion product.

In recent years, fuel cells have emerged as attractive alternatives to energy conversion or storage systems because they are extremely effective and do not produce pollutants such as $S_{O_2}$, $Kn_{O_2}$ and $CO_2$, especially when compared to fossil fuel-driven technologies such as combustion engines (Dodd's et al., 2015; Kim et al., 2015). In addition, it can be argued that the high costs involved in the commercialization of fuel cells may be partially overcome by the development of new materials and fabrication processes. Fuel cells are generally classified according to their operating conditions (e.g. temperature), their structure (e.g. scale of the system and application), and the nature of the electrolyte used in the fuel cell. Five different categories of fuel cells have been identified, each using a different electrolyte (see Table 9.1 and Fig. 9.1). Both the alkaline fuel cells (AFCs) and molten carbonate fuel cells (MCFCs), for example, use only hydrogen ($H_2$) and oxygen ($O_2$) as input fuels, with only water, heat and electrical current being emitted (i.e. no fossil fuels in nor pollutants out).

Kim et al. (2015) prepared a review of polymer-nanocomposite electrolyte membranes for fuel cell applications. From their analysis they concluded that given the design of suitable polymer

Table 9.1. Different types of fuel cells (adapted from Kim et al., 2015).

| Category | Type of electrolyte |
|---|---|
| Alkaline fuel cells (AFCs) | Alkaline solution |
| Phosphoric acid fuel cells (PAFCs) | Acidic solution |
| Proton exchange membrane fuel cells (PEMFCs) | Polymer electrolyte |
| Molten carbonate fuel cells (MCFCs) | Molten carbonate salt electrolyte |
| Solid oxide fuel cells (SOFCs) | Ceramic ion-conducting electrolyte |

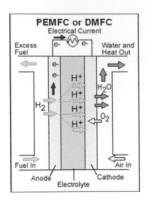
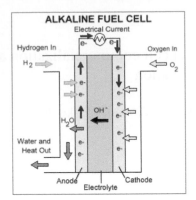
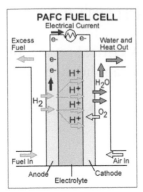
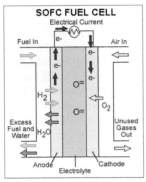
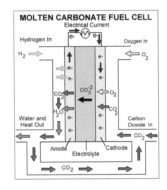

Figure 9.1.   Different types of fuel cells (adapted from DOE, 2012).

composites it is possible to provide long-term durability and better electrical properties, such as current density, of the polymer electrolyte membrane. The working principle of the fuel cell is illustrated schematically in Figure 9.2. Currently, polymer electrolyte membrane fuel cells are based on perfluorosulfonic acid membranes such as Nafion and Flemion, which have shortcomings in terms of proton conductivity, durability, thermal stability, maximum power density and fuel crossover, and cost. In recent years, the hybrid organic-inorganic composite membrane has emerged as an interesting alternative. It provides a unique combination of organic and inorganic properties, and overcomes the limitations of the pure polymeric membranes. Among the various fuel cell systems, solid proton exchange membrane fuel cells (PEMFC) can be applied to small-scale applications such as automotive, portable power generation, and stationary equipment (Fig. 9.1 and Fig. 9.2). The power-generating system in PEMFCs is an electrochemical reaction involving gases such as hydrogen, methanol and ethanol. The reactions occurring in the fuel cell can be described as follows:

$$\text{Anode:} \qquad H_2 \rightarrow 4H^+ + 4e^- \qquad\qquad (9.1)$$

$$\text{Cathode:} \qquad O_2 + 4e^- + 4H^+ \rightarrow 2H_2O \qquad\qquad (9.2)$$

$$\text{Cell reaction:} \quad 2H_2 + O_2 \rightarrow 2H_2O \qquad\qquad (9.3)$$

The commercial and ecological benefits of renewable energies, such as wind, solar, wave and geothermal, and their application to decreasing overall energy consumption have been critically

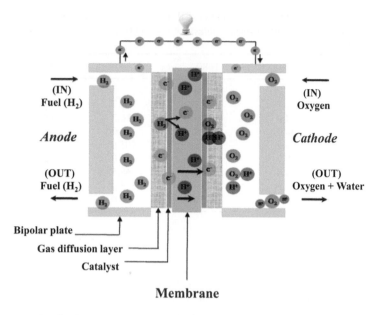

Figure 9.2.    Principle of polymer electrolyte membrane fuel cell (Kim *et al.*, 2015).

assessed in recent articles (Goosen *et al.*, 2014, 2016). Historical recorded accounts show long-standing usage of, for example, geothermal water for bathing, cooking or space heating by the Romans, Japanese, Turks, Icelanders, Central Europeans and the Maori of New Zealand (Lund, 2008). Similarly, early Greeks were the first to use passive solar design in their homes (Butti and Perlin, 1980). Nonetheless, it is a worrying observation that over the past 40 years, renewable energy sources have consistently accounted for only 13% of total energy use (Table 9.2). The global energy structure is still dominated by fossil fuels and we can speculate that fuel cells could, perhaps, better help to bridge the gap between renewable energy and fossil fuels.

Combining renewable energy sources with desalination systems has great potential for water-scarce regions such as the Middle East and North Africa (Ghaffour *et al.*, 2015; Goosen *et al.*, 2014, 2016). Renewable energy-driven desalination systems can be separated into two classes. The first includes distillation processes, such as multi-stage flash, multiple-effect distillation, and thermal vapor-compression, driven by heat produced *directly* by the renewable energy system, while the second embraces *indirect* processes driven by electricity or mechanical energy, such as reverse osmosis membranes and mechanical vapor-compression, produced by renewable energy sources. In addition, there are hybrid systems that also show great promise with, for example, geothermal brine being used directly in membrane distillation technology or to feed a solar still (Bouchekima, 2003; Houcine *et al.*, 1999).

Fuel cells have started to become attractive alternatives because they are extremely effective, and in comparison to fossil fuel-driven technologies do not produce pollutants (Kim *et al.*, 2015). Besides fuel cell CHP systems, microbial fuel cells (MFC) may, for example, be employed directly for water purification and electricity generation. However, these latter are currently only functioning on a small scale and are not yet commercially viable, as far as the authors are aware. In a recent review of the advantages and challenges of MFCs, Choi (2015) argued that the next generation of sustainable energy could come from microorganisms, as indicated by an upsurge in electromicrobiology, the study of microorganisms' electrical properties. He reported that MFCs are gaining acceptance as a future alternative renewable energy technology and energy-efficient wastewater treatment method. In addition, Wang *et al.* (2015) reported in another analysis that the generation of practically usable power is a critical milestone for further

Table 9.2.    Global primary energy use in exajoules [EJ], 1970–2006 (adapted from Moriarty and Honnery, 2009). In describing national or global energy budgets, it is common practice to use large-scale units based upon the joule; $1 \text{ EJ} = 10^{18} \text{ J}$.

| Energy source | 1970 | 1980 | 1990 | 2000 | 2006 |
|---|---|---|---|---|---|
| Fossil fuels | | | | | |
| Coal | 64.2 | 75.7 | 93.7 | 98.2 | 129.4 |
| Oil | 94.4 | 124.6 | 136.2 | 148.9 | 162.9 |
| Natural gas | 38.1 | 54.9 | 75.0 | 91.8 | 107.8 |
| *Total fossil fuels* | *196.7* | *255.1* | *305.0* | *339.0* | *400.1* |
| Nuclear | 0.7 | 6.7 | 19.0 | 24.5 | 26.6 |
| Renewable | 29.4 | 37.6 | 48.5 | 55.6 | 66.2 |
| *All energy* | *216.8* | *299.5* | *372.4* | *419.0* | *492.9* |
| *Renewable [%]* | *13.6* | *12.6* | *13.0* | *13.3* | *13.4* |

MFC growth and application. They concluded that establishing how to effectively and efficiently harvest and utilize MFC energy remains a key challenge. Their review deliberated on the different methods and systems that have been developed for MFC energy extraction and conditions for practical use. It was clear that more work needs to be done to optimize MFC design, improve harvesting efficiency, and reduce the cost. Wang *et al.* (2015) suggested that this was the main bottleneck in MFC application development and should be a new frontier of research. Additional information about fuel cells and how they can be employed as an energy source for desalination may be found in Faisal *et al.* (2017).

The present chapter focuses on the development and commercialization of fuel cells and how this may benefit the desalination industry. After a critical review of energy consumption reduction in desalination, markets and challenges for fuel cell commercialization are discussed. This is followed by a look at fuel cell efficiencies and manufacturers' rated performance versus that in the real world, as well as specific examples of automotive fuel cells and the need to meet cost, performance and durability criteria. The chapter concludes with an analysis of the influence of socioeconomic growth and industrial competition on the desalination industry.

## 9.2   THE NEED TO REDUCE ENERGY CONSUMPTION IN THE DESALINATION SECTOR

There is a strong relationship between water desalination and energy consumption (Alyousef and Abu-ebid, 2012; Goosen *et al.*, 2016). Furthermore, for arid countries such as Saudi Arabia there is an additional drain on energy resources since a large percentage of water consumed is desalinated water, which also has to be conveyed over long distances to reach population centers (Alyousef and Abu-Ebid, 2012). It can be argued that the significant increase in usage of oil and gas for water and electricity production over the next two decades will severely reduce the country's income, thus affecting the government's ability to subsidize electricity and water costs (Bourland and Gamble, 2011). A summary of water production and demand in Saudi Arabia is given in Figure 9.3a. Notably, it reveals that a major problem is the non-sustainability of the water production and consumption cycle, with 81% of the water coming from non-renewable groundwater and 80% of the consumed water being used for irrigation. For example, in 2008, the primary energy production was 1366 TWh year$^{-1}$ (60% coming from gas and 40% from oil), which was used to produce water and electricity (Fig. 9.3b). The demand for energy is expected to double, reaching almost 2600 TWh year$^{-1}$ by 2030, because of an anticipated large increase in population.

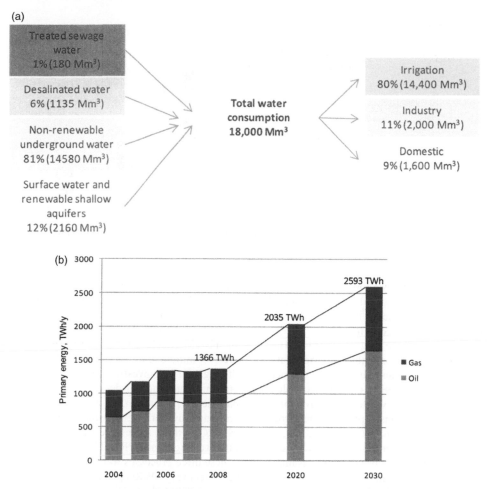

Figure 9.3. (a) Water production and demand in Saudi Arabia; (b) growth of primary energy demand in Saudi Arabia (Alyousef and Abu-ebid, 2012).

The two main desalination methods in general use are multi-stage flash (MSF) distillation and reverse osmosis (RO) (Ghaffour *et al.*, 2015). Power generation produces significant amounts of heat which, if not utilized, will be discarded to the atmosphere. Such heat can be utilized in desalination. Combining power generation with desalination gives higher energy efficiency than generating electricity or desalinating water separately, with improved energy efficiency of 10 to 20%. Thus an effective policy is one which promotes the construction of cogeneration desalination plants in which surplus heat can be recovered and used while at the same time generating electricity. Alyousef and Abu-ebid (2012) maintained that in order to address the predicted growth in water demand in Saudi Arabia, another 20 to 30 desalination plants will be needed by 2030, with a total capital outlay of (US) $50 billion. One effective strategy for reducing energy consumption would be to improve the efficiency of such plants, and an additional area where a contribution may be made is in investigation of the possibility of developing fuel cell technology on a large scale for energy production for desalination plants.

Novel technologies for addressing the problem of high energy consumption in desalination have also recently been reported (Sahin *et al.*, 2016; Wan and Chung, 2016). For example, a potential means of addressing the high energy intensity of traditional desalination plants is the development of pressure-retarded osmosis (PRO) technology. PRO extracts the Gibbs free

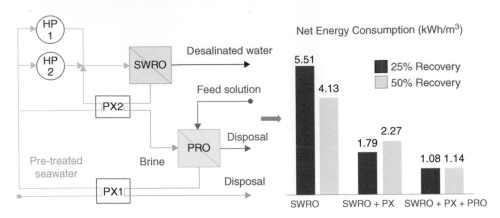

Figure 9.4.   Energy recovery by pressure-retarded osmosis (PRO) in SWRO-PRO integrated processes (Wan and Chung, 2016).

energy of mixing by allowing water to spontaneously flow through a semi-permeable membrane from a low-salinity feed solution to a high-salinity draw solution against a hydraulic pressure. The Gibbs free energy is converted to the hydraulic pressure of the diluted brine, which can be further converted to mechanical energy by a pressure exchanger (PX) or to electrical energy by a hydro-turbine (Fig. 9.4). While Helfer *et al.* (2013) concluded that PRO technology is viable, they mentioned that a barrier to its widespread commercial development is the high cost of PRO membranes. The authors claimed that although further investigations are still needed to ensure the viability of PRO technology, this technical advance in desalination technology is promising as a viable alternative for renewable energy production. In a related study, the specific energy consumption of three processes involving seawater reverse osmosis (SWRO) was scrutinized by Wan and Chung (2016), namely, SWRO without a pressure exchanger, SWRO with a pressure exchanger, and SWRO with pressure exchangers and PRO (Fig. 9.4). The results showed that the specific energy consumptions for these three processes were, respectively, 5.51, 1.79 and 1.08 kWh m$^{-3}$ of desalinated water for a 25% recovery SWRO plant; and 4.13, 2.27 and 1.14 kWh m$^{-3}$ of desalinated water for a 50% recovery SWRO plant, using either freshwater or wastewater as the feed solution in PRO. Whether these results can be translated into large-scale commercial application remains to be seen. However, it does demonstrate the great potential in energy savings available through combination of conventional with novel systems.

## 9.3   DEVELOPMENT AND COMMERCIALIZATION OF FUEL CELLS AND HOW THIS MAY HELP THE DESALINATION INDUSTRY

### 9.3.1   *Markets and challenges in fuel cell commercialization*

As noted in a DOE (2012) article, the major markets for fuel cells are in stationary power, portable power, and auxiliary power units. Around 75,000 fuel cells had been distributed worldwide by the end of 2009 and a further 15,000 fuel cells were distributed in 2010. Companies such as General Motors, Toyota, Honda, Hyundai and Daimler have revealed plans to start commercializing fuel cell vehicles. However, there are a variety of challenges facing the widespread commercialization of fuel cells (Fig. 9.5), ranging from system cost, durability and hydrogen storage, to system integration and market transformation. It can be concluded that success in the application of hydrogen fuel cells as auxiliary power units in the automotive industry, for example, will assist in the translation of this renewable energy technology into desalination applications, particularly in cases where small, remote or residential renewable energy sources are required.

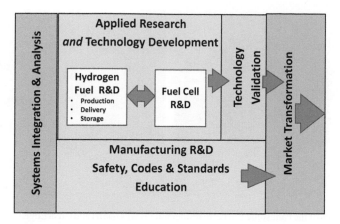

Figure 9.5. Challenges in development and commercialization of fuel cell applications (adapted from DOE, 2012).

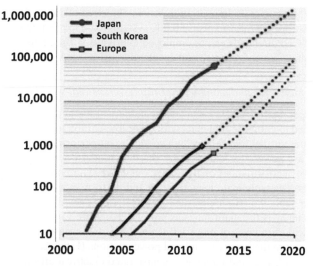

Figure 9.6. Cumulative number of fuel cell combined heat and power (CHP) systems deployed in three major regions (i.e. Japan, South Korea and Europe) since the year 2000, showing historic growth (solid lines) and near-term projections (dotted lines) (Staffell, 2015).

Stationary small-scale combined heat and power (i.e. micro-CHP) currently appears to be the largest and most established market for fuel cells, according to a review by Dodds *et al.* (2015). This scale of fuel cell is applicable to the residential family home or small commercial building in the range up to 50 kW because it involves cogeneration of heat and electrical power. The authors assert that local generation is more efficient since there is no energy loss due to long-distance transport or as a result of heat transfer in local heating networks.

The commercialization of micro-CHP fuel cells has proceeded rapidly since their initial introduction in 2009. As previously indicated, Staffell (2015) reported that nearly 60,000 systems were sold in Japan during a four-year period from 2009 to 2013 and, as shown in Figure 9.6, Japan is leading the way in terms of deployment, almost eight years ahead of South Korea and Europe. However, according to Staffell (2015), all regional markets are roughly doubling in size each year. This remarkable rate of development is expected to continue at least in the proximate future. As an example, Figure 9.7 shows photos of an installed residential PEMFC and a commercial PAFC

Figure 9.7.   Photos of the Panasonic EneFarm residential PEMFC (left) and ClearEdge PureCell commercial PAFC (right) (Staffell, 2015).

(Staffell, 2015). The Japanese administration has an objective of 1.4 million fuel cells installed by 2020, and the European Union expects 50,000 systems to be in position by 2020. The question now is how micro-CHP fuel cells can be employed to drive small-scale reverse osmosis systems. This is one area where further research and development is needed.

### 9.3.2   *Fuel cell efficiencies and manufacturer rating versus real-world performance*

In an excellent paper, Staffell (2015) sets out the numerical methods and evidence of real-world performance needed to compare fuel cells with other low-carbon technologies. It suggested that the efficiency of present-day fuel cells is high enough to outperform the best electric heat pumps, even when these are powered exclusively by the best modern power stations. The equivalent COP (coefficient of performance) of fuel cells was more than comparable to the best ground source heat pumps. The technical characteristics and performance of each fuel cell technology are summarized in Table 9.3. Both electrical and total efficiency are relevant for CHP systems, but electrical efficiency is the main focus as power is the more valuable output. Fuel cells offer the highest electrical efficiency of any CHP technology, and even small, micro-CHP fuel cells are at least equivalent to the best conventional power stations (Hawkes *et al.*, 2009). The leading SOFC systems at residential and larger scales have rated electrical efficiencies of 45–60%, and total efficiencies of 85–90% (Föger, 2011). Fuel processing for PEMFCs incurs greater losses, so electrical efficiencies are lower (with a maximum of 39%) but total efficiencies are higher (i.e. 95%) (Nagata, 2013). European residential systems lag behind the leading Japanese and Australian models, with current SOFCs and PEMFCs limited to 30–35% electrical efficiency (Callux, 2013).

Surprisingly, in residential homes the efficiency of small PEMFC and SOFC systems was found to be lower, by factors of 5 to 10, than in laboratory tests because of electricity consumed by auxiliary systems, reduced part-load efficiency, energy for start-up cycles, and the dumping of excess heat during summer (Staffell, 2015). For example, an SOFC system had a rated specification of 90–96% total efficiency, a field performance of 84% but a real-world performance of just 16% total efficiency. In addition, higher efficiencies were achieved with higher heat demand or power output (Fig. 9.8). For example, in the case of a PEMFC increasing from 40 to 80% power output, electrical efficiency correspondingly increased from 70 to 95%.

### 9.3.3   *Automotive fuel cells and the need to meet cost, performance and durability criteria*

Debe (2012) reported that fuel cells powered by hydrogen from secure and renewable sources are the ideal solution for non-polluting vehicles, and extensive research and development on all

Table 9.3.　Summary of fuel cell performance (adapted from Staffell, 2015).

| | Residential | | Commercial | |
| --- | --- | --- | --- | --- |
| Application | PEMFC | SOFC | PAFC | MCFC |
| Electrical capacity [kW] | 0.75–2 | | 100–400 | 300+ |
| Thermal capacity [kW] | 0.75–2 | | 110–450 | 450+ |
| Electrical efficiency (LHV)[a] [%] | 35–39 | 45–60 | 42 | 47 |
| Thermal efficiency (LHV)[a] [%] | 55 | 30–45 | 48 | 43 |
| System lifetime [years] | 10 | 3–10 | 15–20[c] | 10[c] |
| Annual degradation rate[b] [%] | 1 | 1–2.5 | 0.5 | 1.5 |

[a]Rated specifications when new, which are slightly higher than the averages experienced in practice.
[b]Loss of peak power and efficiency.
[c]Includes overhaul of the fuel cell stack halfway through its life.

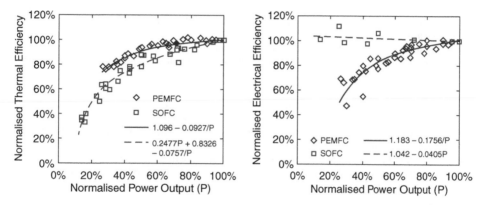

Figure 9.8.　Electrical and thermal efficiency of residential CHP systems versus power output, normalized against each system's efficiency at full power; data from eight PEMFC and six SOFC systems (Staffell, 2015).

aspects of this technology over the past two decades has provided prototype cars. However, the author noted that taking the step towards successful commercialization requires oxygen reduction electrocatalysts, which are crucial components at the heart of fuel cells (as shown in Fig. 9.2). In addition, these catalyst systems will need to be highly durable, fault-tolerant and amenable to high-volume production with high yields and exceptional quality. Debe (2012) reasoned that a fuel cell membrane electrode assembly (MEA) must satisfy three major criteria: cost, performance and durability. The same challenges were noted in the DOE (2012) report. Debe (2012) articulated that while the cathode oxygen reduction reaction (ORR) and anode hydrogen oxidation reaction both occur on the surfaces of platinum-based catalysts, the cathode ORR is more than six orders of magnitude slower than the anode hydrogen oxidation reaction and thus limits performance. This means that virtually all research and development activity is directed towards this aspect. The majority of current MEA catalysts are centered on platinum (Pt) nanoparticles dispersed on carbon black supports. The negative aspect of this approach is the high economic price involved.

　　Minimizing costs by reducing cathode loadings without loss of performance or toughness is the theme of most current electrocatalyst research (Table 9.4). Vehicle operation imposes severe durability and performance constraints on fuel cell cathode electrocatalysts (DOE, 2012), because of serious cathode degradation. Some of the countermeasures being developed are more stable graphitized carbon, the use of catalyst supports that will not electrochemically corrode, and the

Table 9.4.   Development criteria for automotive fuel cell electrocatalysts (Debe, 2012).

*Performance*
- Must meet beginning-of-life performance targets at full and quarter power.
- Must meet end-of-life performance targets after 5000 h or 10 years operation.
- Must meet performance, durability and cost targets and have less than 0.125 mg PGM (platinum-group metal) per $cm^2$.
- Corrosion resistance of both Pt & support must withstand thousands of start-up/shutdown events.
- Must have low sensitivity to wide changes in relative humidity.
- Must withstand hundreds of thousands of load cycles.
- Must have adequate cool start, cold start and freeze tolerance.
- Must enable rapid break-in and conditioning (the period needed to achieve peak performance).

*Materials*
- Must have high robustness, meaning tolerance of off-nominal conditions and extreme-load transient events.
- Must produce minimal $H_2O_2$ production from incomplete ORR.
- Must have high tolerance to external and internal impurities (for example, $Cl_2$) and ability to fully recover.
- Must have statistically significant durability, i.e. individual MEA lifetimes must enable over 99.9% of stacks to reach 5000-hour lifetimes.
- Electrodes must be designed for cost-effective Pt recycling.
- Environmental impact of manufacturing should be minimal at hundreds of millions of square meters per year.

*Process*
- Environmental impact must be low over the total life cycle of the MEAs.
- Manufacturing rates will need to approach several MEAs per second.
- MEA manufacturing quality must achieve over 99.9% failure-free stacks at beginning of life (one faulty MEA in 30,000 for just 1% stack failures).
- Proven high-volume manufacturing methods and infrastructure will be required.
- Catalyst-independent processes will be preferred, to enable easy insertion of new-generation materials.

addition of oxygen evolution catalysts to the mix to clamp the potentials at the start of water oxidation. Table 9.4 summarizes catalyst requirements and manufacturing considerations.

### 9.3.4  *Influence of socioeconomic growth and industrial competition on the desalination industry*

Ziolkowska and Reyes (2016) examined the influence of socioeconomic variables, including population growth, GDP, crude oil price and water abstraction, on advances in the desalination area over a 43-year period (i.e. 1970–2013) using the US as a case study (Fig. 9.9). Only population growth and GDP were found to be statistically significant in defining the quantity of new desalination plants and desalination capacity. The authors performed simple regression for the four variables thought to influence the desalination sector (i.e. population growth, GDP, crude oil price and water abstraction). The independent variables were found to have a higher impact on the number of desalination plants than on the desalination capacity, even though in both cases the statistical significance was given at the level of 99% (Fig. 9.9a to Fig. 9.9d). No significant correlation was found between the desalination capacity and water abstraction or crude oil prices (Fig. 9.9e and Fig. 9.9f). In addition, Ziolkowska and Reyes (2016) maintained that population was found to be a more significant variable than GDP in terms of impact on desalination capacity and number of desalination plants. At the same time, they argued that economic changes and fluctuations in GDP are hard to predict since they depend on numerous internal and global elements.

Gottinger and Goosen (2012), in attempting to gain a better understanding of a nation's fiscal development, assessed strategies of economic growth and industrial catch-up competition and

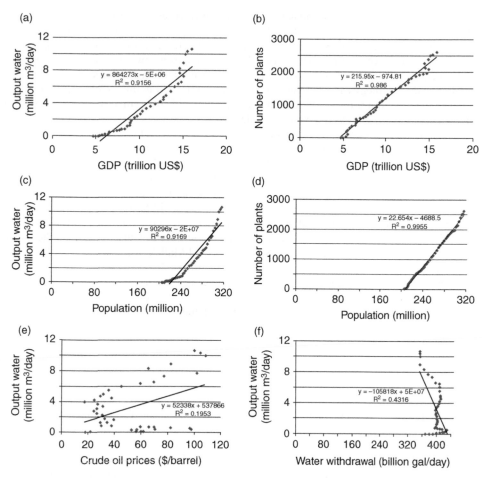

Figure 9.9. Relationships between (a) desalination capacity and GDP; (b) number of desalination plants and GDP; (c) desalination capacity and population; (d) number of desalination plants and population; (e) desalination capacity and crude oil prices; (f) desalination capacity and community water abstraction (Ziolkowska and Reyes, 2016).

their effect on the economic rise and decline of nations. We can speculate that these fiscal aspects may affect the ability of governments to build and operate desalination plants. Gottinger and Goosen (2012) noted that industrial competition among nations or regional economic entities has been an essential driving force for economic growth. Rivalry, for example, can drive a state's standing, prestige, power and economic performance, thus allowing a country or region to get ahead of its competition. Governments can support this competitive process by encouraging entrepreneurship, fostering education and training, and by making it easier for companies to set up new businesses, particularly in the renewable energy area.

A leading mechanism to sustain economic growth among nations lies in the industrial catch-up processes among frontrunners and stragglers (Goosen, 2013). This is born out of a rivalry of enhanced performance, and is comparable to competitions between teams in sporting events. The competitive process may be of benefit in the renewable energy and fuel cell industry through forcing improvements in the areas of technology, scaling up and marketing. Central features in the catch-up process are the use by emerging states, such as the MENA countries, of knowledge learned over the past few hundred years by developed states such as Europe, the USA and Japan, which can help set up or mimic industrial processes in desalination and renewable energy. If executed successfully, this will save time and increase output. The second crucial component described by Gottinger and Goosen (2012) is the need for emerging nations in the catch-up growth stage to invest and save. This provides a fiscal buffer if there is a downturn in the economy. The third element deemed vital is that states must sell what they produce on a global scale. For example, novel fuel cell technology can be developed and commercialized in the Arabian Gulf countries and must then be sold worldwide. If done successfully, this will ensure that a competitive product is created and that it has a sustainable worldwide market.

Finally, significant increases in usage of oil and gas for water and electricity production over the next decades will severely reduce incomes of countries that are located in arid regions and that are dependent on fossil fuels. This will affect a government's ability to subsidize electricity and water costs and may influence regional stability (Goosen *et al.*, 2013). Thus, there is a real need to diversify the economy of such countries. It can be argued that rivalry or competition in the renewable energy and desalination industries should be seen as an opportunity to create new commercial products and process that will enhance GDP. However, all of this must be regarded as a joint effort between the private sector and governments. Writing on the industrialization and economic development of Saudi Arabia, Al Mallakh and El Mallakh (2015) and El Mallakh (1982) noted that governments normally see themselves as partners, not as competitors, to private-sector producers. As in the catch-up process, the aim in the case of Saudi Arabia is to raise production efficiency, with an emphasis on the possible creation of new products and processes in the hydrocarbon-based industries. This, of course, includes fuel cell development and commercialization.

## 9.4   CONCLUDING REMARKS

There is a critical necessity in the desalination industry not only to decrease energy demands but also to make energy production less expensive. It can be reasoned that significant increases in usage of oil and gas for water and electricity production over the next decades will severely reduce the incomes of countries that are dependent on fossil fuels. This will affect such nations' abilities to subsidize electricity and water costs and may affect regional stability. The strategic development and commercialization of large-scale renewable energy technologies is thus crucial in making states less dependent on fossil fuels.

In recent years, fuel cells have emerged as attractive energy-production systems because they are extremely efficient and do not produce pollutants, in contrast to established fossil fuel-driven technologies. The major commercial markets for fuel cells appear to be in stationary power, portable power, and auxiliary power units. It was observed that while hydrogen cars have commanded greater media attention, the fuel cell combined heat and power unit is the largest and most established market, particularly in Japan and the EU. Although hydrogen and fuel cells are not, in the strictest sense of the word, a *renewable* energy resource, they are very environmentally friendly when in operation. The question now is how combined heat and power fuel cells can be employed to drive small-scale reverse osmosis systems, and this is one area where further research and development is needed.

Surprisingly, the efficiency of small commercial fuel cell systems in residential homes was found to be lower by factors of five to ten than in laboratory tests (i.e. manufacturers' ratings), apparently because of electricity consumed by auxiliary systems, reduced part-load efficiency, energy use in start-up cycles, and the dumping of excess heat during summer. In addition, in the case of energy extraction from, and real-world application of, microbial fuel cells, more work needs to be done to optimize the design, improve harvesting efficiency, and reduce the cost. This is a major bottleneck for wider application and should be a new frontier of research. One can argue that a further area where a significant contribution may be made is the investigation into the possibility of developing fuel cell technology on a large scale for energy production for desalination plants.

In closing, more research and development is desirable to close the gap between electricity generation from commercial fuel cells and the power requirements of small- to medium-size desalination units. This will require better coordination between fuel cell manufacturers, the desalination industry and governments. It can be argued that in this partnership the government should improve public utilities and infrastructure, while companies help generate new jobs by making innovative commercial products and creating industries that are competitive both regionally and globally.

REFERENCES

Al Mallakh, R. & El Mallakh, R. (2015) Saudi Arabia: Rush to Development (RLE Economy of Middle East). *Profile of an Energy Economy and Investment* (Vol. 28). Routledge, London.

Alyousef, Y. & Abu-ebid, M. (2012) Energy efficiency initiatives for Saudi Arabia on supply and demand sides. In: Morvaj, Z. (ed) *Energy Efficiency – A Bridge to Low Carbon Economy*. InTech, Rijeka, Croatia. pp. 279–309.

Bouchekima, B. (2003) A small solar desalination plant for the production of drinking water in remote arid areas of southern Algeria. *Desalination*, 159, 197–204.

Bourland, B. & Gamble, P. (2011) Saudi Arabia's coming oil and fiscal challenge. Jadwa Investment, Riyadh.

Butti, K. & Perlin, J. (1980) *Golden Thread: 2500 Years of Solar Architecture and Technology*. Cheshire Books, Palo Alto, CA.

Callux (2013) Field test of residential fuel cells – background & activities. National Organisation Hydrogen and Fuel Cell Technology, Berlin.

Choi, S. (2015) Microscale microbial fuel cells: advances and challenges. *Biosensors and Bioelectronics*, 69, 8–25.

Debe, M.K. (2012) Electrocatalyst approaches and challenges for automotive fuel cells. *Nature*, 486(7401), 43–51.

Dodds, P.E., Staffell, I., Hawkes, A.D., Li, F., Grünewald, P., McDowall, W. & Ekins, P. (2015) Hydrogen and fuel cell technologies for heating: a review. *International Journal of Hydrogen Energy*, 40(5), 2065–2083.

DOE (2012) Multi-year research, development and demonstration plan: technical plan – fuel cells. Office of Energy Efficiency & Renewable Energy, U.S. Department of Energy, Washington, DC. Available from: http://www.eere.energy.gov/hydrogenandfuelcells/mypp/pdfs/fuel_cells.pdf [accessed November 2016].

DOE (2013) US DRIVE Cuel Cell Technical Team Roadmap. Office of Energy Efficiency & Renewable Energy, U.S. Department of Energy, Washington, DC. Available from: http://energy.gov/eere/vehicles/downloads/us-drive-fuel-cell-technical-team-roadmap [accessed November 2016].

El Mallakh, R. (1982) *Saudi Arabia: Rush to Development*. Routledge, London.

Faisal, N., Ahmed, R., Islam, S.Z., Hossain, M., Goosen, M.F.A. & Katikaneni, S.P. (2017) Fuel cells as an energy source for desalination applications. In: Mahmoudi, H., Ghaffour, N., Goosen, M.F.A. & Bundschuh, J. (eds) *Renewable Energy Technologies for Water Desalination*. CRC Press, Boca Raton, FL.

Föger, K. (2011) CFCL: challenges in commercialising an ultra-efficient SOFC residential generator. *4th IPHE Workshop on Stationary Fuel Cells, 1 March 2011, Tokyo, Japan*.

Ghaffour, N., Bundschuh, J., Mahmoudi, H. & Goosen, M.F. (2015) Renewable energy-driven desalination technologies: a comprehensive review on challenges and potential applications of integrated systems. *Desalination*, 356, 94–114.

Goosen, M.F.A. (2013) Institutional aspects of economic growth: assessing the significance of public debt, economic governance & industrial competition. *The Open Business Journal*, 6, 1–13.

Goosen, M.F., Mahmoudi, H. & Ghaffour, N. (2014) Today's and future challenges in applications of renewable energy technologies for desalination. *Critical Reviews in Environmental Science and Technology*, 44(9), 929–999.

Goosen, M.F.A., Mahmoudi, H., Ghaffour, N., Bundschuh, J. & Al Yousef, Y. (2016) A critical evaluation of renewable energy technologies for desalination. *Application of Materials Science and Environmental Materials (AMSEM2015), Proceedings of the 3rd International Conference, 1–3 October 2015, Phuket Island, Thailand*. pp. 233–258.

Gottinger, H. & Goosen, M.F.A. (eds) (2012) *Strategies of Economic Growth and Catch-Up: Industrial Policies and Management*. Nova Science, New York.

Hawkes, A., Staffell, I., Brett, D. & Brandon, N. (2009) Fuel cells for micro-combined heat and power generation. *Energy & Environmental Science*, 2(7), 729–744.

Helfer, F., Sahin, O., Lemckert, C.J. & Anissimov, Y.G. (2013) Salinity gradient energy: a new source of renewable energy in Australia. *Water Utility Journal*, 5, 3–13.

Houcine, I., Benjemaa, F., Chahbani, M.H. & Maalej, M. (1999) Renewable energy sources for water desalting in Tunisia. *Desalination*, 125(1–3), 123–132.

Kim, D.J., Jo, M.J. & Nam, S.Y. (2015) A review of polymer-nanocomposite electrolyte membranes for fuel cell application. *Journal of Industrial and Engineering Chemistry*, 21, 36–52.

Lund, J.W. (2008) Development and utilization of geothermal resources. *Proceedings of ISES World Congress 2007*. Vol. I–Vol. V, pp. 87–95; Springer Berlin Heidelberg.

Moriarty, P. & Honnery, D. (2009) Hydrogen's role in an uncertain energy future. *International Journal of Hydrogen Energy*, 34, 31–39.

Nagata, Y. (2013) Toshiba fuel cell power systems – commercialization of residential FC in Japan. Presented at FCH-JU General Assembly, Brussels.

Sahin, O., Siems, R.S., Stewart, R.A. & Porter, M.G. (2016) Paradigm shift to enhanced water supply planning through augmented grids, scarcity pricing and adaptive factory water: a system dynamics approach. *Environmental Modelling & Software*, 75, 348–361.

Staffell, I. (2015) Zero carbon infinite COP heat from fuel cell CHP. *Applied Energy*, 147, 373–385.

Wan, C.F. & Chung, T.S. (2016) Energy recovery by pressure retarded osmosis (PRO) in SWRO-PRO integrated processes. *Applied Energy*, 162, 687–698.

Wang, H., Park, J.D. & Ren, Z.J. (2015) Practical energy harvesting for microbial fuel cells: a review. *Environmental Science and Technology*, 49(6), 3267–3277.

Ziolkowska, J.R. & Reyes, R. (2016) Impact of socioeconomic growth on desalination in the US. *Journal of Environmental Management*, 167, 15–22.

# CHAPTER 10

## Wireless networks employing renewable energy sources for industrial applications: innovations, trade-offs and operational considerations

Mai Ali & Abd-Elhamid M. Taha

### 10.1 INTRODUCTION

Renewable energy technology can be combined with wireless power transfer to construct sustainable wireless communication networks (Akhtar *et al.*, 2015). Technological improvements have enabled systems to be built in which wireless devices are powered over the air by wireless power transmitters (Bi *et al.*, 2016). While there have been scientific advances in wireless networks, accessible energy still remains a key issue. The amount of available energy has a direct effect on the performance, functionality and lifetime of such systems. The principal reason for the interest in renewable energy is the need to do away with batteries as well as to reduce or eliminate dependence on the electrical grid (Akhtar *et al.*, 2015). As illustrated in Figure 10.1, renewable solar energy can be employed with photovoltaic panels to generate electricity for wireless power and information transfer devices. One can speculate that this technology may eventually be employed to help operate small- to medium-scale desalination units in remote regions.

To put matters into perspective, the extent to which the establishment of telecommunication links depends on wireless networks has continued to increase. In terms of connecting to the internet, Cisco (2016) estimated that traffic from wireless and mobile devices will exceed traffic from wired devices by 2019. Drivers for such growing shifts include enhancement of wireless communications and a reduction in installation costs, as well as easier network topology reconfiguration

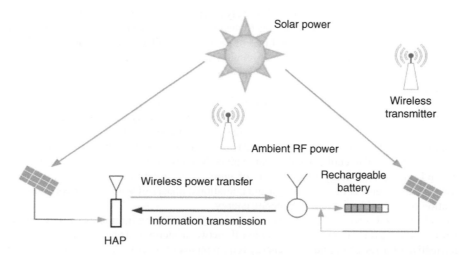

Figure 10.1.   Illustration of a wireless powered communication network with renewable solar photovoltaic energy sources: WD, wireless device; HAP, hybrid information access point; RF, radio frequency. In addition to electricity generation from solar panels, ambient RF power devices use antennas to pick up existing signals and convert them into electricity.

and greater user mobility. There is, however, concern regarding the relative reliability of wireless and mobile links compared to wired ones, especially in the case of autonomous control. There are further concerns regarding the general security of wireless and mobile links, although much effort has been made to address these shortcomings.

The application of wireless networks has also seen a robust upsurge in the context of industrial plants (Sheng *et al.*, 2015). In addition to circuitry for radio frequency (RF) communication, advances in electronics have led to miniaturization of many sensing and actuation devices and all in a cost-efficient manner. Technologies for RF tagging, such as radio frequency identification (RFID), have likewise become popular (e.g. personnel ID, product tagging, and production line and stock monitoring) (DHL, 2016). A sensing/actuation set-up that used dense and costly cabling and installation is now superseded by more economical wireless set-ups.

Estimates of the number of wirelessly connected devices that are expected to exist by 2020 have varied from 25 to 75 billion (Cisco, 2016). The rise in machine sensor/actuator traffic is particularly interesting given its indication of increased automation, as well as the added value from processing such data to enhance industrial output. However, this growth is not without challenges. For one thing, existing telecommunication traffic management techniques were not designed to scale to these projected volumes. Communication technologies and infrastructures in current use will need to be updated and expanded at an unprecedented pace in order to accommodate the data rates to come. One repeatedly voiced issue, for example, concerns the scarcity of the wireless resource (i.e. the spectrum) due to either long-term licensing, spectrum mismanagement or massive contention in unlicensed bands such as Wi-Fi (Song *et al.*, 2012). These concerns have to be addressed.

One particular worry in wireless networks concerns energy. In 2007 (i.e. the year the first iPhone was introduced), a Gartner (2007) report noted that the information and communication technology (ICT) sector contributed 2% to the human $CO_2$ footprint, and that mobile communication networks consumed 0.5% of the global energy supply. More current accounts indicate that ICT's contribution may increase to 4% by 2020. It can be argued that, following the advent of smartphones and tablets, these percentages will probably increase (Fehske *et al.*, 2011).

A typical wireless network consists of three elements: the mobile/wireless device, the access network and the core network. As noted above, the mobile/wireless device can vary from a handheld smartphone or tablet, to a sensor, an appliance or a robot. The access network comprises a configuration of contact points or base stations over a wide geographical area. The arrangement is designed such that the base stations collectively offer a consistent service of wireless coverage. Through this access grid, a device connects to the network core, essentially an interacting infrastructure that handles the access and mobility of devices, as well as relevant contractual logistics such as a monitoring service used for charging. The core also manages linkage of the access system and the devices associated with its coverage to those in other access networks and/or the internet at large.

In 2011, Vodafone presented an extensive assessment of the energy expenses of its mobile network. The most remarkable observation to emerge from the study was that the access network energy consumption comprised 80% of the entire system expenditure (Fehske *et al.*, 2011). Energy is required not only for the communication aspects of the base station, but also includes the air conditioning required to cool down the components. Numerous innovations have thus been made to try and reduce the energy requirements of wireless access networks. These improvements include enhancements at both the design and operational phases of wireless networks, as well as utilization of non-grid resources to generate energy.

This chapter will review innovations in reducing the energy requirements of wireless access networks, as well as discussing the basic trade-offs between energy and operational considerations. Specifically, after reviewing how the energy requirements of a network can be reduced through configuration and operational planning, the core operational functionality of resource allocation will be discussed. It is this functionality that ensures useful connectivity for the wireless devices that matches their communication needs, and which can be used to reduce the energy requirements of both the network and the devices. Wireless network loads that can be offloaded from the grid and onto other sources will be addressed, with a focus on using sustainable renewable sources (i.e. greening). Energy harvesting will be analyzed in the same way, together with the main trade-offs to be considered in greening the operation of wireless networks. The chapter ends with an assessment of the necessity for sensibly designed authentication models, as well as green innovations and smart factories, and how these can be integrated into the desalination industry.

## 10.2   REDUCING ENERGY REQUIREMENTS BY NETWORK CONFIGURATION PLANNING

Countless studies have been directed at decreasing the energy requirements of wireless networks, ranging from addressing basic consideration such as energy metrics and reducing validation mechanisms, to searching for energy efficient network operation solutions that can be applied at the different levels of function (Chen *et al.*, 2011; Hasan *et al.*, 2011; Humar *et al.*, 2011; Mancuso and Alouf, 2011; Marsan and Meo, 2011).

A classification of metrics provided by Hasan *et al.* (2011) helps to illustrate the scope of energy-reduction mechanisms. They divided the metrics into three levels: facility, equipment and network. The standardization of how the metrics in the different levels are defined, though, remains a critical issue. More importantly still, the use of the different metrics to judge qualities of specific networking solutions also remains undefined.

The cataloguing used for metrics can be directly applied to differentiate proposed solutions for energy reduction in wireless networks. For example, the notion of a sleep mode in devices, already a big part of the emerging International Mobile Telecommunications-Advanced (IMT-Advanced) networks (e.g. using LTE Advanced or WiMAX standards), is one that can be considered for application at the user equipment/handset level. The use of similar notions has also been investigated at the network level. For instance, base station sleep mode was discussed by Saker and Elayoubi (2010). There have also been alternative parts of the wireless access network that can be switched off depending on demand usage. Other ideas such as zooming (Ali *et al.*, 2007; Niu *et al.*, 2010) have also been considered where, instead of switching base stations on and off, the coverage of a cell is expanded or contracted. Meanwhile, the introduction of small cells, both in-band and out-of-band, has similarly been shown to offer considerable advantages (Damnjanovic *et al.*, 2011; Khirallah *et al.*, 2011). In small cells, base stations with focused coverage (e.g. Wi-Fi hotspots, femtocells or relay stations) prove to be closer to mobile handsets and thus reduce the uplink energy requirements. However, such benefits of small cells have to be balanced against greater control and coordination requirements.

Attention has been given to the radio level, where the available spectrum is dynamically utilized by either the network or the mobile handset (Lee *et al.*, 2012). With cognitive radio understanding (i.e. sensing), the wireless medium becomes aware of the available spectrum and can exploit specific qualities (e.g. support for reduced signal strength) to the handset's advantage. Another level of intelligence is used in cooperative communications, where data transmission and reception becomes distributed over several network elements, akin to communication using a virtualized antenna array (Lee *et al.*, 2012).

Standardizing or unifying energy-related metrics remains an outstanding issue. One particular challenge of significant impact is how energy expenditure changes in devices at different times. Equipment can vary in terms of settings, and in the number of active interfaces or applications, yielding a wide range of profiles in the way energy can be utilized or batteries can be depleted. There is also the fact that battery capability deteriorates with time: the first 100% charge is thus different from the sixty-first, and so on, a commonly observed effect known as battery degradation or fade (He *et al.*, 2011). Furthermore, the value of the energy becomes subject to variability, where the absolute value of the joule can be different from its relative one. This is especially relevant in cooperative contexts where mechanisms of incentivized sharing of resources are used. Finally, and subject to the same variation and uncertainty associated with depletion, there can be instances where battery energy levels rise due to energy harvesting or generation mechanisms.

It can be noted that much has and is being done to address these problems. For example, a variety of reports continue to be published on the energy profiles of diverse components or entire handsets, which provide further insights. For example, Perrucci *et al.* (2011) investigated the energy profile of smartphones and the impact of multiple radio interfaces on battery consumption, while Baliga *et al.* (2011) offered a broad analysis of energy consumption in access networks, both wired and wireless.

## 10.3    MANAGING THE REQUIREMENTS OF WIRELESS RESOURCE ALLOCATION

When it comes to resource allocation, emphasis on static designs has long given way to adaptive and opportunistic frameworks that are able to make adjustments as time passes. Specifically, effective efforts have traditionally been made to reduce the energy requirements of resource allocation frameworks and these can be categorized into three areas: increasing efficiency, reducing decision complexity and increasing resource/functionality sharing (Humar *et al.*, 2011; Piamrat *et al.*, 2011; Srivastava and Motani, 2005).

Efforts focused on increasing efficiency have examined ways in which the mapping between resources and functionalities can be enhanced. Numerous instances of this approach can be found, for example, within the context of cross-layer design (Srivastava and Motani, 2005) where, given a certain hierarchy and interdependency between resource management functionalities, the necessity of certain dependencies can be revised. Other efforts within this category include some already noted, such as zooming, that control energy consumption in the network according to user demand.

The reduction of decision complexity has always been a major issue in the design of resource allocation frameworks. The impact extends beyond the algorithmic scalability needed to handle a larger number of user connections or a faster rate of variations in the network state. A simplified decision framework reduces the sophistication of the equipment implemented and, in turn, lessens what is called "embodied energy", which is the total energy required to produce the equipment and deliver it to the place of operation. Humar *et al.* (2011) details a critical insight into how ignoring considerations of embodied energy in the design of a resource allocation framework and, more generally, in equipment and network design can render energy-reduction efforts effectively useless.

The simplification of decision complexity has also impacted network design. A case in point concerns the viability of heterogeneous (multi-radio) access, the most popular of which is hybrid cellular/Wi-Fi deployment, which has been driven in large part by a resource allocation thrust that made possible the management of joint resources among diverse systems (Piamrat *et al.*, 2011). Increased autonomy in access-level decisions (e.g. flat architectures, self-optimization) further facilitates more capable network-level cognition (Bokor *et al.*, 2011).

In terms of energy-reduction efforts focused on resource sharing, one can argue that perhaps the most classic example is that of medium-access mechanisms and protocols, throughout their evolution. Here, the energy reductions resulting from the adoption of multi-carrier access techniques, such as orthogonal frequency-division multiple access (OFDMA), in emerging networks should not be lost on the reader. Adopting multi-carrier access techniques has allowed higher bit rates, enhanced and more adaptive communications, reduced handset requirements, more capable multi-antenna techniques, and higher bandwidths through carrier/spectrum aggregation. Although resource sharing is common in various resource allocation operations, the use of certain existing measurement reports, signaling and decisions can be utilized via revised inferences to impact considerations in newer functionalities. These additions are predominantly "non-invasive", in the sense that they do not impose on existing system design nor do they impact the network's existing energy profile.

The emphasis on opportunism continues to surge with the most recent applications, including cognitive radios and cooperative communications. In its basic, implementable from, cognitive radio utilizes extensively processed sensing in order to gain an understanding of the network activity in various bands of interest. Once the radio recognizes a "spectrum hole" with advantageous characteristics, it utilizes it for transmissions. Further cross-layer cognition has been shown to result in other enhancements too, although our particular interest in this work concerns those spectrum holes that reduce the energy requirements for transmission. Meanwhile, cooperation in wireless networks has facilitated reliable and robust communication that overcomes temporary or permanent obstacles, and reduces deployment requirements (Lee *et al.*, 2012).

## 10.4   POWERING THE NETWORK BY UTILISING RENEWABLE ENERGY SOURCES

Renewable energy is sustainable, clean energy harvested from sources that do not deplete with use or are replenished on a human timescale. This is in contrast to non-renewable energy sources, like oil, coal and natural gas, which are derived from a finite resource. Solar and wind power are examples of renewable energy sources: harvesting energy from the sun and wind today will not result in less sunshine or wind tomorrow (CTM Magnetics, 2016; Goosen *et al.*, 2016).

### 10.4.1   *Solar energy*

Solar energy is considered as an essential energy resource due to its availability as well as its clean and inexpensive supply. It is one of the renewable, low-carbon capitals with both the scalability and the technological development to meet fast-growing global demand for electricity. Among solar power technologies, solar photovoltaic (PV) is the most extensively used technology and caters to about 0.87% of the world's electricity demands (Chander *et al.*, 2015). Solar cells are the main module in PV power systems as these convert solar radiation directly into electrical energy. The conversion process is based on the photovoltaic effect. The solar power density at the equator at noon at the equinox is $1000\,\mathrm{W\,m^{-3}}$ or, equivalently, $100\,\mathrm{mW\,cm^{-2}}$ (Le, 2008). Thus, for example, the average energy from sunlight falling on Saudi Arabia is 2200 thermal $\mathrm{kWh\,m^{-2}}$ (Alawaji, 2001), and it is therefore well worthwhile to attempt generation of this clean energy in the country via direct sunlight through PV cells. Based on the Shockley-Queisser efficiency limit, single-junction solar cells can achieve a maximum theoretical conversion efficiency of 33% (Shockley-Queisser, 1961), and Panasonic reported a 25.6% efficiency record in 2014 at the cell level, and 22.5% at the module level in a panel that is suitable for mass production (Clover, 2015). Dimroth *et al.* (2014) reported 44.7% efficiency for multi-junction solar cells but these are proportionately more expensive for mass production.

A PV cell converts a large amount of solar energy into electricity at low temperatures (Patel, 2006). Thorough analysis of the literature reveals a high dependence of solar cell performance on environmental and operational conditions. Experimental results show that the current-voltage (I–V) characteristics, as well as the maximum power point, of solar cells are a function of light intensity and temperature (Patel, 2006).

The maximum power available at lower temperatures is higher than that at higher temperatures, leading to a reduction in output power at high operating temperatures. This effect can be calculated via Equation (10.1) (Almasoud and Gandayh, 2015):

$$P = P_{25^\circ C}[-0.5\% \times (T - 25^\circ C)] \tag{10.1}$$

where $P_{25^\circ C}$ is the manufacturer's rated power output of the PV module, and $T$ is the ambient temperature.

From the plots (Fig. 10.2 and Fig. 10.3), it can be observed that the short-circuit current and maximum power point current of the solar cell is strongly affected by light intensity, while the open circuit voltage and maximum power point voltage is more dependent on the temperature. Given these variations in voltage and current according to temperature and light conditions, a solar power conversion system requires a maximum-power tracking circuit to ensure maximum power transfer between the solar cell and the load. In addition, the performance of PV modules

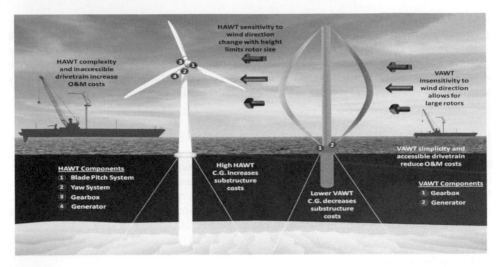

Figure 10.2.    Two types of wind turbine: horizontal-axis (HAWT) and vertical-axis (VAWT) (Excluss Solar, 2016).

is also affected by the presence of dust on their surface. Dust accumulation changes the I–V characteristics of PV cells depending on the amount of dust accumulated per unit area of the module surface [g m$^{-2}$]. Consequently, PV modules must be kept clean in order to maintain PV power plants at maximum efficiency (Almasoud and Gandayh, 2015).

According to REN21 (2016), the solar PV market increased by 25% in 2014 to a record 50 GW, lifting the global total to 227 GW. In 2015, the annual market was nearly ten times the world's cumulative solar PV capacity of a decade earlier. As depicted in Figure 10.4, China, Japan and the United States accounted for the majority of added capacity, but emerging markets on all continents contributed significantly to global growth, driven largely by the increasing cost-competitiveness of solar PV.

### 10.4.2   *Wind energy*

The use of wind power to pump water and grind grain has been practised for centuries (Goosen *et al.*, 2016; GWEC, 2015). In a basic windmill, the force of the wind pushes against the front side of the sails of the windmill causing them to turn. This rotation is mainly caused by drag acting on the sails. The sails are attached to a horizontal wind-shaft so that when the sails turn, the wind-shaft turns. This rotary motion is then converted into mechanical power and can turn the wheel of a grist mill or pump water. Today's wind turbines convert the kinetic energy of air movement into mechanical energy, which is then used to drive a generator that converts this energy into electricity. This is a similar process to the ancient windmill, except that the mechanical power in the rotor shaft is instead used to rotate an electric generator, producing electricity.

One of the world's first wind turbines to generate electricity was built in Denmark by Poul la Cour in 1891 (Masters, 2013). He used the electricity generated by his turbines to electrolyze water, producing hydrogen for gas lights in the local schoolhouse, in which aspect he was considered to be 100 years ahead of his time because the concept he utilized is the main driver behind electrical power generation in fuel cells.

There are two different types of wind turbines, horizontal-axis wind turbines (HAWTs) and vertical-axis wind turbines (VAWTs). In HAWTs, the main rotor shaft is set horizontally and the blades are perpendicular to the ground, as illustrated in Figure 10.2. Most commercial wind turbines that are connected to the electrical grid are HAWTs. Conversely, in VAWTs the main rotor shaft (similar to the wind-shaft of older windmills) is set vertically, or perpendicular to the ground.

In large wind applications, HAWTs are used almost exclusively. However, in small and residential wind use, VAWTs have their place. The advantage of a HAWT is that it is able to produce more electricity from a given amount of wind. So if a person is trying to produce as much wind as possible at all times, the HAWT is the most likely choice. However, the disadvantage of the HAWT is that it is generally heavier and does not perform well in turbulent wind (Windpower Engineering, 2009).

The wind power industry set new records across the world in 2015, when wind power was the leading technology in new power generation, and wind is driving the transformation of the global power system (GWEC, 2016). Overall, 63 GW of wind power capacity was added globally in 2015, reflecting a 23.2% increase on the 51 MW installed in 2014. Much of this growth was driven by China, where booming wind plant construction and underlying wind turbine supplier dynamics show a market split between that country and the rest of the world (see Fig. 10.6b).

## 10.5 ENERGY HARVESTING AS AN ALTERNATIVE IN POWERING THE NETWORK

### 10.5.1 *Radio frequency energy harvesting*

Radio frequency energy is emitted by sources that generate large electromagnetic fields such as TV signals, wireless radio networks and cell phone towers. Through the use of a power harvesting circuit linked to a receiving antenna, this free-flowing energy can be captured and converted into usable DC voltage (Bi *et al.*, 2016; Lazo and Duron, 2007; Tran and Lee, 2010). It is most commonly used in the application of RFID tags, in which the sensing device wirelessly sends a radio frequency to a harvesting device which supplies just enough power to send back identification information specific to the item of interest (Fig. 10.1). The circuits which receive the detected radio frequency from the antenna are made on the scale of a fraction of a micrometer but can convert the propagated electromagnetic waves into low voltage DC power at distances of up to 100 meters. Using energy harvesting to power RFID is an established technique; many examples are available in the literature and it is commercially established as well (Bi *et al.*, 2016; Lazo and Duron, 2007; Tran and Lee, 2010).

The first step in designing a RF energy harvesting system is deciding at which frequencies the power should be harvested. The wireless spectrum is full of signals with different frequency and power levels, ranging from cellular standards to Wireless Local Area Networks (WLANs) and TV signals. The criteria that control the selection of certain frequencies for the purpose of energy harvesting are wide deployment and power level. The world coverage map of the latest deployed mobile telecommunication standard, 4G–LTE (Long Term Evolution) is shown in Figure 10.3. It can be seen that by 2013 LTE covered most of the world except for some parts of Africa and Asia. Thus it is almost guaranteed that 3G/4G/LTE signals would be available wherever the harvesting system is placed. On the other hand, the power level of 4G signals can have a very short reach, with a reference sensitivity of −106 dBm at the receiver (3GPP, 2015). As well as cellular networks, the other widespread wireless technology is TV broadcasting, where digital TV (DTV) standards feature strong transmitter signal power (Fig. 10.4). Local network arrangements also exist that can possess considerable power levels, such as the Wi-Fi signals available in Wi-Fi hotspots and the Bluetooth standard present in almost all mobile handsets. Sensors used in smart homes or inside buildings will be the ones best placed to benefit from these two sources.

A newer frequency range that can also be investigated is that used in road tolling systems. They emit a considerable amount of power over limited time spans to power the RFID tags attached to the vehicles passing through toll gates. Sensors mounted on bridges and streets monitor the road environment, and traffic can benefit from such regimes. Table 10.1 lists examples of standards that could be engaged in an RF energy harvesting system, along with their associated frequency bands.

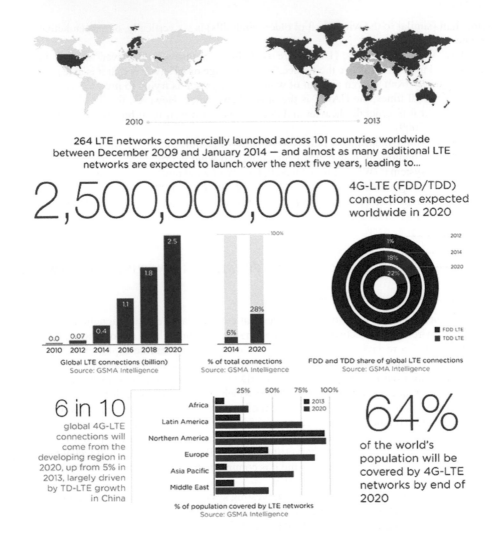

Figure 10.3.   Global 4G-LTE connections forecast: 2010 to 2020 (https://www.gsmaintelligence.com).

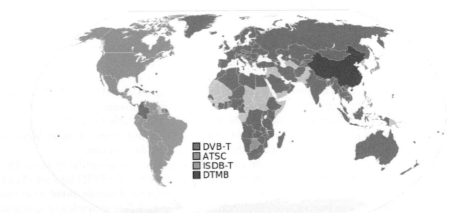

Figure 10.4.   DTV standards coverage map (http://www.dtvstatus.net/map/map.html).

Table 10.1.   Energy harvesting system frequency list (adapted from Bi *et al.*, 2016).

| Technology | Frequency bands | Band of interest |
|---|---|---|
| 4G/LTE | 900, 1800 MHz | 925–960/1805.2–1879.8 MHz |
| DTV | 470–862 MHz | 470–862 MHz |
| Wi-Fi, Bluetooth | 2.4 GHz | 2.4 GHz |
| Tolling system, New Wi-Fi | 5 GHz | 5 GHz |

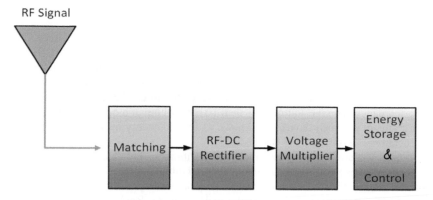

Figure 10.5.   A radio frequency (RF) energy harvesting system.

The main building blocks of an RF energy harvesting system are presented in Figure 10.5. Indeed, this topology can be extended to other harvesting sources with little alteration. Since the key goal of an energy harvesting system is to provide a sustainable energy supply to low-power circuits, it is crucial that the circuits involved in the energy harvesting process perform their tasks efficiently at the minimum possible power.

The interface of the system with the surrounding environment is the antenna. Special care should, therefore, be given to antenna design so that the maximum available energy is picked up and delivered to the harvesting circuitry. The full power available from an antenna $P_{tm}$ is directly proportional to the maximum effective area $A_{em}$. This relation can be written as (Ali *et al.*, 2013):

$$P_{tm} = A_{em}W_i \tag{10.2}$$

where $W_i$ is the power density of the incident wave. The maximum effective area of the antenna is related to the different antenna parameters that need to be taken into consideration when designing a receiving antenna and is given by:

$$A_{em} = e_{cd}(1 - |\Gamma|^2)\left(\frac{\lambda}{4\pi}\right)^2 D_0|\hat{\rho}_w \cdot \hat{\rho}_a|^2 \tag{10.3}$$

where $\lambda$ is the wavelength of the RF source; $e_{cd}$ is the radiation efficiency; $1 - |\Gamma|^2$ and $|\hat{\rho}_w \cdot \hat{\rho}_a|^2$ account for the losses due to impedance and polarization mismatches, respectively; $D_0$ is the maximum directivity, which is determined by the radiation pattern of the antenna. Novel multiband antenna designs are available, in which RF power is harvested from multiple bands using a single antenna (Taghadosi *et al.*, 2015).

Efficient RF-DC conversion dictates the capability of the entire system. This is for two reasons: firstly, this element usually consumes the largest proportion of power in the entire harvesting system and, secondly, it is responsible for conveying the harvested power to the rest of the system, so if the power is not handled efficiently in these first stages the functionality of the entire system will collapse accordingly. The functions reported as accounting for significant power consumption are the forward bias voltages of the diodes employed in rectifier design or the threshold voltages of diode-connected transistors. These voltage values are typically of the order of 0.7 V, which is far from being acceptable for a system dealing with harvested power levels as low as 0.1 $\mu W\ cm^{-2}$. Thus it is of paramount importance to achieve the target of minimum power consumption in the rectifier circuitry by employing novel low-power circuit design techniques, without adding the extra cost of using a non-standard complementary metal-oxide semiconductor (CMOS) process, to allow a wide-ranging deployment of the RF energy harvesting system.

### 10.5.2   Piezoelectric energy harvesting

Mechanical energy is present almost anywhere that wireless sensor networks (WSN) may potentially be deployed, which makes converting mechanical energy from ambient vibration into electrical energy an attractive approach for powering wireless sensors. The source of mechanical energy can be a moving human body or a vibrating structure. The frequency of the mechanical excitation depends on the source: less than 10 Hz for human movements and over 30 Hz for machinery vibrations. Such devices are known as kinetic energy harvesters or vibration-based power generators (Kazmierski and Beeby, 2011).

Piezoelectric materials can convert mechanical vibration into electrical energy with a very simple structure; piezoelectric energy harvesting is highlighted by Kim *et al.* (2011) as a self-powering source for WSN systems. Piezoelectricity describes pressure-generated electricity and is a property of certain crystalline materials, such as quartz, Rochelle salt, tourmaline, barium titanate, and lead zirconate titanate (PZT), that develop electricity when pressure is applied. This is called the direct effect. Furthermore, these crystals undergo deformation when an electric field is applied, which is termed the converse effect and can be used as an actuator, while the direct effect can be used as a sensor or energy transducer. This coupled electromechanical behavior of piezoelectric materials can be modeled by two linearized constitutive equations (Kim *et al.*, 2011):

$$\mathbf{D}_i = e_{ij}^{\sigma}\mathbf{E}_j + d_{im}^d\sigma_m \tag{10.4}$$

$$\varepsilon_k = d_{jk}^c\mathbf{E}_j + \mathbf{S}_{km}^{\mathbf{E}}\sigma_m \tag{10.5}$$

where vector $\mathbf{D}_i$ is the dielectric displacement [$N\ mV^{-1}$ or $C\ m^{-2}$], $\varepsilon_k$ is the strain vector, $\mathbf{E}_j$ is the applied electric field vector [$V\ m^{-1}$], and $\sigma_m$ is the stress vector in $N\ m^{-2}$. The piezoelectric constants are the piezoelectric coefficients $d_{im}^d$ and $d_{jk}^c$ [$m\ V^{-1}$ or $C\ N^{-1}$], the dielectric permittivity $e_{ij}^{\sigma}$ in [$N\ V^{-2}$ or $F\ m^{-1}$], and $\mathbf{S}_{km}^{\mathbf{E}}$ is the elastic compliance matrix [$m^2\ N^{-1}$]. The superscripts c and d refer to the converse and direct effects, respectively, and the superscripts $\sigma$ and $\mathbf{E}$ indicate that the quantity is measured at constant stress and constant electric field, respectively.

Piezoelectric materials can be categorized into piezoceramics and piezopolymers. Piezoceramics have large electromechanical coupling constants and provide high energy conversion rates, but they are too brittle to use in general shape-energy transducers. On the other hand, piezopolymers have smaller electromechanical coupling constants compared to the piezoceramics, but they are very flexible. The mechanical set-up of a piezoelectric generator with a cantilever and proof mass is presented in Figure 10.6. The power generated from this system is proportional to the proof mass, to the square of acceleration, and inversely proportional to the resonant and excitation frequency.

To realize the power supplement of wireless sensor networks, Kong and Ha (2012) presented a low-power design for a piezoelectric energy harvesting system. A discontinuous conduction mode (DCM) flyback converter with constant on-time modulation was used for maximum power point tracking (MPPT) to be implemented with a single current sensor. The proposed system was able to harvest up to 8.4 mW of power under 0.5 g base acceleration using four parallel piezoelectric

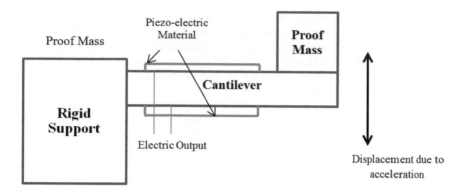

Figure 10.6.   A piezoelectric power generator.

Table 10.2.   Comparison of energy harvesting methods
(Chalasani and Conrad, 2008; Le, 2008).

| Technology | Power density [$\mu W\,cm^{-3}$] |
|---|---|
| Solar cells | 15000 |
| Piezoelectric | 330 |
| Vibration | 116 |
| Thermoelectric | 40 |
| RF @ 900 MHz | 450 |
| RF @ 2.4 GHz | 63.3 |
| Shoe inserts | 330 |
| Acoustic | 0.003 |

cantilevers, and achieved 72% efficiency around the resonant frequency. Table 10.2 shows the power densities of different energy harvesting methods. Clearly, the power harvested from the sun gives the highest power density. Thermal and acoustic methods accounted for the lowest power densities. It is worth noting that RF and piezoelectric harvesting, are two of the most attractive methods for low-power applications.

## 10.6   THE LIMITS AND TRADE-OFFS OF GREEN WIRELESS NETWORKS AND INDUSTRIAL APPLICATIONS

There are two strongly interrelated notions that impact the bounds or limits of *greening* the wireless network operation. The first is a set of trade-offs that limit the overall gains that can be achieved when contemplating an energy-reduction proposal. The second is the validation approach that should be adopted when measuring the impact of that proposal.

A general appreciation of the trade-offs is required to acknowledge the "cost" of any suggested improvement. The following discussion is largely based on the remarkably perceptive effort of Chen *et al.* (2011) who elaborated four trade-offs that govern energy reduction or management:

*Deployment efficiency vs. energy efficiency.* Deployment efficiency (DE) measures the throughput achieved in the network per unit of cost invested in network components, while energy efficiency (EE) captures the network's throughput per unit of energy expended to achieve that throughput level. The relationship is intuitive: higher throughput requires higher energy. Practically speaking, however, network equipment does not consume energy on a proportional basis to throughput level.

*Spectrum efficiency vs. energy efficiency.* Spectrum efficiency (SE) measures how much throughput is achieved per unit of bandwidth. SE plays an important role in measuring network performance, and improving SE has been a principal thrust in the design of IMT-Advanced networks. The emergence of spectrum sensing and dynamic spectrum sensing will also play a major role in enhancing spectrum use. Such SE enhancements typically come at the cost of additional energy consumption.

*Bandwidth vs. power.* This trade-off should be readily discernible from Shannon's theorem. Thus, an increased bandwidth would result in logarithmically reduced power requirements. Consider, for example, the power benefits that come from the bandwidth increase enabled by carrier aggregation. Such power reductions cannot, however, be maintained in non-ideal settings because circuit power may not scale with increase in bandwidth, and because EE may directly impact this trade-off.

*Delay vs. power.* This trade-off can also be discerned from Shannon's theorem, where increased power results in better channel rates and, therefore, lower transmission delays. Combined with the bandwidth-power relationship, these two trade-offs link quality-of-service (QoS) requirements to energy constraints.

With these trade-offs in mind, the notion of high-quality validation becomes more feasible. Our interest in evaluating an energy-reduction proposal should focus on its overall impact. This means that when a specific functionality is adjusted, its evaluation should be concerned with the effect this adjustment has on the complete hierarchy of functionalities. For example, if a proposal is made to reduce the complexity of an admission control module through utilization of additional information, the cost of acquiring such information (e.g. signaling overhead, processing, delay in decision-making, etc.) should be accounted for in some way. If, for some reason, the evaluation isolates the admission metrics (e.g. in terms of packet or call dropping) as the only measure of improvement, the evaluation is rendered inadequate. This applies equally to a more generalized scope, where the energy cost of using the network might be compared with the costs when the network is not used at all.

Let us now consider the applications of wireless network operations in industry. The present trend of automation and data exchange in manufacturing technologies has been called the fourth industrial revolution or "Industrie 4.0" (Li *et al.*, 2015). It comprises cyber-physical systems, the Internet of things (IoT) and cloud computing (Fig. 10.7). According to Li *et al.* (2015), Industrie 4.0, or the creation of "smart factories and cities", has been attracting growing interest from researchers, governments, manufacturers and application developers, as it can offer a reduction in energy consumption and increased economic benefits, as well as facilitating smart production. The aim of this novel approach is to connect and integrate traditional industries, predominantly manufacturing, to permit greater flexibility, adaptability and efficiency, and increase communication between producers and consumers. Examples of accomplishments to date include smart electricity grids equipped with renewable resources, smart water grids, and data acquisition systems applied to decentralized renewable energy plants (Aghaei and Alizadeh, 2013; Hong *et al.*, 2016; Jucá *et al.*, 2011).

Li *et al.* (2015) argued that smart industries would be the next stage of evolution for society. They went on to explain that within the new modular-structured smart factories, cyber-physical systems monitor physical processes, create a virtual copy of the physical world and make decentralized decisions. Cyber-physical systems would communicate and cooperate with each other and with humans in real time over the IoT. It can be argued that industrial wireless networks powered by renewable energies will be one of the key technologies enabling the deployment of smart factories. Of particular interest is how this clever technology can be integrated into the desalination industry in order to improve efficiency, productivity and cost-effectiveness. One conceivable industrial application would be smart small- to medium-sized brackish water desalination plants powered by renewable solar energy and located in remote regions, while the efficiency of current plants could be improved by the incorporation of renewable energy and wireless networks. These are just two areas where further research and development is needed.

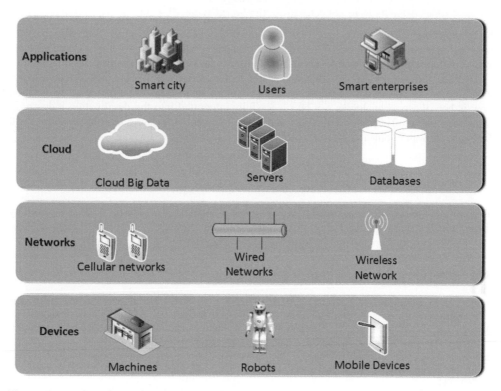

Figure 10.7.   General framework of Industrie 4.0 in a "smart factory or city" with four major layers: physical (devices), networks, cloud and "big data", and applications. Machines, robots, mobile devices, workmen and other intelligent entities constitute the physical layer for acquiring and computing data, and completing mechanical tasks and other primary functions (adapted from Li *et al.*, 2015).

In a related study, Lu *et al.* (2016) reviewed current advances in industrial wireless sensor-actuator networks and cyber-physical design of wireless control systems. The researchers emphasized the significant challenges and opportunities in cyber-physical systems research that cross-cut wireless and control domains. They concluded that interdisciplinary collaboration and partnership among wireless and control researchers and industry communities would be crucial in realizing the dream of wireless industrial control.

## 10.7   CONCLUDING REMARKS

The use of wireless networks is becoming increasingly attractive to manufacturing, including the water desalination industry. As an infrastructure, wireless connectivity foregoes much of the capital and operational cost of wired connectivity, even when considering high-capacity wired alternatives such as fiber optics. However, the core consideration here is the communication requirement of the plant, which must be established prior to any technology considerations of network design.

It has been shown that the use of wireless networks can be aligned to a plant's sustainability objectives. Many innovations have been made in the context of wireless networking that have reduced energy requirements. These improvements include better design approaches to the layout and configuration of base stations, upgraded power management for access networks, and enhanced schemes for wireless resource allocation that reduce the energy requirements of wireless devices. Current advances have also established that access networks can be powered both by renewable sources, such as solar and wind power, as well as by harvesting radio frequencies and mechanical vibrations. The latter, for instance, could be exploited in industrial plants where high operational decibel levels can be utilized to power network components.

To close, industrial wireless networks powered by renewable energies are crucial technologies enabling the large-scale development and deployment of smart factories and cities. The question now is how these technologies can be successfully integrated into the desalination industry in order to improve efficiency, productivity and cost-effectiveness. This is a key challenge that will face not just water companies but also governments and society as a whole.

## ACKNOWLEDGEMENTS

The authors would like to acknowledge the financial support provided by Alfaisal University through its Internal Research Grant, IRG Project Number 16202.

## REFERENCES

3GPP (2015) LTE; Evolved Universal Terrestrial Radio Access (E-UTRA); Base Station (BS) radio transmission and reception (3GPP TS 36.104 version 12.6.0 Release 12). ETSI, Valbonne, France.

Aghaei, J. & Alizadeh, M.I. (2013) Demand response in smart electricity grids equipped with renewable energy sources: a review. *Renewable and Sustainable Energy Reviews*, 18, 64–72.

Akhtar, F. & Rehmani, M.H. (2015) Energy replenishment using renewable and traditional energy resources for sustainable wireless sensor networks: a review. *Renewable and Sustainable Energy Reviews*, 45, 769–784.

Alawaji, H. (2001) Evaluation of solar energy research and its applications in Saudi Arabia – 20 years of experience. *Renewable and Sustainable Energy Reviews*, 5, 59–77.

Ali, K.A., Hassanein, S.H. & Mouftah, H.T. (2007) Directional cell breathing based reactive congestion control in WCDMA cellular networks. *12th IEEE Symposium on Computers and Communications, ISCC'07, 1–4 July 2007, Las Vegas, NV*. IEEE, New York. pp. 685–690.

Ali, M., Albasha, L. & Qaddoumi, N. (2013) RF energy harvesting for autonomous wireless sensor networks. *8th International Conference on Design & Technology of Integrated Systems in Nanoscale Era (DTIS), 26–28 March 2013, Abu Dhabi, UAE*. IEEE, New York. pp. 78–81.

Almasoud, A.H. & Hatim, M.G. (2015) Future of solar energy in Saudi Arabia. *Journal of King Saud University-Engineering Sciences*, 27(2), 153–157.

Baliga, J., Ayre, R., Hinton, K. & Tucker, R.S. (2011) Energy consumption in wired and wireless access networks. *IEEE Communications Magazine*, 49(6), 70–77.

Bi, S., Zeng, Y. & Zhang, R. (2016) Wireless powered communication networks: an overview. *IEEE Wireless Communications*, 23(2), 10–18.

Bokor, L., Faigl, Z. & Imre, S. (2011) Flat architectures: towards scalable future internet mobility. In: Galis, A. & Gavras, A. (eds) *The Future Internet*. Springer, Heidelberg. pp. 35–50.

Chalasani, S. & Conrad, J.M. (2008) A survey of energy harvesting sources for embedded systems. *Proceedings of IEEE Southeastcon, 3–6 April 2008, Huntsville, AL*. IEEE, New York. pp. 442–447.

Chander, S., Purohit, A., Sharma, A., Nehra, S.P. & Dhaka, M.S. (2015) Impact of temperature on performance of series and parallel connected mono-crystalline silicon solar cells. *Energy Reports*, 1, 175–180.

Chen, Y., Zhang, S., Xu, S. & Li, G.Y. (2011) Fundamental trade-offs on green wireless networks. *IEEE Communications Magazine*, 49(6), 30–37.

Cisco (2016) Virtual networking index. Available from: http://goo.gl/H24l6y [accessed November 2016].

CTM Magnetics (2016) How to harvest and store wind energy. Available from: http://www.ctmmagnetics.com/wind-power-basics-how-to-harvest-and-store-wind-energy/ [accessed November 2016].

Damnjanovic, A., Montojo, J., Wei, Y., Ji, T., Luo, T., Vajapeyam, M., Yoo, T., Song, O. & Malladi, D. (2011) A survey on 3GPP heterogeneous networks. *IEEE Wireless Communications*, 18(3), 10–21.

DHL (2016) Trend research. DHL-Cisco report on Internet of things in logistics. Available from: http://goo.gl/iuxbdQ [accessed November 2016].

Dimroth, F., Grave, M., Beutel, P., Fiedeler, U., Karcher, C., Tibbits, T.N.D., Oliva, E., Siefer, G., Schachtner, A., Wekkeli, A.W., Bett, R., Krause, M., Piccin, N., Blanc, C., Drazek, E., Guiot, B., Ghyselen, M., Salvetat, T., Tauzin, A., Signamarcheix, T., Dobrich, A., Hannappel, T. & Schwarzburg, K. (2014) Wafer bonded four-junction GaInP/GaAs//GaInAsP/GaInAs concentrator solar cells with 44.7% efficiency. *Progress in Photovoltaics: Research and Applications*, 22(3), 277–282.

Excluss Solar (2016) HAWT (horizontal) v VAWT (vertical) wind turbines. Available from: http://www.solar.excluss.com/wind-power/wind-tubine-images-gallery.html) [accessed November 2016].

Fehske, A.J., Fettweis, G.P., Malmodin, J. & Biczók, G. (2011) The global footprint of mobile communications: the ecological and economic perspective. *IEEE Communications Magazine*, 49(8), 55–62.

Gartner (2007) Green IT: The New Industry Shockwave. *Gartner Symposium/ITxpo, April 2007, Stamford, CT*.

Goosen, M.F.A., Mahmoudi, H., Ghaffour, N., Bundschuh, J. & Al Yousef, Y. (2016) A critical evaluation of renewable energy technologies for desalination. *Proceedings of the 3rd International Conference on Application of Materials Science and Environmental Materials (AMSEM2015), 1–3 October 2015, Phuket Island, Thailand*. pp. 233–258.

GWEC (2015) Global wind report – annual market update 2014. Global Wind Energy Council, Brussels.

Hasan, Z., Boostanimehr, H. & Bhargava, V.K. (2011) Green cellular networks: a survey, some research issues and challenges. *IEEE Communications Surveys & Tutorials*, 13(4), 524–540.

He, W., Williard, N., Osterman, M. & Pecht, M. (2011) Remaining useful performance analysis of batteries. *2011 IEEE Conference on Prognostics and Health Management (PHM), 20–23 June 2011, Denver, Colorado*. IEEE, New York. pp. 1–6.

Hong, J., Lee, W., Kim, J.H., Kim, J., Park, I. & Har, D. (2016) Smart water grid: desalination water management platform. *Desalination and Water Treatment*, 57(7), 2845–2854.

Humar, I., Ge, X., Xiang, L., Jo, M., Chen, M. & Zhang, J. (2011) Rethinking energy efficiency models of cellular networks with embodied energy. *IEEE Network*, 25(2), 40–49.

Jucá, S., Carvalho, P. & Brito, F.T. (2016) A low cost concept for data acquisition systems applied to decentralized renewable energy plants. *Sensors*, 11(1), 743–756.

Kazmierski, T. & Beeby, S. (eds) (2014) *Energy Harvesting Systems*. Springer, New York.

Khirallah, C., Thompson, J.S. & Rashvand, H. (2011) Energy and cost impacts of relay and femtocell deployments in long-term-evolution advanced. *IET Communications*, 5(18), 2617–2628.

Kim, H.S., Kim, J.H. & Kim, J. (2011) A review of piezoelectric energy harvesting based on vibration. *International Journal of Precision Engineering and Manufacturing*, 12(6), 1129–1141.

Kong, N. & Ha, D.S. (2012) Low-power design of a self-powered piezoelectric energy harvesting system with maximum power point tracking. *IEEE Transactions on Power Electronics*, 27(5), 2298–2308.

Lazo, P. & Duron, M. (2007) Energy harvesting for mobile RFID readers. Patent US 20070200724 A1, 30 August 2007.

Le Triet, T. (2008) *Efficient Power Conversion Interface Circuits for Energy Harvesting Applications*. PhD thesis, Oregon State University, Corvallis, OR. Available from: http://hdl.handle.net/1957/8339 [accessed November 2016].

Lee, D., Seo, H., Clerckx, B., Hardouin, E., Mazzarese, D., Nagata, S. & Sayana, K. (2012) Coordinated multipoint transmission and reception in LTE-advanced: deployment scenarios and operational challenges. *IEEE Communications Magazine*, 50(2), 148–155.

Li, X., Li, D., Wan, J., Vasilakos, A.V., Lai, C.F. & Wang, S. (2015) A review of industrial wireless networks in the context of Industry 4.0. *Wireless Networks*, November 2015, 1–19.

Lu, C., Saifullah, A., Li, B., Sha, M., Gonzalez, H., Gunatilaka, D., Wu, C., Nie, L. & Chen, Y. (2016) Real-time wireless sensor-actuator networks for industrial cyber-physical systems. *Proceedings of the IEEE*, 104(5), 1013–1024.

Mancuso, V. & Alouf, S. (2011) Reducing costs and pollution in cellular networks. *IEEE Communications Magazine*, 49(8), 63–71.

Marsan, M.A. & Meo, M. (2011) Green wireless networking: three questions. *The 10th IFIP Annual Mediterranean Ad Hoc Networking Workshop (Med-Hoc-Net), 12–15 June 2011, Favignana Island, Sicily*. pp. 41–44.

Masters, G.M. (2013) *Renewable and Efficient Electric Power Systems*. John Wiley & Sons, Hoboken, NJ.

Niu, Z., Wu, Y., Gong, J. & Yang, Z. (2010) Cell zooming for cost-efficient green cellular networks. *IEEE Communications Magazine*, 48(11), 74–79.

Patel, M.R. (2005) *Wind and Solar Power Systems: Design, Analysis, and Operation*. CRC Press, Boca Raton, FL.

Perrucci, G.P., Fitzek, F.H.P. & Widmer, J. (2011) Survey on energy consumption entities on the smartphone platform. *Proceedings of the IEEE 73rd Vehicular Technology Conference (VTC Spring), 15–18 May 2011, Yokohama, Japan.* IEEE, New York.

Piamrat, K., Ksentini, A., Bonnin, J.M. & Viho, C. (2011) Radio resource management in emerging heterogeneous wireless networks. *Computer Communications*, 34(9), 1066–1076.

REN 21 (2016) Renewables 2016 global status report. REN21 Network, Paris, France.

Saker, L. & Elayoubi, S.E. (2010) Sleep mode implementation issues in green base stations. *IEEE 21st International Symposium on Personal, Indoor and Mobile Radio Communications (PIMRC), 26–30 September 2010, Istanbul, Turkey.* IEEE, New York. pp. 1683–1688.

Sheng, Z., Mahapatra, C., Zhu, C. & Leung, V.C.M. (2015) Recent advances in industrial wireless sensor networks toward efficient management in IoT. *IEEE Access*, 3, 622–637.

Shockley, W. & Queisser, H.J. (1961) Detailed balance limit of efficiency of p-n junction solar cells. *Journal of Applied Physics*, 32, 510–519.

Song, M., Xin, C., Zhao, Y. & Cheng, X. (2012) Dynamic spectrum access: from cognitive radio to network radio. *IEEE Wireless Communications*, 19(1), 23–29.

Srivastava, V. & Motani, M. (2005) Cross-layer design: a survey and the road ahead. *IEEE Communications Magazine*, 43(12), 112–119.

Taghadosi, M., Albasha, L., Qaddoumi, N. & Ali, M. (2015) Novel miniaturized printed elliptical nested fractal multiband antenna for energy harvesting applications. *IET Microwave, Antennas and Propagation*, 9(10), 1045–1053.

Tran, N. & Lee, J.W. (2010) Simple high-sensitivity voltage multiplier for UHF-band semi-passive radio frequency identification tags using a standard CMOS process. *IET Microwaves, Antennas and Propagation*, 4(11), 1974–1979.

Windpower Engineering (2009) Vertical axis wind turbines vs horizontal axis wind turbine. Available from: http://www.windpowerengineering.com/construction/vertical-axis-wind-turbines-vs-horizontal-axis-wind-turbines/ [accessed November 2016].

WorldTimeZone.com (2016) 4G LTE world coverage map. Available from: http://www.worldtimezone.com/4g.html [accessed November 2016].

# CHAPTER 11

## Fuel-cost apportionment and desalination technology selection based on exergy analysis

M.W. Shahzad, Noreddine Ghaffour & Kim C. Ng

### 11.1 INTRODUCTION AND BACKGROUND

Water and power are two basic resources for the economic development, livelihood and well-being of a country's population. Today's power plant (PP) efficiency, although improved through decades of innovation, has peaked asymptotically in the last decade, ranging from 35% for thermal fuel-fired plants to more than 50% for the advanced, combined-cycle gas turbine (CCGT). Similarly, the practical seawater desalination methods available today are either thermally driven, such as multi-stage flash (MSF) distillation, multi-effect distillation (MED) and adsorption desalination (AD), or pressure- or work-driven, such as reverse osmosis (RO). Today's practical desalination processes's are not only energy intensive but also environment unfriendly and they consume three times higher energy than minimum theoretical required energy limit of $0.85 \, \text{kWh} \, \text{m}^{-3}$.

Conventionally, power and water resources have been generated independently of each other, often due to having separate administrations or authorities in a country. With the amalgamation of power and water authorities in many countries and the search for better process efficiency, these resources are now often produced together using a cogeneration concept and a temperature cascade arrangement for the working fluids involved. In an independent generation of power or water, the fuel input is consumed solely for the creation of a single useful effect, as shown schematically in Figure 11.1. Hence, an enthalpy-based approach is sufficient for system evaluation with respect

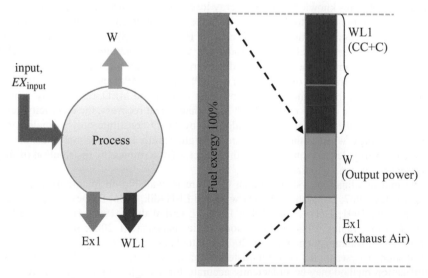

Figure 11.1. A single-purpose power plant configuration where the fuel cost is directly attributed to the enthalpy change or work, $W_A$.

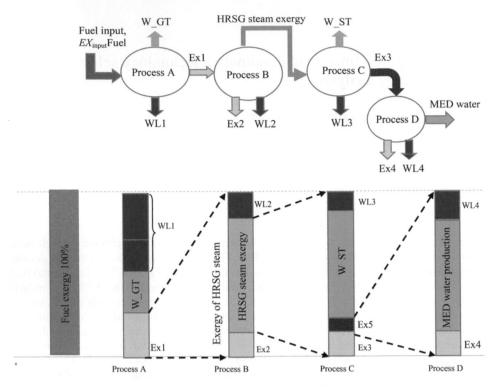

Figure 11.2.   Exergy distribution in a cogeneration plant where both power and water are produced in a cascaded manner.

to the cost of fuel input. The lost work arising from dissipation in compressor, combustor or gas turbine (GT) exhaust is deemed useful even though it has no meaningful role in fuel-cost apportionment. On the other hand, a combined power and water or cogeneration plant, as shown in Figure 11.2, requires knowledge of the exergy loss (lost work) incurred in each process of the cascaded configuration so as to achieve an accurate apportionment of fuel cost. For example, the heat-recovery steam generator (HRSG; process B in Fig. 11.2) extracts the exhaust heat of the GT to produce steam for electricity production via a high-pressure steam turbine (HPST), medium-pressure steam turbine (MPST) and low-pressure steam turbine (LPST) (process C in Fig 11.2). Only small fractions of steam from the MPST (at 1.7 MPa) and the LPST (at 0.27 MPa) are bled off to power the thermal-vapor compressor (TVC) and MED, respectively, where they produce water at a rate of 2813 m$^3$ h$^{-1}$. With this exhaust heat recovery, the electricity generated from the steam turbines and the water production by MED constitute only a small portion of the input fuel exergy, whereas the GT consumes a significantly larger portion. Thus, the relative magnitude of exergy destruction across all the processes determines the distribution of fuel cost in the cogeneration configuration.

For thermal desalination systems, much literature is available on quantitative (i.e. energy) analysis (Burley, 1967; Darwish, 1991; Darwish and El-Hadik, 1986; El-Dessoukey *et al.*, 1995; Hamed and Aly, 1991; Helal *et al.*, 1986; Honburg and Walson, 1993; Hussain *et al.*, 1994; Rasso *et al.*, 1996; Tanios, 1984) and thermodynamic analysis (Ali, 2001; Jiangang *et al.*, 2007; Kafi *et al.*, 2004; Kahraman and Cengel, 2005; Rautenbach and Arzt, 1985; Shih, 2005; Tadros, 1979) but such analyses do not accommodate the quality of energy utilized by the process. Such a quantification method may be sufficiently accurate for a single-output plant where all exergy destruction of working steam is consumed in the production of a single useful effect, that is, either power or water. However, when two or more useful effects such as power and water are produced simultaneously from a single steam expansion, this (energy-based) quantification is unable to capture the energy quality of the working fluid utilized in individual processes. Therefore, an alternative methodology for accounting for the work potentials of the expanding fluids is needed that enables such individual evaluation. For dual-purpose plants, the quality of working fluid is important for primary fuel-cost distribution for the processes where one process may utilize high-quality working fluid (e.g. turbines) and another process uses only low-quality fluid (e.g. desalination).

From the literature, it can be seen that, for dual-purpose plants, two principal methods have been used in the past for fuel-cost apportionment: (i) a method based on cost accounting of steam energy, also called the cost-balance method; (ii) a method based on assigning capital cost and fuel cost directly to the products generated, also called the cost-allocation method. Most authors (Derwish, 1995; Hamed *et al.*, 2006; Helal, 2009; Kamal, 1997; Wade, 1995; Sergio *et al.*, 2002; Syel and Ali, 2011; Valero *et al.*, 2006; Yongqing and Naom, 2007) proposed cost apportionment on the basis of the energetic input of steam to the total output produced.

From a review of this literature, it is noticeable that most of the authors focused on the same energetic analysis for cost allocation in dual-purpose plants, which cannot distinguish the quality of working fluid used at the different levels of the process. Only an exergetic analysis can capture the energy quality of the working fluid as it is used in different processes. Although there are a number of publications available on exergy analysis (Alasfour *et al.*, 2005; Al-Najem *et al.*, 1997; Banat and Jwaied, 2008; Cerci, 2002; Choi *et al.*, 2005; Hamed *et al.*, 1996, 1986; Kahraman and Cengel, 2005; Kotas, 1995; Moran, 1981; Richard, 1983; Sayyaadi and Saffaria, 2010; Spliedler *et al.*, 2001; Sulaiman and Ismail, 1995; Tribus *et al.*, 1966; Yang *et al.*, 2012), they only focus on system performance in cases of single-purpose plants. There are also a number of publications based on exergetic analysis for performance improvement of individual components of dual-purpose plants (Ali *et al.*, 2002; Asam *et al.*, 2005; Chacartegui *et al.*, 2009; Darwish *et al.*, 2007; Kamal, 2005; Mabrouk *et al.*, 2007; Madani, 1996; Mostafa *et al.*, 2011; Nafey *et al.*, 2006, 2008; Sepehr and Saeid, 2013; Yongqing *et al.*, 2006). However, it is noticeable that there is a lack of literature on fuel-cost apportionment based on an exergetic method. Hence, there is good reason to develop a primary fuel-cost apportionment model based on the exergy of working fluid utilized by the processes. Notwithstanding the numerous papers on exergy-based system performance of cogeneration plants, the motivation here is to revisit the fuel-cost analysis using the specification of a practical combined power and MED plant and demonstrate a more equitable method of fuel-cost apportionment using an exergy-based methodology.

## 11.2 AN EXERGETIC ANALYSIS OF A DUAL-PURPOSE PLANT

A practical cogeneration cycle for nominal power and water production of 593 MW and 2,813 m³ h⁻¹ of water production is analyzed, as presented in Figure 11.3. It comprises the following key components: (i) air compressor (C); (ii) combustion chamber (CC); (iii) gas turbines (GT); (iv) heat-recovery steam generator (HRSG); (v) steam turbines (ST); (vi) condenser; (vii) MED-based desalination.

The conventional MED can be coupled with TVC to take advantage of the pressure of the available input steam to pull low-pressure vapors from the last stages of MED and enhance overall MED performance. The incoming steam, called motive steam, is fed into the TVC and its expansion at the throat of the unit allows low-pressure steam from the last stage of MED to enter and mix in the mixing section of the TVC. The mixture is then compressed to the pressure of the first stage through a diffuser section of the TVC. The latent heat of the entrapped vapor is thus recycled in the evaporator and is made available for desalination, leading to energy savings.

The performance of a TVC is expressed by the mass of low-pressure MED steam (in kg) pulled in per kg of motive steam. The higher the motive steam pressure, the better the performance of the TVC. With coupling of TVC in this way, a high gain output ratio (*GOR*) can be obtained for a conventional MED system.

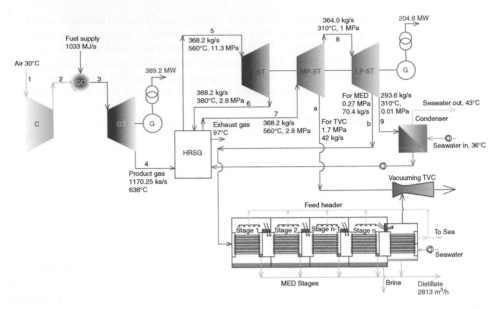

Figure 11.3.   Flow schematic of cogeneration system producing power and water simultaneously (Ihm *et al.*, 2016).

In this cogeneration system, primary fuel, in the form of natural gas, is fed to the combustor at a heat rate of 1033 MJ s$^{-1}$ and burns with air pressurized by the compressor. The product gases enter the GT at high pressure and temperature, producing 389.2 MW of power, before entering the HRSG at 638°C where superheated steam of 224.4 kg s$^{-1}$ and 560°C is generated. The superheated steam enters the HPST and is reheated to the same temperature before entering the MPST; the steam leaving these turbines is at 380°C/2.8 MPa and 310°C/1.0 MPa, respectively. About 4.2 kg s$^{-1}$ of steam is bled off from the MPST at a pressure of 1.7 MPa, which powers the TVC for the de-aeration of the MED unit. After the MPST, steam at 220.2 kg s$^{-1}$ and 310°C enters the LPST and expands to 0.27 MPa before diverting a fraction of 70.2 kg s$^{-1}$ into the top-brine stage of the MED. After partial extraction, the remaining steam expands further into the condenser at 0.01 MPa. The combined electricity output from the steam turbines constitutes 202.4 MW and the combined electricity output of the CCGT system is 594.0 MW. A model for the fuel-cost apportionment of this system, based on exergy analysis, is presented in Table 11.1 and a conceptual model of a methodology for energetic and exergetic calculations is presented in Figure 11.4.

## 11.3   PRIMARY FUEL-COST APPORTIONMENT

Based on this improved model, both the exergy and enthalpy changes across the major components are analyzed and the proportions of exergy and enthalpy utilization of the fuel input are shown in Figure 11.5.

The comparison indicates a stark difference in the proportions with respect to enthalpy and exergy methods: the GT proportion of fuel cost varied from 43.67% by enthalpy to 73.17% by exergy, while the fuel cost for the steam turbine (ST) percentage reduced from 39.58% by enthalpy to 23.43% by exergy. Correspondingly, there was a 393% reduction in the cost of water, that is, from 16.75% by enthalpy to 3.40% by exergy, as shown in Figure 11.6.

Table 11.2 summarizes the percentage ratios that are readily obtained by multiplying their respective ratios of useful effect to useful exergy of process C, that is, {(*electricity of steam*

Table 11.1. Detailed exergy modeling of power plant combined with desalination system.

| State points (ref. Fig. 11.3) | Exergy calculation | Comments |
|---|---|---|
| 1–2 | $DE_C = \dot{m}_1[(h_2 - h_1) - T_0(S_2 - S_1)]$ | Exergy destruction across compressor |
| 2–3 | $\Delta E_{CC} = E_{fuel} - E_3$ | Exergy destruction across combustor |
| | $E_{fuel} = HCV_{CH_4}.\dot{m}_{fuel}$ | |
| 3–4 | $DE_{GT} = \dot{m}_3[(h_3 - h_4) - T_0(S_3 - S_4)]$ | Exergy destruction across GT |
| | $\Delta E_{Brayton\_cycle} = \Delta E_C + \Delta E_{CC} + \Delta E_{GT}$ | Exergy destruction across Brayton cycle |
| 4–4′ | $\Delta E_{HRSG} = \dot{m}_4[(h_4 - h_{4'}) - T_0(S_4 - S_{4'})]$ | Exergy destruction across HRSG |
| | $\dot{m}_5 = m_{product\_gas}Cp(T_4 - T_{4'})\eta_{HRSG}$ | Steam generated, $\eta_{HRSG} = 45\%, m_{product\_gas} = 1170.25\ kg\ s^{-1}$ |
| 5–6 | $\Delta E_{HP-ST} = \dot{m}_5[(h_5 - h_6) - T_0(S_5 - S_6)]$ | Exergy destruction across HPST |
| 7–8 | $\Delta E_{7-a} = \dot{m}_7[(h_7 - h_a) - T_0(S_5 - S_a)]$ | Exergy destruction across MPST |
| | $\Delta E_{a-8} = \dot{m}_8[(h_a - h_8) - T_0(S_a - S_8)]$ | |
| | $\Delta E_{MP-ST} = \Delta E_{7-a} + \Delta E_{a-8}$ | |
| | $\dot{m}_8 = \dot{m}_7 - \dot{m}_a$ | |
| 8–9 | $\Delta E_{8-b} = \dot{m}_8[(h_8 - h_b) - T_0(S_8 - S_b)]$ | Exergy destruction across LPST |
| | $\Delta E_{b-9} = \dot{m}_9[(h_b - h_9) - T_0(S_b - S_9)]$ | |
| | $\Delta E_{LP-ST} = \Delta E_{8-b} + \Delta E_{b-9}$ | |
| | $\dot{m}_9 = \dot{m}_8 - \dot{m}_b$ | |
| | $\Delta E_{ST} = \Delta E_{HP-ST} + \Delta E_{MP-ST} + \Delta E_{LP-ST}$ | Exergy destruction across steam turbines |
| | $\Delta E_{Desal} = \Delta E_{a-MED} + \Delta E_{b-MED}$ | Bleed steam exergy |
| | $\Delta E_{Total} = \Delta E_{brayton\_cycle} + \Delta E_{HRSG}$ | Total exergy |
| | $\%E_{GT} = \dfrac{\Delta E_{brayton\_cycle}}{\Delta E_{Total}}$ | Input exergy proportion to GT |
| | $\%E_{HRSG} = \dfrac{\Delta E_{HRSG}}{\Delta E_{Total}}$ | Input exergy proportion to HRSG |
| | $\%E_{ST} = \dfrac{\Delta E_{ST}}{\Delta E_{ST} + \Delta E_{Desal}}(\%E_{HRSG})$ | HRSG exergy proportion to steam turbines |
| | $\%E_{ST} = \dfrac{\Delta E_{Desal}}{\Delta E_{ST} + \Delta E_{Desal}}(\%E_{HRSG})$ | HRSG exergy proportion to MED |

*turbine/total useful exergy) × (HRSG exergy)*} and {(*water production of MED/total useful exergy) × (HRSG exergy)*}. This implies that for a large-scale seawater desalination plant of one million cubic meters per day, typical in Middle East countries, annual savings of (US) $18 million can be accrued via this exergy method of fuel-cost apportionment (Ghaffour *et al.*, 2013, 2014). The increase in the GT proportion of the fuel cost is attributed to the high level of exergy destruction (lost work) found in the combustor, the air compressor and the turbines. The exergy analysis reduces the fuel-cost percentages of the ST and MED processes, and the enthalpy methodology is shown to be inadequate for cogeneration systems, revealing a significant accounting discrepancy in the fuel costs used hitherto.

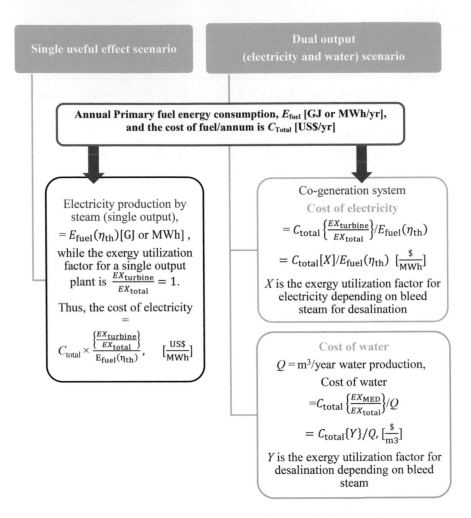

Figure 11.4.    Conceptual model for cost apportionment in single- and dual-purpose plants.

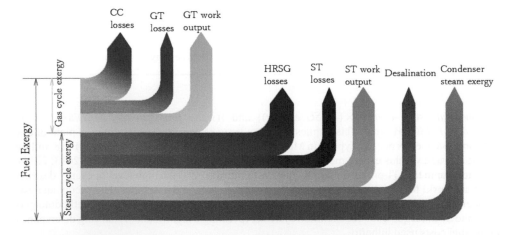

Figure 11.5.    Exergy flow across components in a cogeneration system.

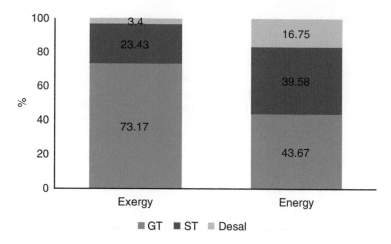

Figure 11.6.   Primary fuel proportions according to energetic and exergetic methods.

Table 11.2.   A summary of enthalpy and exergy methods.

| Process | Conventional energy method (ignoring lost work) | Proposed exergy method |
|---|---|---|
| GT (including combustor, air compressor) | 43.67 | 73.17 |
| ST (HRSG, HP, MP and LP) | 39.58 | 23.43 |
| MED desalination | 16.75 | 3.4 |
| Desalination overcharge by energy method | | 393% |
| Water cost by energy method* | | 1.049 US$ m$^{-3}$ |
| Water cost by exergy method | | 0.213 US$ m$^{-3}$ |
| Water overcharge by energy method | | 0.837 US$ m$^{-3}$ |
| Saving for a 1 000 000 m$^3$ day$^{-1}$ desalination plant | | 26 771 318 US$ year$^{-1}$ |

*Fuel cost used for calculation = (US) \$5 per thousand cu ft or 1055 MJ.

## 11.4   DESALINATION METHOD SELECTION BASED ON SYSTEM EFFICIENCY

In cogeneration plants, desalination technology selection depends on individual systems efficiency. Thermal systems can be compared with RO processes to provide the best choice if there is an option to select between PP+MED or PP+RO, as shown in Figure 11.7. It can be seen that for a conventional low-efficiency MED system ($GOR = 11$) to be combined with a power plant, it needs 70.83 kg s$^{-1}$ steam to produce 2813 m$^3$ h$^{-1}$ water. This extraction steam has about 3.4% of the total input steam exergy and could produce about 11.16 MW electricity if it were to pass through steam turbines. At this point, for the same level of water production, RO consumes 15 MW electricity (red dotted line). As scientists and engineers are working to further improve these technologies, we extended our analysis from a $GOR$ of 11 to a $GOR$ of 30.

Our group developed a hybrid cycle combining MED with an adsorption desalination (AD) cycle, namely a MEDAD cycle, which can achieve a $GOR$ of up to 25. A number of publications are available describing the AD cycle (Ng *et al.*, 2009, 2012, 2013; Saha *et al.*, 2009, 2010) as well as MEDAD hybrid cycles (Ng *et al.*, 2015; Shahzad and Ng, 2016; Shahzad *et al.*, 2013, 2014a, 2014b, 2015a, 2015b, 2016). Figure 11.8 shows the carbon emissions for the same power and water

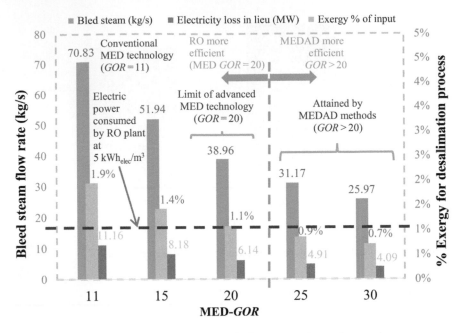

Figure 11.7.    Desalination technology selection criteria for cogeneration cycle operation.

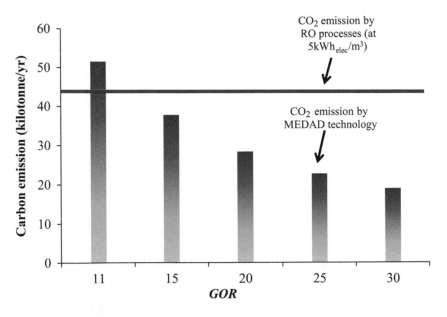

Figure 11.8.    Desalination technologies' $CO_2$ emissions for proposed capacity.

cogeneration system. It can be seen that PP+MED can only be more environmentally friendly than PP+RO if MED *GOR* performance is 15 or more. It can be concluded that for cogeneration systems, a PP+MEDAD system is the only one that is more energetically economical than PP+RO and can address the energy-environment-water nexus.

## 11.5 SUMMARY

In this chapter we have dispelled the misperceptions of primary fuel-cost apportionment for water production in cogeneration plants, which have beset the enthalpy accounting method of seawater desalination. This enthalpy approach to analyzing the cost of water is less accurate and it should be replaced by an exergy-based method such as that described here, which has the potential to lower water charges by as much as 400%, yet has eluded the desalination community thus far.

## NOMENCLATURE

| | |
|---|---|
| AD | adsorption desalination |
| C | compressor |
| CC | combustion chamber |
| CCGT | combined-cycle gas turbine |
| Ex | exergy |
| *GOR* | gain output ratio |
| GT | gas turbine |
| HPST (HP-ST) | high-pressure steam turbine |
| HRSG | heat-recovery steam generator |
| LPST (LP-ST) | low-pressure steam turbine |
| MED | multi-effect desalination |
| MPST (MP-ST) | medium-pressure steam turbine |
| MSF | multi-stage flashing |
| PP | power plant |
| TVC | thermo-vapor compressor |
| W | work |
| WL | work lost |

*Subscripts*

| | |
|---|---|
| A | available work |
| L | lost work |

## REFERENCES

Alasfour, F.N., Darwish, M.A. & Bin Amer, A.O. (2005) Thermal analysis of ME-TVC + MEE desalination system. *Desalination*, 174, 39–61.

Al-Najem, N.M., Darwish, M.A. & Youssef, F.A. (1997) Thermo-vapor compression desalination: energy and availability analysis of single and multi-effect systems. *Desalination*, 110, 223–238.

Al-Sulaiman, F.A. & Ismail, B. (1995) Exergy analysis of major re-circulating multi-stage flash desalting plants in Saudi Arabia. *Desalination*, 103, 265–270.

Banat, F. & Jwaied, N. (2008) Exergy analysis of desalination by solar-powered membrane distillation units. *Desalination*, 230, 27–40.

Burley, M.J. (1967) Analytical comparison of the multistage flash and long tube vertical distillation. *Desalination*, 2, 81–88.

Cerci, Y. (2002) Exergy analysis of a reverse osmosis desalination plant in California. *Desalination*, 142, 257–266.

Chacartegui, R., Sánchez, D., Gregorio, N., Jiménez-Espadafor, F.J., Muñoz, A. & Sánchez, T. (2009) Feasibility analysis of a MED desalination plant in a combined cycle based cogeneration facility. *Applied Thermal Engineering*, 29, 412–417.

Choi, H., Lee, T., Kim, Y. & Song, S. (2005) Performance improvement of multiple-effect distiller with thermal vapor compression system by exergy analysis. *Desalination*, 182, 239–249.

Darwish, M.A. (1991) Thermal analysis of multistage flash desalting systems. *Desalination*, 85, 59–79.

Darwish, M.A. (1995) Fuel cost charged to desalters in co-generation power-desalting plants. *Heat Recovery Systems and CHP*, 15, 357–368.

Darwish, M.A. & El-Hadik, A.A. (1986) The multi effect boiling desalting system and its comparison with the multistage flash system. *Desalination*, 60, 251–265.

Darwish, M.A., Al Otaibi, S. & Khawla, S. (2007) Suggested modifications of power-desalting plants in Kuwait. *Desalination*, 216, 222–231.

El-Dessoukey, H., Shaban, H.I. & Al-Ramadan, H. (1995) Steady state analysis of multistage flash desalination process. *Desalination*, 103, 271–287.

Gaggioli, A.R. (1983) Second law analysis for process and energy engineering. *ACS Symposium Series*. American Chemical Society, Washington, DC. Chapter 1, 235, pp. 3–50.

Ghaffour, N., Missimer, T.M. & Amy, G. (2013) Technical review and evaluation of the economics of water desalination: current and future challenges for better water supply sustainability. *Desalination*, 309, 197–207.

Ghaffour, N., Lattemann, S., Missimer, T.M., Ng, K.C., Sinha, S. & Amy, G. (2014) Renewable energy-driven innovative energy-efficient desalination technologies. *Applied Energy*, 136, 1155–1165.

Hamad, A., Almulla, A. & Gadalla, M. (2005) Integrating hybrid systems with existing thermal desalination plants. *Desalination*, 174, 171–192.

Hamed, O.A. & Aly, S. (1991) Simulation and design of MSF desalination processes. *Desalination*, 10, 1–14.

Hamed, O., Zamamiri, A., Aly, S. & Lior, N. (1996) Thermal performance and exergy analysis of a thermal vapor compression desalination system. *Energy Conversion and Management*, 37, 379–387.

Hamed, O.A., Al-Washmi, H.A. & Al-Otaibi, H.A. (2006) Thermo economic analysis of a power/water cogeneration plant. *Energy*, 31, 2699–2709.

Helal, A.M. (2009) Hybridization – a new trend in desalination. *Desalination and Water Treatment*, 3, 129–135.

Helal, A.M., Medani, M.S. & Soliman, M.A. (1986) A tridiagonal matrix model for multistage flash desalination plants. *Computers and Chemical Engineering*, 10, 327–342.

Honburg, C.D. & Walson, B.M. (1993) Operational optimization of MSF systems. *Desalination*, 10, 331–351.

Hussain, A., Woldai, A., Al-Radif, A., Kesou, A., Borsani, R., Sultan, H. & Deshpandey, P.B. (1994) Modelling and simulation of a multistage flash (MSF) desalination plant. *Desalination*, 10, 555–586.

Ji, J.G., Wang, R.Z., Li, L.X. & Ni, H. (2007) Simulation and analysis of a single-effect, thermal vapor-compression desalination, system at variable operation conditions. *Chemical Engineering & Technology*, 30, 1633–1641.

Kafi, F., Renaudin, V., Alonso, D. & Hornut, J.M. (2004) New MED plate desalination process: thermal performances. *Desalination*, 166, 53–62.

Kahraman, N. & Cengel, Y.A. (2005) Exergy analysis of a MSF distillation plant. *Energy Conversion and Management*, 46, 2625–2636.

Kahraman, N., Cengel, Y.A., Wood, B. & Cerci, Y. (2004) Exergy analysis of a combined RO, NF, and EDR desalination plant. *Desalination*, 171, 217–232.

Kamal, I. (1997) Thermo-economic modeling of dual-purpose power/desalination plants: steam cycles. *Desalination*, 114, 233–240.

Kamal, I. (2005) Integration of seawater desalination with power generation. *Desalination*, 180, 217–229.

Kotas, T.J. (1995) *The Exergy Method of Thermal Plant Analysis*. Krieger, Malabar, FL.

Mabrouk, A.A., Nafey, A.S. & Fath, H.E. (2007) Analysis of a new design of a multi-stage flash-mechanical vapor compression desalination process. *Desalination*, 204, 482–500.

Madani, A.A. (1996) Analysis of a new combined desalination-power generation plant. *Desalination*, 105, 199–205.

Moran, M.J. (1981) *Availability Analysis: A Guide to Efficient Energy Use*. Prentice Hall, Englewood Cliffs, NJ.

Mussati, S., Aguirre, P. & Scenna, N. (2002) Dual-purpose desalination plants. Part II. Optimal configuration. *Desalination*, 153, 185–189.

Nafey, A.S., Fath, H.E. & Mabrouk, A.A. (2006) Exergy and thermo-economic evaluation of MSF process using a new visual package. *Desalination*, 201, 224–240.

Nafey, A.S., Fath, H.E. & Mabrouk, A.A. (2008) Thermo-economic design of a multi effect evaporation mechanical vapor compression (MEE-MVC) desalination process. *Desalination*, 230, 1–15.

Ng, K.C., Thu, K., Chakraborty, A., Saha, B.B. & Chun, W.G. (2009) Solar-assisted dual-effect adsorption cycle for the production of cooling effect and potable water. *International Journal of Low-Carbon Technologies*, 4, 61–67.

Ng, K.C., Thu, K., Chakraborty, A., Saha, B.B. & Chun, W.G. (2012) Study on a waste heat-driven adsorption cooling cum desalination cycle. *International Journal of Refrigeration*, 35, 685–693.

Ng, K.C., Thu, K., Kim, Y.D., Chakraborty, A. & Amy, G. (2013) Adsorption desalination: an emerging low-cost thermal desalination method. *Desalination*, 308, 161–179.

Ng, K.C., Thu, K., Oh, S.J., Li, A., Shahzad, M.W. & Ismail, A.B. (2015) Recent developments in thermally-driven seawater desalination: energy efficiency improvement by hybridization of the MED and AD cycles. *Desalination*, 356, 255–270.

Rasso, M., Beltramini, A., Mazzotti, M. & Morbidelli, M. (1996) Modelling of multistage flash desalination plants. *Desalination*, 108, 335–364.

Rautenbach, R. & Arzt, B. (1985) Gas turbine waste heat utilization for desalination. *Desalination*, 52, 105–122.

Sayyaadi, H. & Saffaria, H. (2010) Thermo-economic optimization of multi effect distillation desalination systems. *Applied Energy*, 87, 1122–1133.

Sepehr, S. & Saeid, A. (2013) Four E analysis and multi-objective optimization of combined cycle power plants integrated with multi-stage flash (MSF) desalination unit. *Desalination*, 320, 105–117.

Shahzad, M.W. & Ng, K.C. (2016) On the road to water sustainability in the Gulf. *Nature Middle East*. doi:10.1038/nmiddleeast.2016.50. Available from: http://www.natureasia.com/en/nmiddleeast/article/10.1038/nmiddleeast.2016.50 [accessed January 2017].

Shahzad, M.W., Myat, A., Chun, W.G. & Ng, K.C. (2013) Bubble-assisted film evaporation correlation for saline water at sub-atmospheric pressures in horizontal-tube evaporator. *Applied Thermal Engineering*, 50, 670–676.

Shahzad, M.W., Ng, K.C., Thu, K., Saha, B.B. & Chun, W.G. (2014a) Multi effect desalination and adsorption desalination (MEDAD): a hybrid desalination method. *Applied Thermal Engineering*, 72, 289–297.

Shahzad, M.W., Thu, K., Saha, B.B. & Ng, K.C. (2014b) An emerging hybrid multi-effect adsorption desalination system. *EVERGREEN Joint Journal of Novel Carbon Resource Sciences & Green Asia Strategy*, 1(2), 30–36.

Shahzad, M.W., Thu, K.T. Kim, Y.D. & Ng, K.C. (2015a) An experimental investigation on MEDAD hybrid desalination cycle. *Applied Energy*, 148, 273–281.

Shahzad, M.W., Ng, K.C., Thu, K. & Chun, W.G. (2015b) Recent development in thermally-activated desalination methods: achieving an energy efficiency less than 2.5 kWh$_{elec}$/m$^3$. *Desalination and Water Treatment*, 57, 7396–7405.

Shahzad, M.W., Thu, K. & Ng, K.C. (2016) Future sustainable desalination using waste heat: kudos to thermodynamic synergy. *Environmental Science: Water Research & Technology*, 2, 206–212.

Sharqawya, M. H., Lienhard, J.H. & Zubair, S.M. (2011) On exergy calculations of seawater with applications in desalination systems. *International Journal of Thermal Sciences*, 50, 187–196.

Shih, H. (2005) Evaluating the technologies of thermal desalination using low-grade heat. *Desalination*, 182, 461–469.

Spliegler, K.S. & El-Sayed, Y.M. (2001) The energetics of desalination processes. *Desalination*, 134, 109–128.

Tadros, S. (1979) A new look at dual purpose, water and power, plants-economy and design features. *Desalination*, 30, 613–630.

Tanios, B.Z. (1984) Marginal operation field of existing MSF distillation plants. *Desalination*, 51, 201–212.

Valero, A., Serra, L. & Uche, J. (2006) Fundamentals of exergy cost accounting and thermodynamics. Part 1. Theory. *Journal of Energy Resources Technology*, 128, 9–15.

Wade, N.M. (1999) Energy and cost allocation in dual-purpose power and desalination plants. *Desalination*, 123, 115–125.

Yang, L., Shen, T., Zhang, B., Shen, S. & Zhang, K. (2012) Exergy analysis of a solar-assisted MED desalination experimental unit. *Desalination and Water Treatment*, 37, 1272–1278.

Yongqing, W. & Noam, L. (2006) Performance analysis of combined humidified gas turbine power generation and multi-effect thermal vapor compression desalination systems. Part 1. The desalination unit and its combination with a steam-injected gas turbine power system. *Desalination*, 196, 84–104.

Yongqing, W. & Noam, L. (2007) Fuel allocation in a combined steam-injected gas turbine and thermal seawater desalination system. *Desalination*, 214, 306–326.

# CHAPTER 12

## Achieving food security in the desalination age

Eanna Farrell & Hassan Arafat

### 12.1 INTRODUCTION

Considerable effort is spent in the quest to understand the natural world and how we as humans interact with it. This quest has been embodied by sustainable development in the recent past, as we strive to live in a world in which our actions will not compromise our future. Our knowledge base is constantly expanding through collection and interpretation of new data, exploring newly appreciated dynamics, and investigating previously misunderstood phenomena. As our understanding grows, so too does our capacity to manage our actions before they cause undesirable consequences in environmental, economic and social spheres. Our knowledge of systems and their actors grows. The dynamics of previously undocumented relationships is explored.

Food security is an issue which is appreciated at varying levels of importance globally. When we view this issue in light of the fact that agriculture is the predominant consumer of water on the planet, the issue warrants further consideration in the context of water security. In water-stressed regions, one such solution to insufficient natural water supply is the use of desalination. Desalination for agricultural use is not yet prominent, accounting only for 1.9% of the total installed worldwide desalination capacity (Zarzo *et al.*, 2012).

Desalination has started to bridge the gap between supply and demand, but groundwater is still the main water source. If groundwater becomes too saline for irrigation, one could propose a change to conventional desalination technologies. However, desalination is not a low-energy process, nor is it cheap, and presently the majority of plants operate on fossil fuels. For example, Abu Dhabi (in United Arab Emirates, UAE) alone spends nearly (US) $3.3 billion a year on desalination (Awerbuch and Walker, 2014). In the UAE, seawater desalination requires about ten times more energy than surface freshwater production, and its costs are projected to increase by 300%. This is not to say that desalination technology cannot serve a purpose, but rather that the economics and potential long-term environmental effects of desalination need to be considered before proposing it as a large-scale solution for agricultural water supply.

Meeting the food needs of increasing populations, overcoming shortfalls in food production, and ensuring that available produce reaches people in need are major challenges for global agriculture (FAO, 2010). By exploring how the issue of food security and desalination are related, the question of whether this technology presents a path to sustainability in the food sector will be more readily understood.

### 12.1.1 *Defining food security*

Food security is defined as a situation where all people, at all times, have physical and economic access to sufficient, safe and nutritious food to meet their dietary needs and food preferences for an active and healthy lifestyle (FAO, 2003). Food security depends on four factors: availability, accessibility, utilization and stability (AFED, 2014; Kotagama *et al.*, 2014). Assessing food security through these factors, it is possible to see variability in different regions and socioeconomic groups, particularly as it relates to accessibility and availability of food for the rural and

211

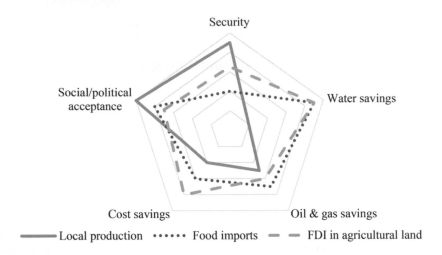

Figure 12.1.    Food security trade-offs in the GCC states (Saif *et al.*, 2014).

urban poor (De Bon *et al.*, 2009). Food security is a mandated objective of several international organizations, most notably the United Nations Food and Agriculture Organization (FAO).

The amount of water needed to meet local food demands depends on diet preferences and how the food is produced. This relationship can be demonstrated by the increasing amount of meat consumption in growing economies. More meat consumption leads to a higher amount of water used per person per day for the purposes of food production. Renault and Wallender (2000) estimated that a vegetarian diet to meet daily calorific needs requires 2000 L of water per person per day, whereas this figure rises to 5000 L for a diet high in grain-fed beef (Renault and Wallender, 2000). Accounting for the water footprint of our food is a core function of the virtual water concept, which we will discuss further in the solutions section of this chapter.

Figure 12.1 shows three possible options for providing food in water-stressed regions such as the Gulf Cooperation Council (GCC) countries. The GCC is composed of Bahrain, Kuwait, Oman, Qatar, Saudi Arabia and the United Arab Emirates, and provides an important context to discuss the relationship between food security and desalination. Each of these countries relies heavily on desalination technology as a means to provide water and each has varying degrees of food security (which will be discussed throughout this chapter). As shown in Figure 12.1, local production is the most favored option, both socially and politically, but comes with a heavy trade-off in water savings, whereas in the option of foreign direct investment (FDI) in agricultural land, water savings are more evenly balanced with other factors in the decision-making process.

One of the key variables for measuring and evaluating the evolution of food systems, and thus food security, is kilojoules [kJ] (or kilocalories, kcal) per person per day. World food supply increased from approximately 10,042 kJ (2400 kcal) per person per day in 1970 to 11,715 kJ (2800 kcal) per person per day in 2000 (FAO, 2010). The production value at the dawn of the millennium is usually taken as a threshold for national food security (Molden, 2007). This accounts for a dietary energy consumption of 7950–10,460 kJ (1900–2500 kcal) per person per day, factored in with inefficiencies in the food chain, inequalities in terms of accessibility, and so on. These figures are averages, and regional differences in food supply can be quite large. For instance, in developed countries, the change between 1970 and 2000 was from 12,761 to 14,435 kJ (3050 to 3450 kcal) per person per day, while in sub-Saharan Africa it increased from 8786 to 9205 kJ (2100 to 2200 kcal) per person per day (Molden, 2007).

A valid question is, therefore: if we already produce enough food in terms of calories, is food security an issue that currently exists? As can be seen from the difference between sub-Saharan Africa and developed countries detailed above, the answer is yes, food insecurity does exist. The fact that enough food is already produced, but does not reach those in most need, is an issue of

Table 12.1.   Summary of water resources and use for GCC countries (adapted from Barghouti, 2010).

| Country | Annual water availability [billion m$^3$ year$^{-1}$] | | | Annual water usage | | Use by sector [%] | | |
| | Natural renewable resources | Desalinated water | Wastewater reuse | [Bm$^3$] | % of total water resources | Domestic | Industry | Agriculture |
| --- | --- | --- | --- | --- | --- | --- | --- | --- |
| Bahrain | 0.11 | 0.14 | Negligible | 0.25 | 170 | 26 | 3 | 71 |
| Kuwait | 0.11 | 0.65 | 0.12 | 0.76 | 87 | 37 | 2 | 60 |
| Oman | 1.6 | 0.12 | 0.02 | 1.22 | 74 | 9 | 1 | 93 |
| Qatar | 0.05 | 0.12 | n.a. | 0.28 | n.a. | 23 | 3 | 74 |
| Saudi Arabia | 2.5 | 2.28 | 0.15 | 17 | 506 | 9 | 1 | 90 |
| UAE | 0.2 | 0.95 | 0.14 | 1.6 | 180 | 24 | 10 | 67 |

the accessibility factor in food security. The purpose of the following discussion is to examine food security in the light of the pressing realities of water security. If food security is already an issue, before future threats such as climate change, population growth and land degradation are taken into consideration, its relationship to and influence on water security needs to be thoroughly embraced by academia, policymakers and industry alike.

### 12.1.2   *Defining water security*

In the same way that food security differs from region to region, so it is for water security. It is possible to quantify a country's level of water security. The FAO defines a water-stressed state as having at least 20% of its renewable water resources consumed by the agriculture sector. Should it rise to 40%, then it is regarded as under a critical degree of stress (FAO, 2010).

To understand the relationship between desalination and food security, we must first look to why desalination is required in the first place. Water security is a current and worsening problem. Between 1960 and 1997, per capita availability of freshwater worldwide declined by about 60%. Another 50% decrease in per capita water supply is projected by the year 2025 (Hinrichsen, 1998). Water demands already exceed supplies in nearly 80 countries, accounting for more than 40% of the world's population (Qadir *et al.*, 2003). By 2025, 1.8 billion people will be living in countries or regions with absolute water scarcity, and two-thirds of the world's population could be living under water-stressed conditions (FAO, 2003).

Taking the GCC countries as an example, the excessive water consumption in the GCC is attributed to the arid environment, wasteful consumer behavior, inefficient water supply systems, economic and population growth, and poor integration of regulatory and management bodies. The presence of subsidies in the GCC water sector has hidden the true cost of water, and has done little to deter consumers from wasteful practices. However, it is not just consumer behavior that leads to unsustainable resource use. Inefficient water supply systems, in conjunction with a subsidized resource, mask the cost of leakages in water networks. A summary of water resources and use is provided in Table 12.1. It is estimated that 22% of water intended for domestic purposes in the GCC is lost to leakages or is not metered (Shepherd, 2014).

The fact that the consumption per capita is so high would not be such a significant problem in itself if the renewable water resource availability was not so low. Having the ability to subsidize the inefficient use of one resource (water) with the exorbitant use of another (energy), serves to provide an unsustainable system. The opportunity cost of dedicating significant energy resources to securing freshwater in countries dependent on desalination is significant. In a region such as the GCC, where fossil fuel resources can contribute significantly to the economy, opportunity costs include long-term energy resources and petrochemical exports. The economics of desalination are also a source of contention. In regions outside the GCC that operate desalination plants on a utility scale, efforts are made to recover the costs via pricing mechanisms. However, within the GCC it is estimated that the gap between the costs (including capital, operating, production and opportunity costs) and revenues of desalination is 92% (Shepherd, 2014).

### 12.1.3    *Defining sustainable desalination*

The world's oceans provide us with an abundant source of water, with 97% of total planetary water housed there in a saline format. The need for desalination is projected to rise quite significantly. In the Middle East and North Africa (MENA) region alone, desalination capacity is expected to nearly double to 15 million $m^3$ $day^{-1}$ by 2030, based on 2011 figures. The historic rise of desalination in GCC countries is shown in Figure 12.2. This is driven in part by rapidly depleting groundwater reserves and in part by an increase in population (Bazilian *et al.*, 2011). When we try to determine what sustainable desalination consists of, environmental impacts are a significant factor. Discharge of concentrate and chemicals to the marine environment, emission of air pollutants and the energy demand of the processes are all factored into impact assessments (Lattemann and Höpner, 2008).

It is important that desalination should also be sustainable in cost as well as resources. In the last 20 years, the cost of producing freshwater from large-scale seawater reverse osmosis (SWRO) has dropped to (US) $0.50 $m^{-3}$, and is below (US) $1.00 $m^{-3}$ for multi-stage flash (MSF) (Ghaffour *et al.*, 2013).

Changes in water availability, quality and abundance are difficult to project. The accuracy of today's models will not be known until we actually experience the modeled environment. What we know for certain is that increases in population and economic activity will require more water resources. We know that there are already areas experiencing water shortages and water-stressed states. We know that agriculture accounts for 65–80% of annual water consumption. So we can conclude that if water security is already an issue, it is likely to affect more regions, more severely, in the future. Therefore, it is imperative that we understand the dynamics between food and water security as a matter of priority.

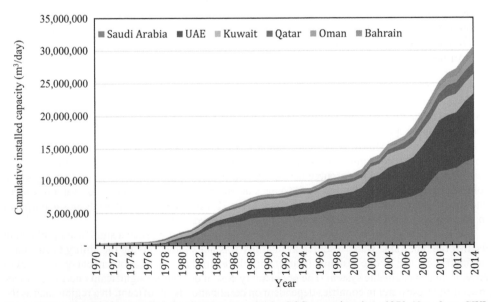

Figure 12.2.    Cumulative installed desalination capacity in GCC countries since 1970 (data from GWI, 2016).

## 12.2   FOOD SECURITY AND DESALINATION IN ACTION – A GCC PERSPECTIVE

Water can be categorized as either potable, meaning it is suitable for drinking, or non-potable, meaning unsuitable for drinking but typically usable for washing clothes, flushing toilets, etc. In deciding how water is allocated, this categorization should be a consideration. Thus, if water is to be desalinated, it should be primarily used for potable uses. The option of desalination should be a last resort in this regard. Once efficient use of freshwater is practiced by consumers and utility operators alike, the focus should then be to maximize the use of secondary water sources, such as treated wastewater and grey water, for non-potable use. The use of non-renewable groundwater should be avoided.

The GCC region in the Arabian Peninsula can serve as an excellent demonstration. The GCC has been dealing with water and food security issues in an increasingly urgent manner. This is due in part to the arid desert climate, low and erratic rainfall patterns, poor-to-marginal land, etc. A perfect storm starts to form for the GCC region when we combine these issues with the reality of a growing population, some of the highest per capita water consumption rates in the world, and some of the lowest per capita renewable water resources. This is why the measures undertaken in the GCC, as they try to mitigate and adapt to the realities of securing water and food, are of great interest to other parties looking to do the same.

On average, 75–80% of the water resources in the region are used for agriculture (Alsharhan *et al.*, 2001), as can be seen from Figure 12.3a. Urban landscapes, public parks and private gardens also consume a large proportion of water resources. The majority of this water comes from groundwater, with estimates for this as high as 85% (Bazza, 2005). This places a large stress on groundwater aquifers in a region where the average renewable water resources can be less than 100 m$^3$ per capita, compared to a global average of 6000 m$^3$ per capita (AFED, 2014). The majority of agricultural water use is for irrigation. The practice of open field irrigation is common and can lead to problems such as salinization of soil, loss of soil fertility and structure, and non-optimal fertilizer uptake when irrigation is not properly managed. These negative environmental impacts can occur in parallel with damage to groundwater resources in the form of increasing salinity.

One reason for this depletion of groundwater reserves in GCC countries has been the provision of direct and indirect subsidies for well excavation, pumps, fuel and other inputs as well as price support programs and trade protection (Saif and Mushtaque, 2014). This mismatch of priorities is central to the problem of food security. For example, in Qatar, agriculture accounts for only 1% of GDP, but consumes 74% of its freshwater. Comparable GDP figures exist for other GCC countries, as shown in Figure 12.4. This seemingly disproportionate dedication of water resources

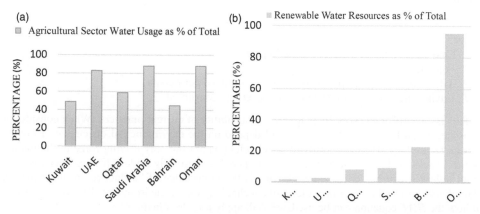

Figure 12.3.   (a) Water consumption by agriculture in GCC countries (2013) (Alpen Capital, 2015); (b) renewable water consumed in same GCC countries (2013) (data from Strategy&, 2014).

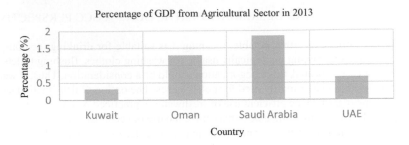

Figure 12.4.   Agriculture as a percentage of GDP (GaStat, 2015).

is such that the usage rate is approximately six times the natural renewal rate, with 50% of the water used coming from desalination or wastewater treatment plants (Spiess, 2008). And herein lies the dilemma between food and water security, destabilized by economics. In regions with poor water resources, why focus on aiding food security at the expense of water security? An alternative is to seek food security directly with monetary resources, and improve water security in so doing. For example, refer to the earlier trade-off discussion surrounding local production, food imports, and FDI in agricultural land. As food and water security issues become more widespread throughout the world, this dilemma will become even more pertinent.

## 12.3   DRIVERS AFFECTING FOOD SECURITY

Food and water are two of three pillars, the other being energy, that form the food-water-energy nexus, which most commonly describes relationships between food security and desalination. Vulnerabilities in water supplies and growing competition from other sectors have fuelled the insecurities of this paradigm. In order to meet global demand in 2050, the International Renewable Energy Agency (IRENA) have reported that food production will need to increase by 60%, water availability by 55% and energy generation by 80% (IRENA, 2015). This sizeable increase in food production would also need to occur in regions already suffering from water scarcity, where competition for water resources will come from manufacturing, electricity production and drinking water. Figure 12.5 illustrates the nexus dynamics for regions utilizing desalination technology.

The stresses placed on resources due to increased demand by a population are not new. Ehrlich and Holdren (1971) designed an equation to illustrate this relationship known as the "*IPAT* equation":

$$I = PAT \qquad\qquad (12.1)$$

where:
  $I$: Total anthropogenic environmental impact
  $P$: Population
  $A$: Affluence: number of products/services consumed per person
  $T$: Environmental impact per unit product/service consumed

This is a useful framework to gauge the sustainability of current practices. When considering an issue, it can be seen that "impact" will depend on the environmental efficiency ($T$) of the chosen technology. The availability of a technology is not the only driving force in the impact, as the benefit of having an ultra-efficient rather than a grossly under-efficient technology is moot unless it can be consumed/used. This is the importance of "affluence" in the equation. "Population" multiplies the impact level of a technology on the environment. To give an example of how the *IPAT* equation can be used, we shall apply it to desalination.

Currently, the dominant techniques in desalination are multi-stage flash (MSF) distillation or reverse osmosis (RO). MSF plants are usually coupled with power plants to cogenerate

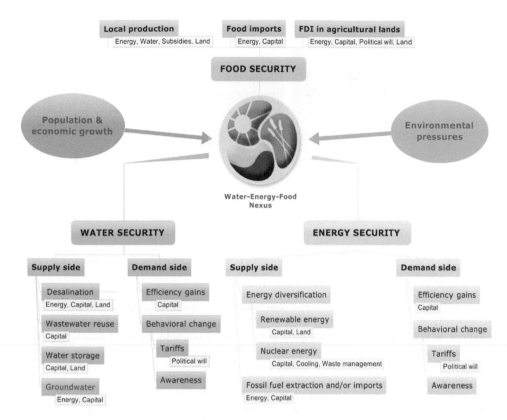

Figure 12.5.   Food-water-energy nexus dynamics in regions utilizing desalination technology (Saif *et al.*, 2014).

freshwater and electricity, with waste heat from power production being utilized by the desalination plant. RO requires less energy per cubic meter of freshwater produced than MSF, but requires electricity within its process, rather than utilizing a power generation byproduct as in MSF.

As the efficiency of these desalination technologies has improved over time ($T$ is lowered), their environmental impact has reduced ($I$ is lowered). However, the regions demanding the current best practice in cogeneration have grown in number (climate change, scarcity issues, unsustainable practices, etc.) and population ($P$ is raised), and almost certainly in affluence as well ($A$ is raised). Therefore, a neat solution is to limit the environmental impact at the start, and approach neutrality. The largest proportion of environmental impacts from desalination plants occur during plant operation, for example, the discharge of concentrate and chemicals to the marine environment, and air pollution from fossil fuel consumption. This is why several projects are pioneering the use of renewable energy, brine management strategies, and closed-loop designs to achieve the end goal of a reduced environmental impact. Figure 12.6 shows the environmental impact of three of the most common desalination technologies over their respective lifecycles. A reduced environmental impact has relevance to the pillars of sustainability. The economic and social constituents will benefit from a more efficient use of resources too. Minimizing waste and moving toward renewable energy options provide avenues for improving the production cost of desalination. Similarly, the less waste entering our ecosystems the better, so that society can continue to function without propagating detrimental consequences for our existence.

Technology was at the core of the desalination impact example above. However, the population factor is important for revealing the sources of change. This is denoted by terms such as socio-technical change, wherein the dichotomy between society and technology should ideally

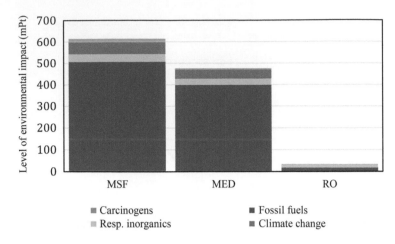

Figure 12.6.    Lifecycle analysis of various desalination technologies (Raluy *et al.*, 2006; Saif *et al.*, 2014).

be attuned to both social processes and consequences of using a technology (Mulder, 2006). For example, it is estimated that we are currently consuming one-and-a-half planet's worth of resources, with the figure expected to grow to two planets' worth by 2030 (GFN, 2015). Another stark reality, which can be viewed through the "affluence" factor in the *IPAT* equation, is the unequal distribution, or rather consumption, of resources. The United Nations Development Program reports that 20% of the world's population is consuming 80% of its resources; this leaves 20% for the remaining 80% of people on the planet (UNDP, 2003).

Cassils (1996) conveys an interesting argument for the "profound shift of paradigms" that needs to occur. This shift would be from the currently accepted zeitgeist that economic growth brings prosperity and the opportunity for civilization to govern and provide human rights, to the reality that we exist on a finite planet and that a lower population of about two billion would allow us a better chance of sustainability; that our human rights are actually dependent on the Earth and its survival (Cassils, 1996). This is a logical and well-structured interpretation of what impact we actually have on our ecosystem, and mirrors the sentiments of other measures that are discussed in the food-water nexus, particularly resource control. Either we become more efficient with what we have, or we use more, to the point where it no longer exists. The well-known spiritual writer Thomas Merton more or less encapsulated this dynamic in 1980 when he said, "The biggest human temptation is to settle for too little" (Fisher, 2002).

In the 1960s, 66% of the world's population lived in rural areas, and 60% of employment was in agriculture. Today, 46% of the world's population lives in rural areas, a proportion that is expected to decrease to 34% by 2050 (UN, 2014). The agricultural sector now accounts for just 40% of employment (Molden, 2007); a modal change can be seen from the last century to this one. This urbanization is a "phenomenon" which provides a challenge to food and water security, and is at odds with the fundamental attributes which sustain large urban centers – the availability and access to food and water.

Food security is not an issue that is addressed in a static system. One of the considerable forces affecting the strategies for food security, particularly in arid regions, is desertification. Desertification represents a loss of land and potential for food production. It intertwines with climate change to threaten production of traditional food crops. Predicted increases in frequency and duration of drought events are expected to accelerate the onset of desertification (UNEP, 2006). Figure 12.7 provides an overview of global land degradation. Rises in global temperatures will exacerbate the severity of water shortages in regions already prone, or potentially susceptible, to such environmental sensitivity (Al Kolibi, 2002). Causes of desertification and long-term drought include natural climate variability, the non-equilibrium dynamics of arid ecosystems

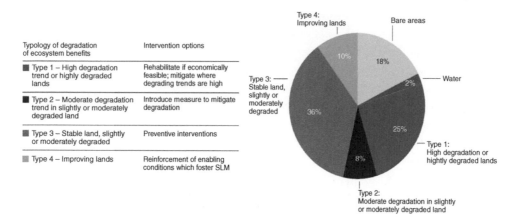

| Typology of degradation of ecosystem benefits | Intervention options |
|---|---|
| ■ Type 1 – High degradation trend or highly degraded lands | Rehabilitate if economically feasible; mitigate where degrading trends are high |
| ■ Type 2 – Moderate degradation trend in slightly or moderately degraded land | Introduce measure to mitigate degradation |
| ■ Type 3 – Stable land, slightly or moderately degraded | Preventive interventions |
| ■ Type 4 – Improving lands | Reinforcement of enabling conditions which foster SLM |

Figure 12.7.    Status and trends in global land degradation (FAO, 2011).

and the exacerbating effects of extractive land uses such as over-cultivation, overgrazing and deforestation (Hermann and Hutchinson, 2006).

When vegetation cover is incorporated into agricultural systems, either as ground cover or an agroforestry layer, it can reduce the impacts of minor to moderate drought on agricultural production (Scheer and McNeely, 2007). This type of land management is arguably the most important consideration when it comes to food security, especially when water stress is an issue. Intelligent mechanisms to bring forth these advantages are key to capturing this value. For example, Polgreen (2007) discusses a case where a change in policy for farmers in Niger led to an increase in tree cover of 7.4 million acres. The policy initiative placed the ownership of the newly planted trees with the farmers. This resulted in reduced wind-based topsoil erosion and led to an increase in the ability of soil to retain water, while also increasing nitrogen fixation within the soil and improving fertility (Polgreen, 2007).

Rumbaitis del Rio (2012) discusses the importance of resilience in our ecosystems, in contrast to specific and anticipatory adaptation measures. This is an important delineation. Rumbaitis del Rio (2012) cites the example of agricultural development, demonstrating that vulnerability to climate change may increase in areas with higher yields by comparison to areas with marginal yields. The primary driver of this greater vulnerability is higher susceptibility to temperature-related yield decreases, which in turn causes soil nutrient depletion (Schlenker and Lobell, 2010). Resilience would create a buffer against such shocks. Building resilience into our ecosystems, and thus benefitting our food and water security, will come from developing appropriate research initiatives and continuing examination and evaluation of present-day ecosystems. For example, arid ecosystems provide a reservoir of the biodiversity needed to buffer against the impacts of drought, temperature fluctuations and salinity.

## 12.4   HUMAN HEALTH – WHY WE NEED FOOD SECURITY

Any examination of food security as an issue should take into account why food is important in the first place. We use it as a source of fuel, and without it we grow weak, leading to compromised health and eventually death. Consuming the wrong types of food leads to the same results. Therefore, when we look to areas where food security is an emergent issue, we see more clearly the effects that this issue of food quality can have. Populations struggle with being underweight, stunted growth, wasting and acute protein energy malnutrition, as well as poor eyesight and anemia, due to the absence of key vitamins and nutrients in their diet (Pierre-Louis *et al.*, 2004). The World Health Organization estimates that over 1.62 billion people suffer from anemia,

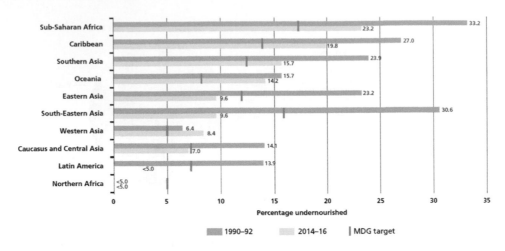

Figure 12.8.    Undernourishment trends (FAO, 2015).

caused in over half by deficiencies in essential minerals and micronutrients (de Benoist *et al.*, 2008). Figure 12.8 shows undernourishment trends in relation to WHO's associated Millennium Development Goal (MDG).

These health problems are often thought to be primarily present in the least developed countries. However, there is also a rise in malnutrition in developing and developed countries, represented by non-communicable diseases such as obesity. For example, in Morocco there has been a shift toward an energy-dense diet, resulting in an increased level of obesity that is inversely associated with education and positively associated with income (Pierre-Louis *et al.*, 2004). Similarly, in Egypt, instances of hypertension and diabetes are on the rise due to the effects of obesity. A primary reason behind this unfortunate trend is the reliance on cereal crops for calorie intake. Historically, cereals have been the main import food for arid regions as they are energy-dense and do not perish as quickly as other foodstuffs. Economic growth is often associated with overconsumption of resources, and food mirrors this trait amongst changing demographics. Diets mainly composed of carbohydrates, with an absence of key nutrients for a balanced diet, are an important issue to combat to ensure the health of societies in these regions.

The Royal Society stated, in a report on the sustainable intensification of agriculture, that the preferred approach to eliminating hidden hunger will always involve strategies to increase access to fruit and vegetables and diversify diets (Royal Society, 2009). In practice, implementation of such strategies is not always as acceptable to society in general as to the solution-finders. Take, for example, the case of Golden Rice. Rice is a staple crop of many developing and least developed nations. The absence of key nutrients from the diets of vulnerable populations, whilst reaching an acceptable dietary calorific intake, leads to instances of hidden hunger. To combat this, Golden Rice was genetically engineered to provide vitamin A in such diets, and so offset the health effects and loss of life attributed to the absence of this micronutrient. However, several regulatory and developmental roadblocks to Golden Rice have ignored the urgency of this health crisis and prolonged the impacts on vulnerable populations which otherwise have no access to this vital dietary component (Potrykus, 2010).

Health problems are a possibility even when the ideal food groups are included in dietary consumption. For example, whey protein has long been considered an excellent source of nutrition, used in consumer food products ranging from baby formula to fitness supplements. In 2013, the New Zealand dairy industry was forced to issue warnings and recall large quantities of exported whey protein when it was discovered that it contained traces of bacteria that can cause botulism (Tajitsu, 2013). This echoed previous recalls in 2008 when contaminated dairy products in China were designated as the cause behind the death of six children, rendering hundreds more seriously ill. This was a tragic event, and further underscores the importance of not only food availability, but also food safety.

The raising of awareness and education are important parts of ongoing efforts to ensure population health and vitality. Keatinge *et al.* (2010) and Lutaladio *et al.* (2010) report strategies that are of particular importance to developing and the least developed countries, as these strategies aim to combat the issue of malnutrition and health effects from poor diets. The first goal is to promote a general increase in fruit and vegetable consumption, complemented by the second goal of intelligent diet diversification.

Viewing the issue of food security from its roots is useful in avoiding issues before they arise. For example, when deciding how higher quantities of macronutrients and protein could be introduced in the populace's diet (FAO, 2010), several scenarios might be pursued. One such scenario could be to assign land for the specific purpose of growing this food, which could require a degree of land clearance. However, some case studies have shown that land clearance activity indirectly caused health problems amongst the population. Cases of yellow fever in African countries have been linked to deforestation (Kawachi and Wamala, 2006). Land clearance of wild areas is more of a grey area. During times of food scarcity, additional food sources may be found in these wild areas. Unmanaged extraction of these food sources will result in loss of ecological function, as would land clearance measures. However, the spread of zoonitic diseases, such as those transmitted by bushmeat consumption, increases with the onset of wild food-source extraction (Rumbaitis del Rio, 2012). New habitats and greater food availability can attract disease-carrying insects and animals, with closer proximity to humans than existed prior to land clearance.

By examining the issue of food security in terms of its effects and relationship dynamics, the solution matrix becomes much more intricate. These health trends represent significant problems, to be tackled at national levels. Designing and implementing strategies to combat them requires a concerted effort in a highly complex system. It is therefore important to consider the "human" element of any such solution in a food security strategy, in order to avoid the negative health, societal and economic consequences that can arise. Diets are often linked to culture and represent personal and societal preferences. Implementing changes to improve population health via dietary elements has been shown to be more effective at a local level than a national one (Institute of Medicine, 2014).

## 12.5 SOLUTIONS – WORKING TOWARDS FOOD SECURITY

### 12.5.1 *Biodiversity approaches*

Now that the constructs have been defined for what is meant by food and water security, as well as what drives them, and ultimately the issues they pose, we will begin to review available solutions and their practicalities. While it is not the intention of this chapter to focus on specific and detailed agronomy practices, it is important to provide an overview of the characteristics that are present in biodiversity strategies to alleviate food and water security concerns.

The first solution topic to be examined is biodiversity. Essentially, the solution here culminates in risk management. By spreading reliance over numerous food sources, we are less likely to be adversely affected by single-source supply disruptions. As an example, the Irish Potato Famine happened due to blight, over-reliance on monoculture, and climate change. Crop and dietary diversity have been linked with enhanced food security during periods of stress, including natural disasters (Kahane *et al.*, 2013).

Studies have shown that, globally, we rely on only 82 crop species to provide 90% of our dietary energy consumption, which has been described as unwise due to the ramifications of a potential sudden disturbance to this norm (Prescott-Allen and Prescott-Allen, 1990). Increased mechanization and market demands lead farmers to concentrate on fewer and fewer crops. The result is a steady loss of biodiversity (Smale *et al.*, 2009). Furthermore, diversifying agricultural ecosystems will improve their adaptability to extreme climatic conditions, resilience to biotic and abiotic stresses, and produce harvestable yields where major crops may fail (Padulosi *et al.*, 2002).

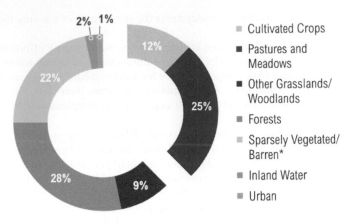

* Permanent ice cover, desert, etc.

Figure 12.9.    Distribution of earth's land mass by use, with "Cultivated Crops" and "Pastures and Meadows" occupying 37% of the total (excludes Antarctica) (FAO, 2011).

The World Bank estimates that 37.7% of total land area is used for agriculture (defined as arable, under permanent crops and under permanent pasture), which is illustrated in Figure 12.9. With demand for volume set to increase due to population growth, and demand for variety expected to change due to socioeconomic changes in developing nations, the finite resource of land is the base point for this change. Accessing more land, through forest clearance for example, has already been shown to be a health concern, and is also regarded as a risk to current ecosystems and biodiversity (Rockstrom et al., 2009). Therefore, it is through enhanced, well-managed productivity that food security will be achieved, while mitigating the burden on water resources.

Kahane et al. (2013) argue that diversification with highly valuable crops and species is an essential element for sustainability amongst smallholder agriculture (Kahane et al., 2013). Development efforts represent an important consideration for ongoing work in developing countries to ensure food and water security. As of 2011, there were just over one billion people living beneath the poverty line (UN, 2011), for whom agriculture is either the primary or a tributary source of sustenance/income. Similarly, appreciation of local realities can also assist in harnessing synergies that may otherwise go untapped. For example, the UN Standing Committee for Nutrition highlighted the value of local food production in food security efforts (UNSCN, 2010). In African countries, where local realities such as topography can negate widespread cultivation of crops, appreciation of current food sources as a basis for developing food security has been successful.

Internationally, awareness is growing about new food products, partially due to globalization. The demand generated by this awareness is something which can be capitalized on by biodiversity strategies to facilitate investment and policy. The key is therefore for consumers to become more interested in the plants, land, supply chains and farming communities that produce and deliver what they eat (Kahane et al., 2013).

### 12.5.2    Policy mechanisms

Spiess (2008) discusses the importance of a harmonized approach to designing and implementing environmental and sustainable development policy initiatives. Creating mechanisms by which the value of resources is appreciated instead of undermined is key. In other words, administrative and institutional failures should be avoided, as they will lead to wastage and lack of conservation of the resource (Spiess, 2008). Monitoring and tracking protocols are vital components of these successful mechanisms.

We have already shown how changes in diet to address food security issues can encounter cultural resistance, and the same is true for water policy. For example, water allocation and pricing policies in agriculture are often designed to support local tribal or traditional economic activities or special interest groups (Elhadj, 2006). This is partially due to the risk that can be borne by these stakeholders in the higher pursuit of what is seen as the best path forward, and is why education and training initiatives should be aligned with such moves.

The land, labor and capital embodied in agricultural imports and exports must also be considered in countries where one or more of those resources are limited, or where reducing unemployment is an important policy goal (Wichelns, 2001). In GCC countries, employment generation is not an objective of agricultural policy. New policy measures in Saudi Arabia mean that by 2016 the government will no longer buy wheat from national producers. This policy is part of an effort to conserve non-renewable water resources, hoping that the non-support of wheat will provide the impetus for cultivation of less water-intensive crops (Shepherd, 2014).

Baumgärtner and Quaas (2010) looked into the effects of agro-environmental policies and insurances in situations where risk-averse farmers invest in on-farm biodiversity and found that insurances can be detrimental to efforts to increase biodiversity (UNSCN, 2010). In other words, the benefits of diversity were motivation enough to pursue change. Kahane *et al.* (2013) write of the damage that unfit national economic policies can cause, where negative externalities like damage to the environment and loss of diversity for future use go unaccounted for. They argue that these losses should have an artificial price attributed to them (Kahane *et al.*, 2013).

Restrictive tariff and non-tariff barriers such as the European Union 'Novel Food Regulation', restrict the access of 'new' foods into member states (Hermann, 2009). This regulation results in high premiums being paid for niche products by consumers in order to cover the technically complex, time-consuming and financially burdensome nature of certification procedures (Hermann, 2009).

In contrast to some of the drivers of food and water security issues, namely climate change, the effects and repercussions of these problems are more universally accepted and understood. Agriculture is responsible for 14% of total global greenhouse gas emissions (Wright, 2010). As a result, its problems are subject to less ambivalence when they and their solutions are debated. This is important, as the greater the level of consensus around such serious issues, the more likely action will be taken in terms of prevention, adoption and mitigation. By contrast, the notion of climate change was so strongly contested that those charged with taking action did so without being in agreement on whether the problem existed or not, not to mention what action should be taken (Weart, 2003). This called for practice of the "precautionary principle", which also has value in the continuing work on food and water security issues (Mulder, 2006). Suffice to say, lack of consensus should not be used as an excuse for inaction, as this will lead to paralysis within the decision-making community. The analysis and modeling efforts of the research community should focus on bridging this inaction time-gap to the best of their ability. In such ways, scrutinized elements can be defined as either potential strengths, weaknesses, opportunities or threats.

### 12.5.3 *Leadership, coordination and strategy*

As with all decision-making activities, a certain level of information is required or sought to inform the best outcome from any imposed action. With this in mind, the relationship between food security and desalination is perhaps best aligned as part of the effort made in modeling scenarios within the food-water-energy nexus. Decision makers are aiming to navigate the inherently complex dynamics between these different issues.

With an array of starting points for such models, it can often be quite difficult to decide which problem(s) require most attention when optimizing a solution. Considering this, Bazilian *et al.* (2011) suggest that inequalities arising from lack of access, caused by economic and security-related issues, might take precedence over environmental concerns (Bazilian *et al.*, 2011). The political realities of adopting such an approach surround decision-making, particularly when there are several parties involved with disparate priorities. Coupling this consensus-building effort with an institutional capacity to design and deliver effective solutions is an important step. As has been

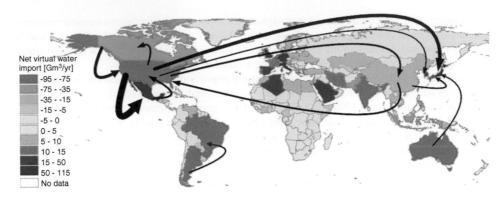

Figure 12.10.    Virtual water balance per country and direction of gross virtual water flows related to trade in agricultural and industrial products over the period 1996–2005. Only the biggest gross flows ($>15\,\mathrm{Gm^3}$ year$^{-1}$) are shown; the fatter the arrow, the bigger the virtual water flow (Mekonnen and Hoekstra, 2011).

mentioned previously, the dynamics of the relationships involved in these systems requires a cross-cutting approach and an avoidance of "silo thinking" (Bazilian *et al.*, 2011).

Synergy must exist amongst the research community in pursuing unified approaches to solving local and national problems in food security (Kahane *et al.*, 2013). There is little research into the economic benefit of enhanced agro-biodiversity (Wojtkowski, 2008), despite the documented advantages that such strategies can bring in terms of enhanced food security.

In regions suffering from these problems, advantage should be taken of existing network components to instigate greater levels of food and water security. The absence of a few key elements, such as a strategic body or dedicated research facilities, can be added as critical needs to an agenda for addressing local problems with local people. This, in turn, helps to build the human capital and capacity of regions to effectively solve their own issues, further contributing to food security.

### 12.5.4    *Virtual water*

In the same way that lifecycle analyses provide an insight into how much energy and carbon is utilized by a product (or service), the concept of virtual water provides a means to account for the various stresses placed on water resources. Virtual water represents the embedded water content of an agricultural product. It has served as a useful tool to base decisions on, to gain insight as to how water-intensive agricultural practices and products are, and at what point it may be better to import foodstuffs rather than consume domestic water resources in their local production.

The concept of virtual water is founded on the reality that regions throughout the world possess unique attributes and resources, of which water is one, that makes some of them better suited for producing agricultural products (Hoekstra, 2010). What this translates into is an advantage, which can be leveraged by the "producer" through trade with a "buyer". The buyer has access to product from the producer that would otherwise not be available in their region. Locally producing the same may even have led to compromising local sustainability. In this way, the virtual water trade generates water savings for the buyer, albeit as a physical saving and not necessarily a monetary one. The producer bears the opportunity cost of consuming water for this trade purpose, in the pursuit of an economic and/or political gain, though seldom an environmental one (Qadir *et al.*, 2003). Figure 12.10 shows the extent of the virtual water trade at the turn of the century.

To illustrate the concept further, and give an idea of just how much water can be virtually imported, we can examine the case of wheat. For every kilogram of wheat imported, the buyer will also get about 1 m$^3$ of virtual water at much less cost than the price or value of the same quantity of water from local water resources, which may not even be available in the country itself (Allan, 1999). This means that each metric ton of wheat imported has about 1000 m$^3$ of virtual water embedded in it. It is therefore easier and less ecologically destructive to import grain than to produce the same commodity locally (Turton, 2000).

Virtual water is not just a means of accounting for the embedded water content of agricultural products, but also demonstrates where the majority of virtual water flows exist. Mekonnen and Hoekstra (2011) estimate that crops and crop-derived products account for 76% of virtual water flows, with animal and industrial products representing 12% each. Allan (1996) discusses how virtual water is also present in services such as electricity generated from hydroelectric schemes.

When addressing food security by leveraging the concept of virtual water, a certain level of convenience is extolled. This convenience exists to the degree that food needs are met in the food-insecure region, but the import-dependent country is only food-secure as long as this trade relationship remains viable. Self-sufficiency is not always advisable, or indeed possible, and is dependent on a country's resources. The goal, therefore, is to aim for the optimum solution that accounts for these dynamics, that is, increasing imports when prices are low and using local water resources when prices are high (Qadir *et al.*, 2003).

Debate surrounds certain elements of the virtual water concept. For instance, the attribution of equal value to initial sources, whether they be rain or groundwater, does not take into account long-term sustainability. Also, virtual water does not account for environmental effects of usage practices in "producer" countries where, for example, improper resource use may lead to increased instances of soil salinization and loss of fertility. There may also be a transferral of the same problems that instigated the need for food imports in the buyer region. Leveraging the trade advantage may be such that the water extraction rates begin to exceed the renewal rates, and may propagate the decline of both the buyer and the producer (Frontier Economics, 2008).

### 12.5.5 *Trade*

As can be seen from examination of the virtual water concept, trade offers an avenue to moderate food and water security issues. Another way in which trade is used as a means to reduce these problems is through land leases, dubbed by some as "land-grabbing". The basic premise of a land-grab is that a buyer/investor will acquire the usage rights of agricultural land in a foreign country in order to produce food that will be imported back to the investor country. Deals are often brokered at the intergovernmental level, and may also be done through the function of a specific corporate entity, with key governmental interactions during negotiations. An important dynamic within this potential solution to food security is the fact that the producer nations are often the least developed or developing countries. This can distort the true value of the producer's land, as access to financial resources from foreign parties may be of more immediate benefit than developing and/or preserving their own land rights. Table 12.2 summarizes GCC countries' investments in foreign agricultural lands.

Shepherd (2014) documents situations in which food insecurity is transferred from the buyer country to the producer country. Governments in some producer countries pursue eviction orders against their own people for the purpose of securing agricultural land to sell to foreign buyers (Shepherd, 2014). In some cases, prior to negotiations between producer and buyer countries, farmers are pre-emptively evicted so that the producer government can claim lands granted to buyers are "unused". Such inequality is due in part to the absence of clear regulation surrounding foreign agricultural land-grabs. This not only damages communities in producer countries, but also serves to destabilize the investment environment for buyers, and thus destabilizes their food security. Shepherd (2014) suggests that reform measures could focus on investment in the host countries' farmers instead of in their farmlands. Targeted financial and technical investment would increase the capability of farmers to boost the productivity from their land, allowing for export of surplus food to the buyer/investor countries, thus transitioning the solution from a land-centric to a productivity-centric approach.

The impact that land-grabs have on communities in producer countries should not be underestimated. Davis *et al.* (2014) discuss the fact that rural households in agriculture-based economies are limited in their opportunities for non-farm employment, and that the income lost from

Table 12.2.    Investments by GCC countries in foreign agricultural land (Agrimoney, 2013; Dorondel, 2012; GRAIN, 2012; Land Coalition, 2015; Malek, 2012; National Aquaculture Group, 2013; Nepomuceno, 2012; Saif *et al.*, 2014).

| Investor country (both private and public entities) | Producer country receiving the investment | Land acquired through purchase or lease [ha] |
|---|---|---|
| Bahrain | Philippines | 10000 |
| | India | 4000 |
| Kuwait | Sudan | 62136 |
| | Philippines | 500 |
| Qatar | Australia | 750000 |
| | Argentina | 113460 |
| | Sudan | 106382 |
| | Kenya | 40000 |
| | Brazil | 33790 |
| | Angola | 25000 |
| Saudi Arabia | Morocco | 700000 |
| | Ethiopia | 561248 |
| | Tanzania | 500000 |
| | Argentina | 223000 |
| | Sudan | 177239 |
| | South Sudan | 105000 |
| | Mali | 100000 |
| | Egypt | 73000 |
| | Mauritania | 52000 |
| | Philippines | 50000 |
| | Ukraine | 33000 |
| | Senegal | 20000 |
| | Turkey | 20000 |
| | Zambia | 5000 |
| | Poland | 2700 |
| | Nigeria | 1000 |
| United Arab Emirates | Sudan | 1643100 |
| | Morocco | 700000 |
| | Pakistan | 334100 |
| | Egypt | 100837 |
| | Romania | 50000 |
| | Algeria | 31000 |
| | Serbia | 20000 |
| | Ghana | 10000 |
| | Ethiopia | 5600 |
| | Spain | 5050 |
| | South Africa | 4046 |
| | Namibia | 220 |

land-grabs represents a reduced ability of the area to support a given number of people. It is estimated that over 12 million people have lost their main source of income due to the practice of land-grabs (Davis *et al.*, 2014). Tacoli (2011) further describes the issues of fragile regions, and how instances of land-grabs can lead to urbanization and rural community migration. Land acquisitions are not only sought for the purposes of obtaining agricultural production. The fuel versus food debate, characterized by biofuels, also emerges in the context of land-grabs, as land is primarily sought for the purpose of cultivating energy feedstocks and curbing $CO_2$ emissions rather than producing food (EU, 2009). Similarly, forested areas are purchased by industrial-ized countries using the Kyoto Protocol's Clean Development Mechanisms for the purpose of climate change mitigation, while also operating as a resource for new revenue stream generation (Fairhead *et al.*, 2012). In contrast to this, the least developed countries are expected to experience a disproportionate amount of the consequences that arise from climate change (Parry *et al.*, 2004).

This brief introduction to land leases may appear to portray this potential food security solution in a negative light. However, research has clearly demonstrated these inequalities in the early years of this practice. Nevertheless, the need for further evidence and quantifiable results is needed to even out and manage this solution such that it can benefit both buyer *and* producer countries. McCarthy (2010) shows that the options presented to farmers and the ways in which they choose to interact with commercial agriculture ultimately help to determine whether the change is positive or negative. While this remains an important consideration, marginalization of rural communities should not be excused for now because of the absence of consensus surrounding the available data. Development of appropriate regulations is an immediate priority, as the opportunity for wide-scale adoption of any proposed standard is likely to diminish with time. As it stands, countries that have unstable legal systems, issues in enforcement, and land tenure systems, and who are attracted to large and sudden investment, are vulnerable to buyer abuse (Anseeuw *et al.*, 2013).

Having discussed biodiversity and its relation to food security, it is important to note a relationship that applies to the principle of trade. The biodiversity paradox states that as new species of flora and fauna are introduced to an ecosystem, there is an increase in local biodiversity, but a global decrease. Thus, mixing of species makes the world locally more varied, but globally more uniform (Mulder, 2006). The same is true for globalization, by which the variety of products has increased locally, while global variety has decreased.

There are several drivers that can exacerbate food security issues, as has been described above. The distribution of natural resources can lead to regional disparities in supply and demand. For instance, China and India are home to 35% of the world's population, but both have exploited most of the land and water within their borders, rendering the prospect of supplying all food domestically unlikely (Bazilian *et al.*, 2011). Even countries such as South Africa, that were once net food exporters, are starting to become import-dependent due to productivity losses and an increasing population. When the potential for South American and sub-Saharan African countries to use untapped resources and increase food production is considered without considering gains in local productivity, the idea of trading our way to food security becomes more attractive.

Education about correct dietary behavior will play a part in tackling the issue of food security. However, this measure also highlights the impacts that imported food can have on population health when a newly imported food is not beneficial to the health of the local population. For example, the widespread availability of soft drinks such as cola has increased the spread of food diversity, but has not contributed essential nutrients to local diets.

Although trade may seem like an attractive proposition to align food and water security initiatives, it is not a perfect solution. A systems-based approach should be part of the solution process, and "silo thinking" should be avoided. This is pertinent to the design of a trade policy which aims to alleviate food and water security concerns, as the factors involved in the decision-making process must include consumption patterns, market complexities, policy priorities and wealth endowment (IRENA, 2015).

The underlying importance of the discussion surrounding trade policy, virtual water, biodiversity and so on, is that sustainability must not be compromised, and should remain as the end goal. Failure to reach this or to act sustainably will ultimately result in the collapse of systems. In this context, the system is society, or a regional embodiment of same. Diamond (2005) discusses the choices made by societies that determine their failure or success, and provides a framework, culminating in five factors which guide the outcome:

- Human impacts on the environment.
- Climate change.
- Friendly neighbors leaving.
- Advent of hostile society.
- Dysfunctional political, economic, social and cultural practices.

By using this framework, a "solution", such as a trade policy to remedy food and water security, can be more readily assessed for how it may affect a country or region. For example, poor water management may lead to reduced water reserves, affecting crop productivity. Couple this with a potentially reduced market demand and the effects of climate change on crop productivity, and a warning to our current systems emerges from the failed societies of the past.

### 12.5.6   *Efficiency*

The overarching fact that will always pervade agriculture is that it needs water. Hopefully, innovations in technology and policy, coupled with behavioral changes, will lead to greater efficiency in how we use water to produce food. The question to ask then is how we measure efficiency. When we consider the water used in producing food, its use depends on numerous factors depending on the type of food, that is, vegetables, dairy, meat, etc. A lifecycle analysis can account for the water used within specified boundaries, but as plants represent the largest constituent of most people's diets, it is best to perhaps look to them for an example. The amount of water transpired by the plant is the first parameter that must be accounted for.

Efficiencies can be improved in many parts of the food production cycle. Brown (2006) showed that 30–40% of food is lost to waste in developing and developed countries. An interesting point of note regarding where these losses occur is that in developing countries, the majority are borne on-farm, during transportation, and in processing. In developed countries, most losses occur after the retail stage. Cuellar and Weber (2010) estimated that the energy involved in the wastage from farm-to-plate in the USA accounted for 2% of total annual energy consumption. This is a significant waste of resources, and should warrant a serious level of concern when addressing food security issues. There are many natural and uncontrollable causes that lead to food losses, such as outbreaks of pestilence and disease, but the level of human involvement, and control, in the wastage described is worrisome. As with all problems, the very act of identifying them is the beginning of a solution. Harnessing the benefits of leaner production cycles in developing countries, coupled with a certain level of citizenship in "owning" the problem in developed countries, will have to be part of the solution hierarchy if the end goal is to progress to food security.

Allan (1997) cites five factors which have affected, and may continue to affect, the availability of good-quality irrigation water. These consist of an inherited shortage of water, increased cropping intensities, cultivation of crops on new land, increased competition for water from other sectors and, finally, contamination of water resources. With this in mind, one of the key reasons for using greenhouse technology is the provision of a controlled environment. This allows for the possibility of higher levels of resource efficiency and a greater level of control in the cultivation process, which in turn leads to reduced risk compared to alternative agronomy practices. The following example provides an example of the opportunities for increased efficiency. A common method of cooling greenhouses in arid environments, such as those in GCC countries, is evaporative cooling. This method is well-suited to arid regions (MAF and ICARDA, 2011; Sabeh *et al.*, 2011), but the gain in water efficiency in a greenhouse can be lost due to the presence of an open-cycle evaporative cooler, as these use 67% of the water consumed in greenhouses that require cooling in arid environments (Al-Mulla, 2006). This essentially puts the efficient greenhouse on a resource efficiency level that is comparable with open field cultivation. Therefore, with careful design, the potential risk mitigation can be developed in the context of the economic, social and environmental pillars of sustainability. Greenhouses are already a recognized solution for food production, with some 5.4 million hectares of greenhouse area installed globally, currently producing 60% of vegetables consumed (Tester, 2015). While irrigation is not a new technology, its controlled use can be an effective means of ensuring food productivity. Currently, 17% of global arable land is irrigated, from which 40% of total food calories are derived (Strategy&, 2014).

Practicing the merits of industrial ecology in achieving more efficient closed-loop systems is important. This is most readily represented by "waste" water, treated municipally or otherwise

Figure 12.11. Greenhouse technology provides a means to achieve greater efficiency and control of resources in food production. Pictured is a hydroponic system in use for a tomato crop (Agra Tech, 2015).

ignored, that can go on to serve additional purposes within broader and better-connected systems. For example, in Kuwait, the cost of treated effluent is estimated at (US) $0.66 per m$^3$ compared with (US) $2.30 per m$^3$ for desalinated water (Strategy&, 2014). A growing awareness of the importance of using resources efficiently will further increase the likelihood that such avenues are utilized and adopted as norms.

## 12.6 CONCLUSION AND RECOMMENDATIONS

Food security is a diverse issue with many different drivers contributing or detracting from its realization. The purpose of this chapter was to provide the reader with an introduction to food security, and how it is related to desalination. This is of particular importance in the context of concerns about sustainability and the food-water-energy nexus, as regions that operate desalination systems often face food security concerns too. The question to ask oneself now is whether or not desalination can contribute to the achievement of food security in arid regions. Having dealt with the predominant solution themes in this area, the scope of answers should have informed opinion on trade, policy, strategy, efficiency, virtual water etc.

In conclusion, the authors would like to offer their own recommendations to ensure the security of food and water for generations to come:

- Improve self-sufficiency.
- Boost crop productivity.
- Improve water productivity of crops.
- Increase use of treated wastewater.

- Reduce post-harvest losses.
- Develop regional co-operation and relationships.
- Practice sustainable development of livestock and fishery holdings.
- Improve education on food and water security initiatives.
- Emphasize a resilience-based approach.
- Expand competency and best practice in sustainable desalination technologies.

## REFERENCES

AFED (2014) Arab Environment 7: Food Security – Challenges and Prospects. Arab Forum for Environment and Development, Beirut, Lebanon.

Agra Tech (2015) Hydroponic system in use with a tomato crop [photo]. Available from: http://www.commercial-hydroponic-farming.com/wp-content/uploads/2014/04/tomato-greenhouse-nft-production.jpg [accessed November 2015].

Agrimoney (2013) Saudi extends foreign land spree with CFG takeover. Available from: http://www.agrimoney.com/news/saudi-extends-foreign-land-spree-with-cfg-takeover–5673.html [accessed November 2016].

Al Kolibi, F.M. (2002) Possible effects of global warming on agriculture and water resources in Saudi Arabia – impacts & responses. *Climatic Change*, 54, 225–245.

Allan, J.A. (1996) Policy responses to the closure of water resources: regional and global issues. In: Howsam, P. & Carter, R.C. (eds) *Water Policy: Allocation and Management in Practice*. E. & F.N. Spon, London. pp. 3–13.

Allan, J.A. (1997) Virtual water: a long term solution for water short Middle Eastern economies? Paper presented at the *British Association Festival of Sciences, Water and Development Session, 9 September 1997, University of Leeds, UK*.

Allan, J.A. (1999) Water in international systems: a risk society analysis of regional problem sheds & global hydrologies. *Conference on Water Resources and Risk, Department of Geography, Oxford University, March 1999*. Occasional Paper Number 22, SOAS Water Issues Group.

Al-Mulla, Y.A. (2006) Cooling greenhouses in the Arabian Peninsula. *International Symposium on Greenhouse Cooling, 24–27 April 2006, Almeria, Spain*. pp. 499–506.

Alpen Capital (2015) GCC food industry 2015. Available from: http://futuredirections.org.au/wp-content/uploads/2015/07/GCC_Food_Industry_Report_April_20 15.pdf [accessed November 2016].

Alsharhan, A.S., Rizk, Z.A., Nairn, A.E.M., Bakhit, D.W. & Alhajari, S.A. (2001) *Hydrogeology of an Arid Region: the Arabian Gulf and Adjoining Areas*. Elsevier Science, Amsterdam.

Anseeuw, W., Lay, J., Messerli, P., Giger, M. & Taylor, M. (2013) Creating a public tool to assess and promote transparency in global land deals: the experience of the land matrix. *Journal of Peasant Studies*, 40, 521–530.

Awerbuch, L. & Walker, A. (2014) A Webinar on Water Sustainability in Arid Regions. 12 December 2014, Abu Dhabi, UAE. International Desalination Association (IDA).

Barghouti, S. (2010) Sustainable Management of a Scarce Resource: Water Sector Overview. Arab Forum for Environment and Development (AFED), Beirut, Lebanon.

Baumgärtner, S. & Quaas, M.F. (2010) Managing increasing environmental risks through agrobiodiversity and agrienvironmental policies. *Agricultural Economics*, 41(5), 483–496.

Bazilian, M., Rogner, H., Howells, M., Hermann, S., Arent, D., Gielen, D. & Yumkella, K.K. (2011) Considering the energy, water and food nexus: towards an integrated modeling approach. *Energy Policy*, 39(12), 7896–7906.

Bazza, M. (2005) Policies for water management and food security under water-scarcity conditions: the case of GCC countries. *7th Gulf Water Conference, 19–23 Nov 2005, Kuwait, UAE*.

Brown, L. (2006) Plan B 2.0: Rescuing a planet under stress and a civilization in trouble. Earth Policy Institute, Washington, DC.

Cassils, J.A. (1996) Overpopulation, sustainable development, and security: developing an integrated Strategy. *Population and Environment*, 25(3), 171–194.

Cuellar, A.D. & Weber, M.E. (2010) Wasted food, wasted energy: the embedded energy in food waste in the United States. *Environmental Science and Technology*, 44(16), 6464–6469.

Davis, K.F., D'Odorico, P. & Rulli, M.C. (2014) Land grabbing: a preliminary quantification of economic impacts on rural livelihoods. *Population and Environment*, 36(2), 180–192.

de Benoist, B., McLean, E., Egli, I. & Cogswell, M. (eds) (2008) Worldwide prevalence of anemia 1993–2005: WHO Global Database on Anaemia. World Health Organization, Geneva, Switzerland.

De Bon, H., Parrot, L. & Moustier, P. (2009) Sustainable urban agriculture in developing countries: a review. *Agronomy for Sustainable Agriculture*, 6, 619–633.

Diamond, J.M. (2005) *Collapse: How Societies Choose to Fail or Succeed.* Viking, New York.

Dorondel, S. (2012) Romanian agriculture between subsistence, large associations and land grabbing. *European Agricultural Policy Conference, 16 October 2012, North Rhine-Westphalia Representative Office to the European Union, Brussels.* Available from: http://www.demeter.de/sites/default/files/Presentation%20S%20Dorondel.pdf [accessed November 2016].

Ehrlich, P. & Holdren, J. (1971) Impact of population growth: complacency concerning this component of man's predicament is unjustified and counterproductive. *Science*, 171, 211–217.

Elhadj, E. (2006) *Experiments in Achieving Water and Food Self-Sufficiency in the Middle East: The Consequences of Contrasting Endowments, Ideologies, and Investment Policies in Saudi Arabia and Syria.* Dissertation.Com, Boca Raton, FL.

EU (2009) Directive 2009/28/EC of the European Parliament and of the Council of 23 April 2009 on the promotion of the use of energy from renewable sources. *Official Journal of the European Union* (L 140), 52, 16–62.

Fairhead, J., Leach, M. & Scoones, I. (2012) Green grabbing: a new appropriation of nature? *Journal of Peasant Studies*, 39, 237–261.

FAO (2003) Trade reforms and food security, conceptualizing the linkages. In: Food security: concept and measurement. Economic and Social Development Department, Food and Agriculture Organization of the United Nations, Rome, Italy.

FAO (2010) The state of the food insecurity in the world 2010: addressing food insecurity in protracted crises. Food and Agriculture Organization of the United Nations, Rome, Italy.

FAO (2011) Distribution of earth's land mass by use. Food and Agriculture Organization of the United Nations, Rome, Italy.

FAO (2015) Undernourishment trends: progress made in almost all regions, but at very different rates. Food and Agriculture Organization of the United Nations, Rome, Italy.

Fisher, J. (2002) *Crowded Greenhouse: Population, Climate Change, and Creating a Sustainable World.* Yale University Press, New Haven, CT.

Frontier Economics (2008) The concept of 'Virtual Water – a critical review. A report prepared for the Victorian Department of Primary Industries, Australia. Available from: http://agriculture.vic.gov.au/__data/assets/pdf_file/0006/233529/Virtual-Water-The-Concept-of-Virtual-Water.pdf [accessed November 2016].

GaStat (2015) Saudi Arabia GDP from agriculture forecast. General Authority for Statistics, Riyadh.

GFN (2015) World footprint; do we fit on the planet? Global Footprint Network. Available from: http://www.footprintnetwork.org/en/index.php/GFN/page/world_footprint/ [accessed November 2016].

Ghaffour, N., Missimer, T.M. & Amy, G.L. (2013) Technical review and evaluation of the economics of water desalination: current and future challenges for better water supply sustainability. *Desalination*, 309, 197–207.

GRAIN (2012) GRAIN releases data set with over 400 global land grabs. Available from: https://www.grain.org/article/entries/4479-grain-releases-data-set-with-over-400-global-land-grabs [accessed November 2016].

GWI (2016) Cost Estimator. Global Water Intelligence, Oxford, UK.

Hermann, M. (2009) The impact of the European novel food regulation on trade and food innovation based on traditional plant foods from developing countries. *Food Policy*, 34, 499–507.

Hermann, S.M. & Hutchinson, C.F. (2006) The scientific basis: links between land degradation, drought and desertification. In: Johnson, P.M, Mayrand, K. & Paquin, M. (eds) *Governing Global Desertification: Linking Environmental Degradation, Poverty and Participation.* Ashgate, Aldershot, UK. pp. 11–26.

Hinrichsen, D. (1998) Feeding a future world. *People and the Planet*, 7, 6–9.

Hoekstra, A.Y. (2010) The relation between international trade and freshwater scarcity. Staff Working Paper ERSD-2010-05, Economic Research and Statistics Division, World Trade Organization, Geneva.

Institute of Medicine (2014) *Including Health in Global Frameworks for Development, Wealth, and Climate Change: Workshop Summary.* Board on Population Health and Public Health Practice, and Roundtable on Environmental Health Sciences, Research, and Medicine, National Academies Press, Washington, DC.

IRENA (2015) *Renewable energy in the water, energy & food nexus.* International Renewable Energy Agency, Abu Dhabi, UAE.

Kahane, R., Hodgkin, T., Jaenicke, H., Hoogendoorn, C., Hermann, M., Hughes, J.D.A. & Looney, N. (2013) Agrobiodiversity for food security, health and income. *Agronomy for Sustainable Development*, 33(4), 671–693.

Kawachi, I. & Wamala, S. (eds) (2006) *Globalization and Health*. Oxford University Press, New York.

Keatinge, J.D.H., Waliyar, F., Jamnadass, R.H., Moustafa, A., Andrade, M., Drechsel, P., Hughes, D.A.J., Palchamy, K. & Luther, K. (2010) Re-learning old lessons for the future of food: by bread alone no longer diversifying diets with fruit and vegetables. *Crop Science*, 50, 51–62.

Kotagama, H., Al Jabri, S.A.N., Boughanmi, H. & Guizani, N. (2014) Impact of food prices, income and income distribution on food security in Oman. In: Shahid, S.A. & Ahmed, M. (eds) *Environmental Cost and Face of Agriculture in the Gulf Cooperation Council Countries*. Springer International, Cham, Switzerland. pp. 145–162.

Land Coalition (2013) LAND MATRIX. Available from: http://www.landmatrix.org/en/get-the-idea/global-map-investments/ [accessed March 2017].

Lattemann, S. & Höpner, T. (2008) Environmental impact and impact assessment of seawater desalination, *Desalination*, 220, 1–15.

Lutaladio, N., Burlingame, B. & Crews, J. (2010) Horticulture, biodiversity and nutrition. *Journal of Food Composition and Analysis*, 23, 481–485.

McCarthy, J.F. (2010) Processes of inclusion and adverse incorporation: oil palm and agrarian change in Sumatra, Indonesia. *Journal of Peasant Studies*, 37, 821–850.

MAF & ICARDA (2011) Assessing returns to support investments in the two agricultural development projects (protected agriculture and modern irrigation systems) in the Sultanate of Oman, Final report. Ministry of Agriculture and Fisheries, Muscat, Oman; International Center for Agricultural Research in the Dry Areas, Beirut, Lebanon.

Malek, C. (2012) Serbian farmland to provide substantial amount of UAE's food. The National, Abu Dhabi, UAE. Available from: http://www.thenational.ae/news/uae-news/serbian-farmland-to-provide-substantial-amount-of-uaes-food [accessed November 2016].

Mekonnen, M. & Hoekstra, A. (2011) National water footprint accounts: the green, blue and grey water footprint of production and consumption. *Value of Water Research Report Series* No. 50. UNESCO-IHE, Institute for Water Education, Delft, The Netherlands. Available from: http://www.waterfootprint.org/Reports/Report50-NationalWaterFootprints-Vol1.pdf [accessed November 2016].

Molden, D. (ed.) (2007) *Water for Food, Water for Life: A Comprehensive Assessment of Water Management in Agriculture*. Earthscan, London; International Water Management Institute, Colombo, Sri Lanka.

Mulder, K. (2006) *Sustainable Development for Engineers*. Greenleaf Publishing, Sheffield.

National Aquaculture Group (2013) Mauritania and Al-Rajhi sign MOU for food security. Available from: http://www.naqua.com.sa/Mauritania-news.php [accessed November 2015].

Nepomuceno, J. (2012) State visit of Amir to boost country's economic relations with Kuwait. Zambo Times (Philippines), 24 March 2012. Available from: http://www.kglinvest.com/index.php/2012-03-14-03-28-45/285-state-visit-of-amir-to-boost-countrys-economic-relations-with-kuwait [accessed November 2016].

Padulosi, S., Hodgkin, T., Williams, J.T. & Haq, N. (2002) Underutilized crops: trends, challenges and opportunities in the 21st century. In: Engels, J.M.M, Rao V.R., Brown, A.H.D. & Jackson, M.T. (eds) *Managing Plant Genetic Diversity*. CAB International, Wallingford. pp. 323–338.

Parry, M.L., Rosenzweig, C., Iglesias, A., Livermore, M. & Fisher, G. (2004) Effects of climate change on global food production under SRES emissions and socioeconomic scenarios. *Global Environment Change*, 14, 53–67.

Pierre-Louis, A.M., Akala, F.A. & Karam, H.S. (eds) (2004) Public health in the Middle East and North Africa: meeting the challenges of the 21st century. World Bank Publications, Washington, DC. Available from: http://elibrary.worldbank.org/doi/abs/10.1596/0-8213-5790-5 [accessed November 2016].

Polgreen, L. (2007) In Niger, trees and crops turn back the desert. New York Times, 11 February 2007.

Potrykus, I. (2010) Lessons from the humanitarian golden rice project: regulation prevents development of public good genetically engineered crop products. *New Biotechnology* 27(5), 466–472.

Prescott-Allen, R. & Prescott-Allen, C. (1990) How many plants feed the world? *Conservation Biology*, 4(4), 365–374.

Qadir, M., Boers, T.M., Schubert, S., Ghafoor, A. & Murtaza, G. (2003) Agricultural water management in water-starved countries: challenges and opportunities. *Agricultural Water Management*, 62(3), 165–185.

Raluy, G., Serra, L. & Uche, J. (2006) Life cycle assessment of MSF, MED and RO desalination technologies. *Energy*, 31(13), 2361–2372.

Renault, D. & Wallender, W.W. (2000) Nutritional water productivity and diets. *Agricultural Water Management*, 45(3), 275–296.

Rockström, J., Steffen, W., Noone, K., Persson, Å., Chapin, F.S. 3rd, Lambin, E.F., Lenton, T.M., Scheffer, M., Folke, C., Schellnhuber, H.J., Nykvist, B., de Wit, C.A., Hughes, T., van der Leeuw, S., Rodhe, H., Sörlin, S., Snyder, P.K., Costanza, R., Svedin, U., Falkenmark, M., Karlberg, L., Corell, R.W., Fabry, V.J., Hansen, J., Walker, B., Liverman, D., Richardson, K., Crutzen, P. & Foley, J.A. (2009) A safe operating space for humanity. *Nature*, 461, 472–475.

Royal Society (2009) Reaping the benefits: science and the sustainable intensification of global agriculture. The Royal Society, London, UK. Available from: https://royalsociety.org/~/media/Royal_Society_Content/policy/publications/2009/4294967719.pdf [accessed November 2016].

Rumbaitis del Rio, C. (2012) The role of ecosystems in building climate change resilience and reducing greenhouse gases. In: Ingram, J.C., DeClerck, F. & Rumbaitis del Rio, C. (eds) *Integrating Ecology and Poverty Reduction*. Springer, New York. pp. 331–352.

Sabeh, N.C., Giacomelli, G.A. & Kubota, C. (2011) Water use in a greenhouse in a semi-arid climate. *Transactions of the ASABE*, 54(3), 1069–1077.

Saif, A. & Mushtaque, A. (2014) Opportunities and challenges of using treated wastewater in agriculture. In: Shahid, S.A. & Ahmed, M. (eds) *Environmental Cost and Face of Agriculture in the Gulf Cooperation Council Countries*. Springer International, Cham, Switzerland. pp. 109–123.

Saif, O., Mezher, T. & Arafat, H. (2014) Water security in the GCC countries: challenges and opportunities. *Journal of Environmental Studies and Sciences*, 4(4), 329–346.

Scheer, S. & McNeely, J. (2007) *Farming with Nature: The Science and Practice of Ecoagriculture*. Island Press, Washington, DC.

Schlenker, W. & Lobell, D.B. (2010) Robust negative impacts of climate change on African agriculture. *Environmental Research Letters*, 5, 014010.

Shepherd, B. (2014) Investments in foreign agriculture as a Gulf state food security strategy: towards better policy. In: Shahid, S.A. & Ahmed, M. (eds) *Environmental Cost and Face of Agriculture in the Gulf Cooperation Council Countries*. Springer International, Cham, Switzerland. pp. 125–144.

Smale, M., Hazell, P., Hodgkin, T. & Fowler, C. (2009) Do we have an adequate global strategy for securing the biodiversity of major food crops? In: Kontoleon, A., Pascual, U. & Smale, M. (eds) *Agrobiodiversity Conservation and Economic Development*. Routledge, Abingdon. pp. 40–50.

Spiess, A. (2008) Developing adaptive capacity for responding to environmental change in the Arab Gulf states: uncertainties to linking ecosystem conservation, sustainable development and society in authoritarian rentier economies. *Global and Planetary Change*, 64(3–4), 244–252.

Strategy& (2014) Achieving a sustainable water sector in the CC: managing supply and demand, building institutions. Strategy & Beirut, Lebanon. Available from: http://www.strategyand.pwc.com/reports/achieving-sustainable-water-sector-gcc [accessed November 2016].

Tacoli, C. (2011) Not only climate change: mobility, vulnerability and socioeconomic transformations in environmentally fragile areas in Bolivia, Senegal and Tanzania. International Institute for Environment and Development (IIIED), London.

Tajitsu, N. (2013) China bans NZ milk powder imports on botulism scare. *Reuters*, 4 August 2013. Available from: http://www.reuters.com/article/2013/08/04/newzealand-milk-idUSL4N0G500D20130804 [accessed November 2016].

Tester, M. (2015) Greenhouse business. *Global Forum for Innovations in Agriculture, 9–11 March 2015, Abu Dhabi, UAE*.

Turton, A.R. (2000) Precipitation, people, pipelines and power: towards a 'virtual water' based political ecology discourse. In: Stott, P. & Sullivan, S. (eds) *Political Ecology: Science, Myth and Power*. Arnold, London. pp. 132–156.

UN (2011) Water Policy Brief no. 6, 2011. Africa's progress towards the MDGs. Available from: http://www.un.org/en/africa/osaa/pdf/policybriefs/2011_towards_mdgs.pdf [accessed March 2017].

UN (2014) World's population increasingly urban with more than half living in urban areas. United Nations, 10 July 2014. Available from: http://www.un.org/en/development/desa/news/population/world-urbanization-prospects-2014.html [accessed November 2016].

UNDP (2003) Human development report 2003. Millennium Development Goals: a compact among nations to end human poverty. United Nations Development Programme, New York.

UNEP (2006) Global deserts outlook. United Nations Environment Programme, Nairobi, Kenya.

UNSCN (2010) Progress on nutrition. 6th Report on the World Nutrition Situation of the United Nations Standing Committee on Nutrition, Geneva, Switzerland. Available from: http://www.unscn.org/files/Publications/RWNS6/report/SCN_report.pdf [accessed November 2016].

Weart, S.R. (2003) *The Discovery of Global Warming*. Harvard University Press, Cambridge, MA.

Wichelns, D. (2011) The role of virtual water in efforts to achieve food security and other national goals, with examples from Egypt. *Agricultural Water Management*, 49, 131–151.

Wojtkowski, P. (2008) *Agroecological Economics: Sustainability and Biodiversity*. Academic Press, Amsterdam.

Wright, J. (2010) Feeding nine billion in a low emissions economy: challenging, but possible – a review of the literature for the Overseas Development Institute. Available from: https://www.odi.org/sites/odi. org.uk/files/odi-assets/publications-opinion-files/6389.pdf [accessed November 2016].

Zarzo, D., Campos, E. & Terrero, P. (2013) Spanish experience in desalination for agriculture. *Desalination and Water Treatment*, 51(1–3), 53–66.

# CHAPTER 13

## Renewable energy policy and mitigating the risks for investment

Nadejda Komendantova

### 13.1 INTRODUCTION

There is a consensus that in order to lessen the impact of climate change, global warming has to be limited to a maximum of 2°C above the pre-industrial average temperature (IPCC, 2014). Furthermore, the stabilization of global warming will only be possible through a transition towards low carbon energy generation energy, since 60% of all greenhouse gas (GHG) emissions lead to climate alteration (Riahi *et al.*, 2012). This transition can be accomplished by using, for example, renewable energy sources (RES) that would help to mitigate climate change and also provide energy security (Francés *et al.*, 2013).

In the Middle East and North Africa (MENA) region energy demand and GHG emissions have been growing steadily over the last few decades (Brand, 2015). By 2030 GHG emissions are projected to double and the demand for electricity will triple. Brand (2015) noted that the MENA region also has extensive renewable energy sources (RES), which would be sufficient to satisfy a growing demand for electricity while at the same time satisfying the need for low carbon energy sources. Furthermore, the solar irradiation in the region is three times higher than in Europe, where most of the reported solar power extension projects have taken place (IRENA, 2013). However, despite the available RES, the transition to a low carbon economy in the MENA region has been slow. There are many barriers for the deployment of renewable energy projects; among them are high upfront costs, complex administrative and regulatory procedures, and insufficient legal regulations. It can be argued that, for the effective deployment of RES on a large scale, there is an acute need for special policies and programs to help overcome these barriers.

Existing literature disclosed the technical feasibility of the deployment of renewable energy sources in the MENA region, not only for local use but also for exporting electricity to Europe (DLR, 2006; Schellekens *et al.*, 2010). Similarly, several studies exist on the costs of energy transition, the subsidies needed and the levelized costs of electricity. These studies mainly used the energy modeling and engineering approach, however, only a few of them focused on human factors and the decision-making processes in the deployment of renewable energy sources (RES) (Sovacool and Brown, 2010).

In a recent Intergovernmental Panel on Climate Change (IPCC) assessment report Kunreuther *et al.* (2014) argued that, for the successful large-scale deployment of renewable energy sources, significantly more attention needs to be given to the decision-making processes; in particular how different stakeholders involved in the decision-making processes perceive existing risks. The concept of risk perception refers to people's subjective judgements of the characteristics and severity of a risk and is related to how much risk people are willing to accept. Furthermore, the literature argues that the realization of renewable energy projects and their economic efficiency depends on project financing (UNDP, 2013) where the weighted average cost of capital (WACC) is influenced by the perceived risks that investors associate with a specific technology or geographical location (Komendantova *et al.*, 2012; Schmidt, 2014).

Risk perceptions are also closely connected with an investor's decision on whether or not to invest in a certain technology (Lüthi and Prassler, 2011), thus leading to a situation where renewable energy projects may be delayed or not realized at all. Additionally, renewable energy projects may face implementational and operational barriers and risks along the whole deployment process. This would likewise depend on different stakeholders and their respective perceptions of these risks. At one end of the spectrum, renewable energy projects might face barriers and risks in the project planning and financing phase, while at the other end of the spectrum, there might appear to be problems associated with a society's acceptance of these new technologies. Therefore, understanding the risks as well as the decision-making process and governance structures is crucial.

Lüthi and Prassler (2011) maintained that human factors must be considered. These include support or opposition from the public and other stakeholders, such as established actors in the policy debate, as well as coalitions or civil movements that might emerge in direct response to renewable energy issues related to the deployment of projects in local communities. This is also associated with the question of how different stakeholders perceive risks, such as the costs versus the benefits of renewable energy technology. Is there consumer willingness to pay and is there a political will to shoulder the economic burden of technology deployment? Perceptions of equity and fairness also need to be considered regarding who will carry the costs of technological deployment and what will be the temporal, generational and spatial distribution of the benefits.

Prevailing scientific evidence shows that the efficiency of policy mechanisms for the deployment of renewable energy technology depends on several factors, such as public acceptance and consumers' voluntary renewable energy purchases. To address the decision-making process about the willingness to use renewable energies not only at the individual level, which involves public acceptance and risk perceptions, but also for a large-scale deployment of technology, it is necessary to understand which institutional and regulatory structures are needed for the proper governance of the energy transition process. In this chapter the following aspects shall be considered: technological lock-ins and inflexibility, as well as the influence of stakeholder groups, such as energy companies, and the interactions of the private and public sectors. The major question is if and how participatory multi-level governance of energy transition can involve the voices, opinions, perceptions and views of the different stakeholder groups concerning the placement of renewable energy sources on a large scale.

The aim of this chapter is to contribute to the understanding of how human factors affect energy transition in the MENA region. The innovative aspect of this approach is in the integrated and holistic assessment associated with the deployment of renewable energy technologies and how the positioning of renewable energies can contribute to the socioeconomic development of the area. The economic feasibility of renewable energy projects, for instance, is assessed by taking into account the risk perceptions of different stakeholders. The interdisciplinary aspect of this work is in the combination and the quantification of actual risks and perceptions; thus, bringing different social views into the determination of actual energy costs.

## 13.2   OBSTACLES TO DEPLOYMENT AND ASSOCIATED INVESTMENT RISKS

### 13.2.1   *Barriers for deployment of renewable energy projects in the Middle East & North African region*

Van der Hoeven (2013), the executive director of the International Energy Agency, recently noted that energy demand has been growing steadily in the Middle East and North African (MENA) region and that it is also connected to a growing level of greenhouse gas (GHG) emissions. The demand for power is projected to grow at around 7% per year over the next 10 years, caused mainly by population growth and economic development. According to the World Energy Outlook, as reported by van der Hoeven (2013), the share of renewable energy sources in the MENA region is expected to increase from 2% in 2010 to 12% by 2035. The major drivers of this increase are population growth, increasing energy activity due to industrialization and urbanization, as well as increasing standards of living. Needless to say, it can be argued that such growth is based on subsidized energy prices. The greater share of electricity consumption falls on air conditioning and seawater desalination, which is also expected to increase energy demand significantly (Brand and Zingerle, 2011).

The countries of the MENA region have very diverse levels of availability of non-renewable energy sources, ranging from the exporters of oil and gas, such as Egypt and Algeria, to countries that import almost all the energy they need, such as Jordan and Morocco (Brand and Zingerle, 2011). Today the dominating source of power generation in the region is natural gas, whose share in the energy mix has been growing steadily, pushing oil based generation in to second place. While natural gas makes up the biggest share in electricity generation in Algeria, Egypt and Tunisia, coal dominates in Morocco.

Currently renewable energy contributes only 4% to MENA's primary energy consumption. This is significantly below the 17% average for the rest of the world. The existing level of energy consumption from renewable energy sources also stands in strong contrast to the availability of resources as, according to the World Bank estimations, the MENA region receives between 22% and 26% of all solar energy striking the earth. The region has particularly good potential for solar energy as it is situated in the Earth's Sunbelt with especially high direct normal irradiation ranging between 4000 and 8000 Wh m$^{-2}$ day$^{-1}$ (Mason and Kumetat, 2011). In addition to solar energy, MENA also has wind resources. Nonetheless, solar energy is considered to have by far the largest potential. In comparison to the North of the Mediterranean region, where the principal capacities of solar energy technologies are currently deployed, the South Mediterranean region has much higher direct solar irradiance. For example, the prospect for concentrated solar power (CSP) is about 150,000 TWh year$^{-1}$, which is sufficient to satisfy the entire global primary energy demand in 2012 (DLR, 2006).

Recognizing the mounting energy demand, growing GHG emissions and the potential of RES, the governments of the MENA countries have set ambitious national targets to deploy RES technology. Algeria, for example, foresees the installation of 12 GW of RES by 2030 (MEM, 2011). By 2030 renewable energy is expected to supply 40% of its domestic electricity demand. In addition, solar power should satisfy more than 37% of domestic electricity demand by 2030 (MEM, 2011). Egypt plans to increase the share of RES in its energy mix from 9% to 20% by 2020 (GIZ, 2014) due to its high import dependency on fossil fuels, which amounted to 97% in 2011 (IEA, 2013a). Furthermore, Morocco plans to deploy 6 GW of RES capacities by 2020. Its strategy to foster the promotion of renewable energy foresees 42% of its installed power generation capacity to be based on renewable energies by 2020. Tunisia plans to achieve 25% of RES in electricity production by 2030 and a 10% share of renewables in total electricity production and a 20% reduction in energy demand by 2020 (IEA, 2013b).

Deployment of renewable energy technology on a large scale requires not only political will but also financial resources. According to the IEA's new policy scenario, non-OECD countries have to invest on average US$ 1200 billion annually in the energy supply infrastructure (IEA, 2014). Compared to historical investments of US$ 708 billion annually, there remains a financing gap of almost US$ 500 billion annually. To reach the necessary investment, it is essential to create attractive conditions for private or foreign direct investment in renewable energy projects and to use available public funds as a means of leverage for private investment.

The involvement of private capital in renewable energy projects in the MENA region can be a serious challenge. Currently the region is attracting the lowest volumes of foreign direct investment (FDI) in the world and the share of this investment going into renewables is minimal. The area currently receives only 15% of the European FDI, with France, the UK and Spain being the most important European investors. In Libya, Algeria and Egypt the FDI flows primarily to the hydrocarbon sector, in Tunisia to the energy and manufacturing sector and in Morocco to tourism, real estate and the industrial sector (REN21, 2013).

### 13.2.2  *Risk for investment in renewable energy technology*

Risk, in relation to investments in renewable energy projects, can be described by a negative impact that uncertain future events may have on the financial value of a project. A serious issue in the development of renewable energy ventures is how imminent happenings may upset the value of the project and what threats are involved with the planned investment. Thus dealing with risk is a crucial element when it comes to valuing a new project and deciding whether or not to invest in it.

The key to determining the effectiveness of any technology's deployment is its comparative cost advantage vis-à-vis existing technologies (Coenraads *et al.*, 2006). Renewables are facing more difficulties because of the lengthy life cycles required to recover their high initial capital costs, the absence of any meaningful way to consider externalities in the calculation of the full cost of renewable and non-renewable energy sources, and the additional cost of integrating intermittent energy sources into stable power systems. Therefore, addressing barriers and risks for investment in renewables from the point of view of investors is one of the conceivable drivers for the positioning of this technology. Correspondingly, not only economic efficiency but the entire realization of an investment project depends on affordable financing and the cost of capital itself depends on perceived risks by investors associated with concrete projects.

Several studies have identified risks observed by project developers as a barrier for investment in renewable energy sources under existing European conditions, such as regulatory and political risks of financial support for renewables, available resources, and technological risks and planning. Investors perceive the majority of these risks as predictable and therefore possible to deal with. In a survey, the regulatory and political risks are seen as unpredictable, and therefore, much more difficult to deal with and more risky for investment (REN21, 2013). Another survey identified that senior executives in renewable energy industries in the European Union identify financial risks as the most significant danger for renewable energy projects (i.e. 76% of all respondents). Financing risks were perceived as being serious because of the capital intensity of renewable energy projects and the high initial costs for infrastructure. Due to the large volumes of capital required for large-scale installations, the capital for such investment is typically leveraged, with up to 70–80% of a total project being financed through debts (Coenraads *et al.*, 2006).

Renewable energy investments generally require greater financing for the same capacity. Lower fuel and operating outlays may make renewable energy cost competitive on a life cycle basis, in spite of a higher initial capital. Depending on the circumstances, capital markets may demand a premium in lending rates for financing renewable energy projects because more capital is being risked upfront than in conventional energy projects. Renewable energy technologies may also face higher taxes and import duties. These duties may exacerbate the high first-cost considerations relative to other technologies and fuels.

The behavior of most investors can be complex. On the one side they are hoping to earn profits, particularly from investment, which promises high benefits. On the other side they hope to minimize risk. Thus, even if a given investment promises expected profits, if the possibility of losing money is too large, investors will stay away. In economic theory, such behavior is known as risk aversion, and is a long-accepted feature of economic behavior (Khattab and Davies, 2008). Many financial instruments allow investors to reduce their risk at some cost to expected profits. Often, however, the markets for such instruments fail due to information asymmetries and other barriers, and investors are left with significant amounts of risk (Mendonça, 2007).

Miranda and Glauber (1997) have identified that developers of renewable projects have difficulties in obtaining financing for projects that banks perceive as risky. Other studies by del Rio and Gual (2007), which examined the effectiveness of the feed-in tariff system for renewable energy support, identified that this system is successful, relative to other policy instruments, because it reduces the downside risk associated with project development, and makes it easier to attract finance. Likewise, human issues, such as perception of risks, were considered much less frequently than economic and technical factors. From limited existing studies, the risks perceived by renewable energy investors in developing countries were mainly addressed qualitatively (Shrimali *et al.*, 2013). Only very few studies investigated how these risk perceptions translate into higher costs of capital or what impact they have on financial de-risking strategies of renewable energy investment (Komendantova *et al.*, 2012; Schmidt, 2014; UNDP, 2013).

## 13.3   THE CASE STUDY METHOD FOR ASSESSING BARRIERS & RISKS

The results reported in this chapter are based on the case study method, which is frequently used in social sciences as a research technique for the in-depth examination of a subject and its related contextual conditions. The situation can be defined as follows: "case studies are analyses of persons, events, decisions, periods, projects, policies, institutions, or other systems that are considered holistically by one or more means. The case that is the subject of the inquiry will be an instance of a class of phenomena that provides an analytical frame, an object, within which the study is conducted and which the case illuminates and explicates" (Thomas, 2011). One of the most famous examples of application of the case study method is Galilei's rejection of Aristotle's law of gravity. The rejection was based on information-oriented sampling and not by random sampling (Flyvbjerg, 2006).

As opposed to random sampling, which is also often used in social sciences, cases are selected based on information-oriented sampling and often include extremes, such as, in our research, a country that not only has rich fossil fuel reserves for local consumption but is also one of the largest exporters of fossil fuel energy, and a country that does not have such reserves, either for export or for local consumption, and is covering its energy demands by imports. The majority of existing studies on public acceptance and willingness to pay or to use renewable energy, apply quantitative methods of research using random sampling and large-scale surveys of public opinion. Research methods include hedonic analysis of actual price premiums charged for green electricity in deregulated markets (Roe *et al.*, 2001), the elicitation technique to develop survey design (Bollino and Polinori, 2006), contingent valuation and double bound dichotomous choice format to collect people's elicitations (Zografakis *et al.*, 2010) and other methods based on large-scale surveys of stakeholders' opinions and views. We argue here that a case study approach can create additional benefits through detailed and more in-depth evaluation.

To identify barriers to the deployment of renewable energies in the region several approaches were employed. To identify risk perceptions by stakeholders a number of rounds of semi-structured interviews with stakeholders were conducted (Komendantova *et al.*, 2011). Komendantova *et al.* (2011) modeled how risk perceptions were reflected in higher costs for renewable energy projects and for photovoltaics (PV) (Ondraczek *et al.*, 2015) as well as what impact investments into renewable energies have on the socioeconomic development of MENA countries, focusing in particular on Egypt (Farag and Komendantova, 2014) and Morocco (Komendantova and Patt, 2014). By taking the regional approach, additional costs were modeled of renewable energy projects, taking into account each risk, reflected in the perceptions of investors (Schinko and Komendantova, 2016). The methodology of this research is described in the relevant papers.

## 13.4   CASE STUDY RESULTS ON MITIGATING INVESTMENT RISKS

### 13.4.1   *Risk perceptions of investors*

Komendantova *et al.* (2011, 2012) interviewed stakeholders in the business of renewable energy development in MENA in order to identify risks that were perceived by investors as being serious and likely to happen. The majority of respondents identified the complexity of bureaucratic procedures and corruption as significant barriers. Other risks identified as significant were the instability of national regulations, the absence of guarantees from national governments and the international community on invested capital and revenues from projects, a low level of political stability, and the lack of support from local governments, including commitment and co-operation, due to a low level of awareness about the advantages of renewable energy sources. Furthermore, stakeholders were also asked to evaluate how serious and likely investment risks were for renewable energy projects. The resultant list of threats contained nine classes: technical, construction, operating, revenues, financial, force majeure, regulatory, environmental, and political (Fig. 13.1). The seriousness of each was ranked according to the following scores: 3 as high importance, 2 as medium importance, 1 as low importance. The likelihood of happening was evaluated based on the

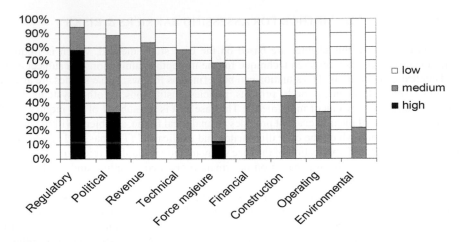

Figure 13.1.    Risks perceived as being most serious by investors ($n = 18$) (Komendantova *et al.*, 2012).

following percentages: very likely (>90% chance of occurring), medium likelihood (33–89%), and unlikely (<33%).

As shown in Figure 13.1, only three types of risks were perceived as being serious by stakeholders. These were regulatory risks (which included complexity and corruption of bureaucratic procedures and instability of national regulations), political risks (including a low level of political stability in a country and the lack of support from local governments) and force majeure risks (which include natural and human-made disasters, including terrorism). All of the other six classes, such as technical, construction, operation, financial, revenue and environmental, were evaluated as of low or medium importance.

### 13.4.2    *How do risk perceptions reflect the financing of renewable energy projects?*

To understand if risk perceptions were associated with the financing of renewable energy projects, a number of deployment scenarios were constructed for concentrated solar power in the MENA region using the Mediterranean Area Renewable Generation Estimator (MARGE), which was developed by Williges *et al.* (2010). The model allows for the examination of alternative assumptions about perceived risks by changing the internal rate of return (IRR) of a project. The effects of varying the IRR were examined for new concentrated solar power (CSP) plants in North Africa, with IRR ranging between 0% and 20%. It was assumed that CSP, at least in the short to medium run, would remain the most attractive renewable energy source for large-scale applications in the North African region, mainly because of its longer track record and its ability to provide dispatchable clean energy by storing energy in thermal storages, which is already technically and economically feasible today (World Bank, 2011).

The results showed that the levelized costs of electricity varied by a factor of three, depending on the IRR (Fig. 13.2). Changes in IRR also have a large impact on the year when the LCOE for CSP will be equal to the LCOE for coal, which is 5€ cents per kWh. If the rate of return to investors was close to 0% (potentially feasible for a purely public investment) then CSP would become competitive with privately developed coal power in about 5 years, about the time when the currently planned projects will have been completed. At a 5% rate (not unreasonable for a public/private partnership with very low perceived risk) CSP would become competitive by 2020. At higher IRRs, the year in which CSP would become cost competitive with coal would be pushed back to beyond 2025 (Fig. 13.2).

If the LCOE of CSP is compared with the LCOE of the cheapest energy generation option, which is coal, it is seen that it will take years for CSP to become cost competitive with coal. Currently the LCOE of coal stands at 5€ cents per kWh.

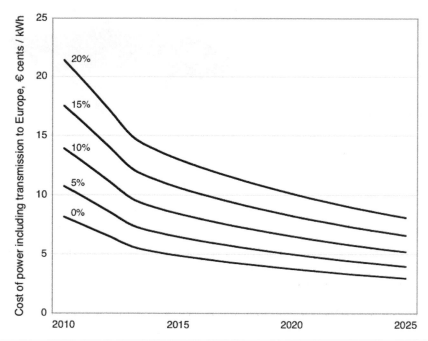

Figure 13.2.   Levelized electricity cost curves for CSP assuming different internal rates of return. Each curve represents a different rate that project developers must pay their investors (Komendantova *et al.*, 2011).

The cost of photovoltaic (PV) cells has decreased significantly during the last few years (Ondraczek *et al.*, 2015). This has led to an annual growth rate of PV electricity generation of about 47% over the period 2000–2011. The growth of CSP was not so significant and amounted to about 20% between 2000 and 2011. If we look at the deployment of PV, it can be seen that there is a strong dependency not only between the levelized costs of electricity and solar irradiance but also with financing costs. In their analysis, Ondraczek *et al.* (2015) mapped the costs of solar PV globally, taking into account the differences in solar resources and financing costs in order to calculate levelized costs of electricity (LCOE) from solar PV systems in 143 countries, including the MENA region (Figs. 13.3a and 13.3b) (Ondraczek *et al.*, 2015). Figure 13.3a shows the effects of solar irradiance on LCOE. In countries where GHI is above the global average ($1862 \, \text{kWh m}^{-2}$ $\text{year}^{-1}$), LCOE is lower (dark green and blue colors) and vice versa for countries with below average global horizontal irradiance (GHI) (orange and red colors). It also shows that LCOE is generally lower along the equator and greater in high latitude countries. The effect of differences in the cost of capital on LCOE is also demonstrated. In countries where the weighted average cost of capital (WACC) is above the uniform rate of 6.4% (Breyer and Gerlach, 2010), LCOE is higher (orange and red colors) and vice versa for countries with the WACC below this rate (dark green and blue) (Fig. 13.3b). In contrast, LCOE is higher along the equator and lower in high latitude countries.

Schinko and Komendantova (2016) employed the financing cost waterfall approach, developed by UNDP (2013), to quantify the contribution of different categories of investment risk in financing the cost gap between developing and developed countries (Fig. 13.4). Furthermore, the weighted average cost of capital (WACC) was compared for the Euro area and four North African countries. Due to the limited availability of data on financing costs for concentrated solar power (CSP) projects in the MENA region the researchers relied on available data for two CSP projects, Hassi R'Mel in Algeria and Ouazazate in Morocco. Since the credit ratings of Algeria and Egypt, as well as of Morocco and Tunisia, are very similar (Wikirating, 2014), it was assumed that financing costs for CSP projects were also similar in Algeria and Egypt and respectively in

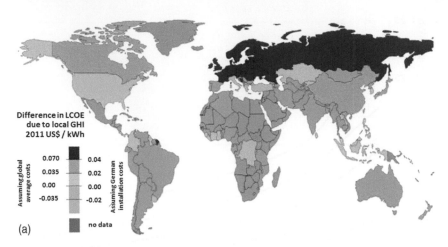

Figure 13.3a.   Effect of the global horizontal irradiance (GHI) on levelized costs of electricity (LCOE) per country [US$2011/kWh] (Ondraczek *et al.*, 2015).

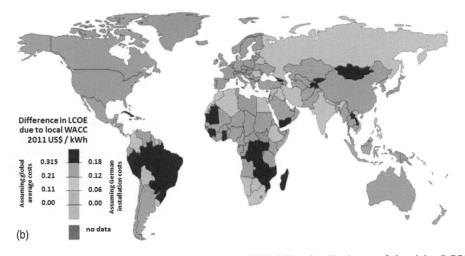

Figure 13.3b.   Effect of weighted average cost of capital (WACC) on levelized costs of electricity (LCOE) per country [US$2011 kWh$^{-1}$] (Ondraczek *et al.*, 2015).

Morocco and Tunisia. Results showed that the WACC in Algeria and Egypt was 8.3%, compared to only 4% in Europe. The WACC in Morocco and Tunisia was even higher and amounted to 9.2%. Figure 13.4 shows the different types of risks in North Africa, which are in addition to the financing conditions applicable in Europe.

By comparing the levelized costs of electricity (LCOE) for CSP electricity generation between North Africa and Europe (Fig. 13.5), Schinko and Komendantova (2016) concluded that even though North Africa has a substantially higher solar potential than Europe, the resulting LCOE for Europe (0.25 US$ kWh$^{-1}$) was not dramatically higher than the mean for North Africa (0.21 US$ kWh$^{-1}$). It can be argued that this was due to substantially lower financing costs in Europe compared to the North African region. Furthermore, they concluded that if a CSP investor in North Africa could get project financing under similar conditions and costs to those in Europe, then the LCOE could be reduced from 0.21 to 0.15 US$ kWh$^{-1}$, which is a 32% reduction. At the same time, if a CSP investor in Europe needed to realize a CSP project with the costs and conditions available in North Africa, the LCOE for a CSP project in Europe would increase from 0.25 to 0.37 US$ kWh$^{-1}$, which is a 51% increase, as shown in Figure 13.5.

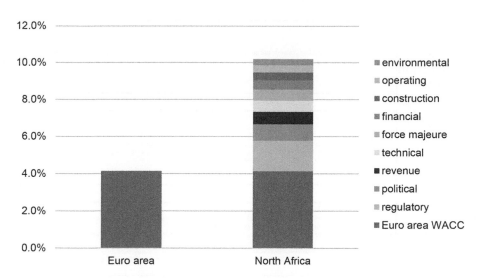

Figure 13.4.   The weighted average cost of capital (WACC) for concentrated solar power (CSP) investment in North Africa compared to the Euro region (Schinko and Komendantova, 2016).

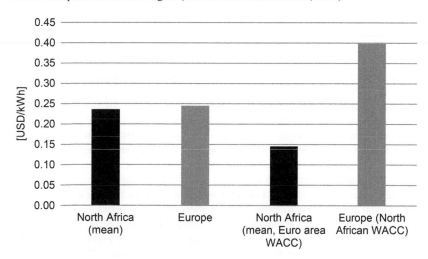

Figure 13.5.   The levelized costs of electricity (LCOE) for concentrated solar power (CSP) investments in Europe and North Africa with alternative financing costs (Schinko and Komendantova, 2016).

### 13.4.3   *Impacts of investment on socioeconomic development*

Komendantova and Patt (2014) estimated the impact of investment into renewable energy projects on socioeconomic development in the MENA region. Impacts were estimated in terms of created direct, indirect and induced job-years. They concluded that the deployment of renewable energy sources in the MENA region will have an impact in terms of the number of jobs created or multiplier effects on local economies only if a considerable share, not below 40%, of the needed power station components are manufactured locally. This study focused on the transfer of renewable energy technologies (RET), which is an important objective in industrialization by limiting greenhouse gas (GHG) emissions in these countries (Romer, 1990). Technology transfer can have several forms, such as joint ventures, foreign direct investment, governance assistance programs, direct purchases, joint research and development programs, franchising and sale of turnkey plants (Karakosta *et al.*, 2010). In order to limit the impact of climate change and to diversify the energy

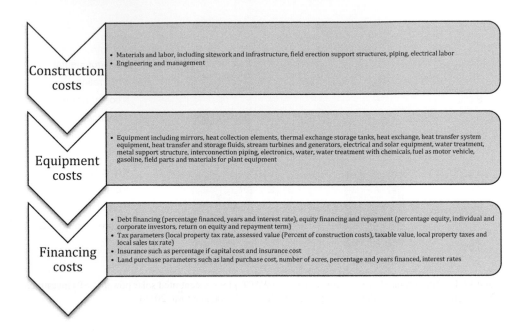

Figure 13.6.    Financing parameters (Komendantova and Patt, 2014).

supply, it is crucial to accelerate the spread of renewable energy technology around the world (IEA, 2010).

An analysis by Komendantova and Patt (2014) was conducted for two types of technology transfer, horizontal and vertical. Vertical technology transfer occurs, for example, when large multinational corporations establish factories in developing countries to bring down production costs. The managers and technical staff for these new factories are nationals of the developed countries, while the general workforce is from the developing country. Transfer of know-how to local manufacturers and technological spillovers is limited. On the other hand, horizontal technology transfer includes the formation of joint ventures between foreign and local companies, including technical and business training. This is a more lengthy process but it allows the embedding of technology within a local population and economy (Kammen *et al.*, 2004). It is preferable for local economies as skills and knowledge are transferred, but it makes it more difficult for foreign companies to protect their design and to control quality (Sáinz, 2008).

To be able to estimate the impact of technology transfer, Komendantova and Patt (2014) collected data on several investment parameters such as construction, equipment and financing costs (Fig. 13.6). These data were collected from the World Bank, IEA and SolarPaces. The results showed that MENA countries would benefit from technology transfer and the deployment of renewable energy projects only if the share of components manufactured locally was higher than 40% (Fig. 13.7).

Komendantova and Patt (2014) indicated that the number of induced jobs would be significantly higher than the number of direct jobs. Furthermore, comparison with existing estimations of ESTELA showed that their estimations for the number of created job-years were much higher and were based on a qualitative assessment (European Solar Thermal Electricity Association, 2009). The comparison with estimations of NREL show that their numbers of created job-years were much lower as they were based on wages typical for California.

Farag and Komendantova (2014) assessed the impact of investment in renewable energy projects on the economies of the hosting countries. For instance, Egypt predicts the deployment of renewable energies due to national targets and international incentives. The Desertec industrial initiative (DII) forecasts a large share of electricity being exported to Europe. By applying the

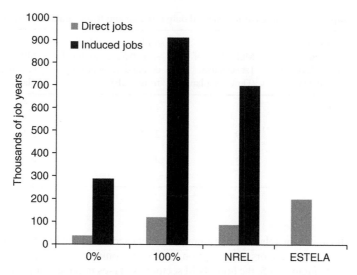

Figure 13.7.   Expectations on growth and jobs can be realized but with the deployment of manufacturing RES components industries in 0 and 100% scenarios as well as in the estimations of the National Renewable Energy Laboratory (NREL) and the European Solar Thermal Electricity Association (Komendantova and Patt, 2014).

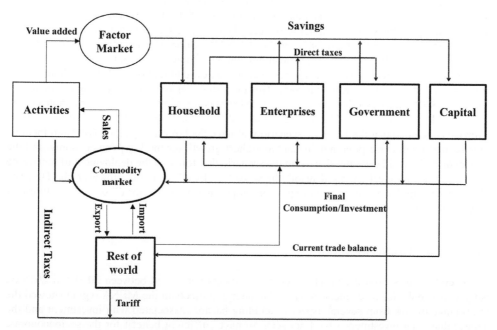

Figure 13.8.   Economic representation of the flow of income in the Egyptian economy (Farag and Komendantova, 2014).

method of the Leontief input-output model and social accounting matrix of Egypt, Farag and Komendantova (2014) evaluated the impact from Desertec, business as usual and renewable energy deployment according to national targets scenarios in terms of their impact on income, output and GDP (Fig. 13.8).

Table 13.1.   Multiplier effects on GDP, income and output (Farag and Komendantova, 2014).

| Type of multiplier | Multiplier of base scenario (current level of investment) | Multipliers of 1st scenario (Desertec plan) | Multipliers of 2nd scenario (secure local demand of electricity from CSP) | Multipliers of 3rd scenario (government plan till 2020) |
|---|---|---|---|---|
| GDP multiplier | 1.62 | 2.12 | 1.67 | 1.72 |
| Income multiplier | 2.15 | 2.19 | 2.04 | 2.16 |
| Output multiplier | 4.04 | 4.32 | 4.46 | 4.21 |

Figure 13.8 illustrates the relationships among different sectors of the Egyptian economy, such as household, government, factors of production, activities, commodity market, and foreign trade sector. According to Figure 13.8, the household sector pays taxes to the government, on one side, and buys products at the market, on the other side. The market includes both domestic activities and international imports. The commodity market contributes to the export sector, which, in its turn, pays tariffs, going directly to the government as revenues. The reserves of the government, which are collected through revenues, are used for investment. The results showed that from three investigated scenarios, investment under the Desertec scenario will have the largest impact on the GDP of Egypt. Under this situation the growth of the Egyptian economy will be the highest. Applying the GDP multiplier to the real data of the 2012 Egyptian economy, the GDP was predicted to be US $500 billion instead of US $236 billion. However, investment under the Desertec scenario will have a less significant impact on income and output.

The scenario where renewable energy projects are deployed to satisfy local energy demand will have the lowest impact in terms of income. However, this scenario will have a higher impact on output. The outcomes also showed that multiplier impacts from investment into renewable energy sources will be different in all three scenarios (Table 13.1). Farag and Komendantova (2014) likewise concluded that the Desertec scenario will have the largest impact on GDP and income. This scenario will correspondingly mean the highest growth for the Egyptian economy, and the impact on GDP will be greater than from the scenario that foresees the deployment of renewable energies to secure local energy demand, the so called "local" scenario. However, the Desertec scenario will not have a significant impact on output and here the "local" scenario will have a much higher impact on output.

## 13.5   CONCLUDING REMARKS

The overall results from the different research projects conducted between 2009 and 2015 on barriers and risks for investment in renewable energy projects in the MENA region showed the importance of risk from several aspects, including hazards associated with investment and the danger that such investment would not seem to have sufficient benefit for the socioeconomic development of the hosting countries. Furthermore, the innovative character of this research approach is in its multidisciplinary character by quantifying risk perceptions in human reasoning and decision-making and modeling how these hazard perceptions increase the costs of capital for renewable energy projects. The results also provide a holistic view on risk and human factors, such as perceptions, through the entire cycle of the renewable energy project deployment.

Several specific conclusions were achieved. The first is that human reasoning, such as the perception of risks affecting investment, can be a significant driver or barrier for the deployment of renewable energy sources. Risk perceptions can make investment into renewable energies more expensive due to required hazard premiums or prevent investment from happening at all. These threat perceptions also drive the weighted average costs of capital (WACC). Taking into account the available costs of capital, it might indeed make more sense to deploy highly investment sensitive renewable energy projects, like solar projects, in places with lower WACC rather than available solar resources, such as global horizontal irradiance (GHI). Addressing the risk perceptions of investors will allow for reducing the levelized costs of electricity significantly. The second set of results deals with the impacts from investment into renewable energies on socioeconomic development of MENA countries. Analysis indicated that there will be significant impacts on socioeconomic development only if up to 40% of all components are manufactured locally. Also investment in a scenario which foresees electricity exports to Europe will have higher impacts on GDP and income. Investment in a scenario that covers local energy needs first, will have higher impacts on output.

## REFERENCES

Bollino, C. & Polinori, P. (2006) An assessment of consumer willingness to pay for renewable energy sources use in Italy: a payment card approach. *26th USAEE/IAEE North American Conference Energy in a World of Changing Costs and Technologies, 24–27 September 2006, Ann Arbor, MI.*

Brand, B. (2015) *The integration of renewable energies into the electricity systems of North Africa.* Verlag Dr. Kovač, Hamburg.

Brand, B. & Zingerle, J. (2011) The renewable energy targets of the Maghreb countries: impact on electricity supply and conventional power markets. *Energy Policy*, 39(8), 4411–4419.

Breyer, C. & Gerlach, A. (2010) Global overview on grid-parity event dynamics. *25th European Photovoltaic Solar Energy Conference and Exhibition / 5th World Conference on Photovoltaic Energy Conversion, 6–10 September 2010, Valencia, Spain.*

Coenraads, R., Voogt, M. & Morotz, A. (2006) Analysis of barriers for the development of electricity generation from renewable energy sources in the EU-25. ECOFYS, Utrecht, The Netherlands.

Del Rio, P. & Gual, M.A. (2007) An integrated assessment of the feed-in tariff system in Spain. *Energy Policy*, 35(2), 994–1012.

DLR (2006) Trans-Mediterranean interconnection for concentrating solar power. German Aerospace Center Institute of Technical Thermodynamics, Section Systems Analysis and Technology Assessment, Stuttgart.

European Solar Thermal Electricity Association (2009) Solar power from the sun belt: the solar thermal electricity industry's proposal for the Mediterranean Solar Plan. A programme of the union for the Mediterranean, Brussels.

Farag, N. & Komendantova, N. (2014) Multiplier effects on socioeconomic development from investment in renewable energy projects in Egypt: Desertec versus energy for local consumption scenarios. *International Journal of Renewable Energy Research*, 4, 1108–1118.

Flyvbjerg, B. (2006) Five misunderstandings about case-study research. *Qualitative Inquiry*, 12(2), 219–245.

Francés, G.E., Marín-Quemada, J.M. & San Martín González, E. (2013) RES and risk: renewable energy's contribution to energy security: a portfolio-based approach. *Renewable and Sustainable Energy Reviews*, 26, 549–559.

GIZ (2014) Egyptian-German Committee on Renewable Energy, Energy Efficiency and Environmental Protection. Deutsche Gesellschaft für Internationale Zusammenarbeit (GIZ) GmbH, Eschborn, Germany; Ministry of Electricity and Renewable Energy (MoRE) and New and Renewable Energy Authority (NREA), Egypt. Available from http://www.giz.de/en/worldwide/16274.html [accessed November 2016].

IEA (2010) Projected costs of generating electricity. International Energy Agency, Paris.

IEA (2013a) Tracking clean energy progress 2013: IEA input to the clean energy ministerial. International Energy Agency, Paris.

IEA (2013b) Energy balances of non-OECD countries, 2013 edition. International Energy Agency. Paris.

IEA (2014) World energy investment outlook: special report. International Energy Agency, Paris.

IPCC (2014) Summary for Policymakers, In: Edenhofer, O., Pichs-Madruga, R., Sokona, Y., Farahani, E., Kadner, S., Seyboth, K., Adler, A., Baum, I., Brunner, S., Eickemeier, P., Kriemann, B., Savolainen, J., Schlömer, S., von Stechow, C., Zwickel, T., & Minx, J.C. (eds) *Climate Change 2014, Mitigation of Climate Change. Contribution of Working Group III to the Fifth Assessment Report of the Intergovernmental Panel on Climate Change.* Cambridge University Press, Cambridge and New York, NY.

IRENA (2013) Renewable energy technology: cost analysis series. Volume 1: Power sector issue 2/5. Concentrating solar power. IRENA working paper, International Renewable Energy Agency, UEA.

Kammen, D., Kapadia, K. & Fripp, M. (2004) Putting renewables to work: how many jobs can the clean energy industry generate? Report of the renewable and appropriate energy laboratory, Energy and Resources Group, Goldman School of Public Policy, University of Berkeley, CA.

Karakosta, C., Doukas, H. & Psarras, J. (2010) Technology transfer through climate change: setting a sustainable energy pattern. *Renewable and Sustainable Energy Reviews*, 14(6), 1546–1557.

Khattab, A. & Davies, A. (2008) Managerial practices of political risk assessment in Jordanian. *International Business Risk Management: an International Journal*, 10, 135–152.

Komendantova, N. & Patt, A. (2014) Employment under vertical and horizontal transfer of concentrated solar power technology to North African countries. *Renewable and Sustainable Energy Reviews*, 40, 1192–1201.

Komendantova, N., Patt, A. & Williges, K. (2011) Solar power investment in North Africa: reducing perceived risks. *Renewable and Sustainable Energy Reviews*, 15(9), 4829–4835.

Komendantova, N., Patt, A., Barras, L. & Battaglini, A. (2012) Perception of risks in renewable energy projects: the case of concentrated solar power in North Africa. *Energy Policy*, 40, 103–109.

Kunreuther, H., Gupta, S., Bosetti, V., Cooke, R., Dutt, V., Ha-Duong, M., Held, H., Llanes-Regueiro, J., Patt, A. & Shittu, E. (2014) Integrated risk and uncertainty assessment of climate change response policies. In: Edenhofer, O., Pichs-Madruga, R., Sokona, Y., Farahani, E., Kadner, S., Seyboth, K., Adler, A., Baum, I., Brunner, S., Eickemeier, P., Kriemann, B., Savolainen, J., Schlömer, S., von Stechow, C., Zwickel, T. & Minx, J.C. (eds) *Climate Change 2014: Mitigation of Climate Change. Contribution of Working Group III to the Fifth Assessment Report of the Intergovernmental Panel on Climate Change.* Cambridge University Press, Cambridge and New York.

Lüthi, S. & Prassler, T. (2011) Analyzing policy support instruments and regulatory risk factors for wind energy deployment: a developers' perspective. *Energy Policy*, 39(9), 4876–4892.

Mason, M. & Kumetat, D. (2011) At the crossroads: energy futures for North Africa. *Energy Policy*, 39(8), 4407–4410.

MEM (2011) Renewable energy & energy efficiency program. Algerian Ministry of Energy and Mines. Available from: http://portail.cder. dz/IMG/pdf/Renewable_Energy_and_Energy_Efficiency_Algerian_ Program_EN.pdf [accessed November 2016].

Mendonça, M. (2007) *Feed-in Tariffs: Accelerating the Deployment of Renewable Energy*. Earthscan, London.

Miranda, M. & Glauber, J. (1997) Systemic risk, reinsurance, and the failure of crop insurance markets. *American Journal of Agricultural Economics*, 79(1), 206–215.

Ondraczek, J., Komendantova, N. & Patt, A. (2015) WACC the dog: the effect of financing costs on the levelized cost of solar PV power. *Renewable Energy*, 75, 888–898.

REN21 (2013) MENA renewables status report. Renewable Energy Policy Network for the 21st century, Paris.

Riahi, K.F., Dentener, D., Gielen, A., Grubler, J., Jewell, Z., Klimont, V., Krey, D., McCollum, S., Pachauri, S., Rao, B., van Ruijven, D.P. & Van Vuuren, C. (2012) Energy pathways for sustainable development. In: Johansson, T.B. (ed) *Global Energy Assessment: Toward a Sustainable Future*. Cambridge University Press, Cambridge and New York; International Institute for Applied Systems Analysis, Laxenburg, Austria.

Roe, B., Teisl, M., Levy, A. & Russell, M. (2001) US consumers' willingness to pay for green electricity. *Energy Policy*, 29(11), 917–925.

Romer, P. (1990) Endogenous technological change. *Journal of Political Economy*, 98(5), 71–102.

Sáinz, J. (2008) Employment estimates for the renewable energy industry. Instituto Sindical de Trabajo Ambiente y Salud (ISTAS) and Comisiones Obreras, Madrid.

Schellekens, G., Battaglini, A., Lillestam, J., McDonnell, J. & Patt, A.G. (2010) 100% renewable electricity: a roadmap to 2050 for Europe and North Africa. PricewaterhouseCoopers, London.

Schinko, T. & Komendantova, N. (2016) De-risking investment into concentrated solar power in North Africa. *Renewable Energy*, 92, 262–272.

Schmidt, T.S. (2014) Low-carbon investment risks and de-risking: commentary. *Nature Climate Change*, 4, 237–239.

Shrimali, G., Nelson, D., Goel, S., Konda, C. & Kumar, R. (2013) Renewable deployment in India: financing costs and implications for policy. *Energy Policy*, 62, 28–43.

Sovacool, B. & Brown, M. (2010) Competing dimensions of energy security: an international perspective. *Annual Review of Environment and Resources*, 35, 77–108.

Thomas, G. (2011) A typology for the case study in social science following a review of definition, discourse and structure. *Qualitative Inquiry*, 17(6), 511–521.

UNDP (2013) De-risking renewable energy investment: a framework to support policymakers in selecting public instruments to promote renewable energy investment in developing countries. United Nations Development Programme, New York.

Van der Hoeven, M. (2013) World energy outlook 2013. International Energy Agency, Paris.

Wikirating (2014) List of countries by credit rating. Available from: http://www.wikirating.org/wiki/List_of_countries_by_credit_rating_%28comparisons%29 [accessed November 2016].

Williges, K., Lilliestam, J. & Patt, A. (2010) Making concentrated solar power competitive with coal: the costs of a European feed-in tariff. *Energy Policy*, 38(6), 3089–3097.

World Bank (2011) Middle East and North Africa region assessment of the local manufacturing potential for concentrated solar power (CSP) projects. January 2011, Washington, DC.

Zografakis, N., Sifaki, E., Pagalou, M., Nikitaki, G., Psarakis, V. & Konstantinos, V. (2010) Assessment of public acceptance and willingness to pay for renewable energy sources in Crete. *Renewable and Sustainable Energy Reviews*, 14(2010), 1088–1095.

# CHAPTER 14

## Integrating renewable energy sources into smart grids: opportunities and challenges

Muhammad Anan

### 14.1 INTRODUCTION

The growing number of renewable energy sources in the world has resulted in a more dispersed power generation energy sector (Fuchs *et al.*, 2012). The replacement of centralized power generation by decentralized renewable energy generation is the principal goal of smart grids everywhere. A smart grid has sensors throughout to enable it to monitor itself. The sensors will send information back to a control station where it will be analyzed and, if necessary, corrective actions can be taken. If, for example, a transformer in a substation were to malfunction, the control system will know from the information sent back by the sensors. Once the smart grid is aware of the failure it can supply customers with electricity through another route in the grid while the transformer is being repaired or replaced. While a regular power grid only allows electricity to flow from the producer to the customer, a smart grid, also known as an intelligent grid, intelligrid, future grid, intergrid, or intragrid, allows for electricity to flow in both directions. It will also have more distributed power generation locations compared to a regular power grid, be self-monitoring and self-correcting.

A smart grid is designed with three major systems in mind (Fand *et al.*, 2012). The first is the smart infrastructure system. This is the backbone of the smart grid. It is responsible for the two-way energy and information flow. The smart infrastructure is further divided into three more categories: smart energy subsystem, smart information subsystem and smart communication subsystem. The smart energy subsystem is responsible for advanced electricity generation, delivery and consumption. The smart information subsystem is responsible for advanced information metering, monitoring and management. The smart communication subsystem is responsible for the communication connectivity and information transmission among systems, devices and applications. The smart information and communication subsystems will need to work together more often than anything else.

The second system is the smart management system. This structure provides advanced management, control services and functionalities. With the development of new management applications and services, the smart grid can keep getting "smarter". This is like a nervous system. It is responsible for controlling the entire grid to make sure that everything runs smoothly and nothing harmful happens. Finally, the third scheme is the smart protection system, which provides advanced grid reliability analysis, failure protection, and security and privacy protection services. By taking advantage of the smart infrastructure, the smart grid must not only realize a smarter management system, but also provide a smarter protection system, which can more effectively and efficiently support failure protection mechanisms, address cybersecurity issues and preserve privacy.

A high-level architecture of the smart grid can be seen in Figure 14.1, which illustrates all the major systems and their interactions. It shows how the systems communicate in a similar way to the power flow. Everything must go through the substation, whether it is the power coming from the generators or signals being sent from smart meters to the control centre. Figure 14.1 also shows the interactions between the smart grid and electric vehicles and smart appliances. Furthermore, Table 14.1 gives a simple comparison between the old power grid and the future smart grid. Some

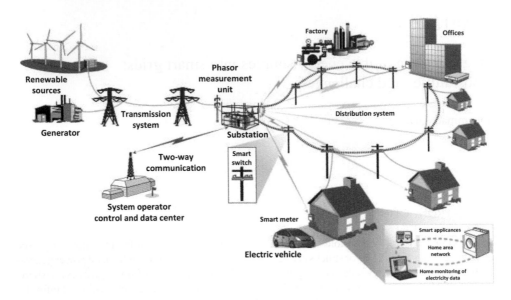

Figure 14.1.    Smart grid architecture (GAO, 2011).

Table 14.1.    Comparison between a conventional grid and a smart grid.

| Properties | Conventional power grid | Smart grid |
|---|---|---|
| Power flow | Unidirectional | Bidirectional |
| Power generation | Centralized generation | Distributed generation |
| Corrections | None | Self-correcting |
| Sensors | Few | Throughout |
| Customer choices | Few | Many |

of the main differences between them are the utilization of sensors and the system control center. Power flow is also bidirectional, while the power generation will be distributed in an ideal smart grid. Also, the future smart grid entities have the ability to communicate. The smart grid will also be able to correct itself while giving customers many different choices for rates and power supply.

Renewable energy resources are best when used to generate smaller amounts of energy as well as being located in many areas. The idea is to have more distributed power generation points. When a large group of distributed generators are connected and controlled by one central control system, it is known as a virtual power plant (VPP). A VPP will normally have a capacity similar to conventional power plants (Molderink *et al.*, 2010).

There are many advantages to having a smart grid, with the main reason being for cleaner energy. With all the renewable energy sources available, the existing conventional grid is being transformed. Due to all the different technologies and power levels of each renewable energy source the current power grid cannot keep up. Furthermore, with the ageing infrastructure of the current grid, there are many problems with utilities and customers, such as low power quality and increasing costs (Rao *et al.*, 2013). Thus there is a need for further research and development into the integration of renewable sources with the smart grid. While there are many ways to integrate different types of renewable sources, the main problem is inconsistency. Solar power, for example, depends on the sun's rays being able to reach the solar panels. If there are clouds then the rays will not reach the panel, leaving it powerless. Since the weather is so inconsistent and relatively unpredictable, different techniques have been developed to help ease the integration.

Similar difficulties can be found with wind energy (Molderink *et al.*, 2010). Wind energy, for example, cannot produce energy on calm days when there is no wind. In this case location is most important.

## 14.2   OPPORTUNITIES

### 14.2.1   *Seamless integration of renewable energy sources and ushering in a new era of consumer choice*

The increasing amount of electrical energy generated from distributed energy sources requires their proper integration into the grid. The renewable power sources are mainly feeding in the power through power electronic converter systems and can be well-controlled (Fuchs *et al.*, 2012).

One thing that should be taken into account when optimizing renewable energy resources is forecasting and scheduling. Because of the variations in weather there are many differences to the energy being produced. Therefore, knowing the relevant long-term weather patterns is a big boost for optimization (Potter *et al.*, 2009).

A new area of opportunity is consumer choice. With the current grid there are many ways for a power generation company to charge and bill customers. The most common rate used is the flat rate. The company will charge the customer based on the total amount of energy used and add to that a demand charge. Demand is the peak of energy usage. The reason demand is so important is because someone has to be able to provide that peak amount of energy (Roozbehani *et al.*, 2010).

The newest form of rates is the real-time rate. The idea behind real-time rates is that the price rises as more energy is being used, not only by the individual customer but by the amount being used by the community. As the energy being used increases, so does the price of energy. The price around the peak time, which usually occurs in the middle of the day, is higher than the price at night-time (Kempton and Tomić, 2005a). This will give customers not only another type of rate but another style of energy usage. This rate makes non-peak times more attractive because of the low pricing, which will most probably make customers change the time of their energy usage to either late night or early morning. Customers with a big energy profile will look for savings by changing the time of their operations to more affordable times. This will be a big help to energy producers, as they will have less of a load during peak times. The lower the demand and the more stable the energy needed, the more efficient the energy producers can be.

### 14.2.2   *Exploiting the use of green building standards to help "lighten the load" and enabling nationwide use of plug-in hybrid electric vehicles*

Another opportunity is the building or redesigning of green buildings. Approximately 20–30% of a building's energy consumption can be saved through optimization of the operations without changing the building structure (Guan *et al.*, 2010). As mentioned earlier, the less energy needed at a certain point in time, the more efficient are the operations. One way to need less energy is by reducing the load. The most common way to optimize the operations of a building is by undergoing an energy audit. An energy audit is the analysis of the energy flow through a building and finding a way to minimize the energy input to the building without harming the functionality. The way the smart grid can help with such things is through the smart sensors. With better sensors, customers will know how much energy they are consuming with more accuracy (Fand *et al.*, 2012). Therefore, making a building follow green standards will help to improve efficiency.

Enabling nationwide use of electric vehicles (EV) is also an enormous step in improving efficiency. The combination of both renewable energy sources and EVs can significantly reduce the amount of greenhouse gases being emitted. As of January 2014, electric power generators produce 41% of the world's greenhouse gases, while transportation systems produce another 23% (Tushar *et al.*, 2014). That is a combined total of 64% of the world's greenhouse gases. The complete replacement of all electric generators and transportation systems may never happen, but there will be a significant decrease in the future (Tushar *et al.*, 2014).

An EV can also be used as a battery. In the smart grid, bidirectional power flow is a must. Therefore, an EV can be charged when the cost of electricity is cheap and used as a battery to power buildings when the cost is high. Such implementations of an EV can be very efficient and effective (Kempton and Tomić, 2005b).

### 14.2.3   *Making large-scale energy storage a reality*

Energy storage is a very important aspect of the smart grid. It was in fact used in the early days of the power industry. Energy storage can be useful for many different applications. Battery energy storage systems (BESS) are being used in many projects to help integrate wind power to the grid (Such and Hill, 2012). They are capable of providing voltage and frequency support when the generation is fluctuating.

A BESS consists of batteries, power electronics and the control system. The batteries are used to store electrical energy as chemical energy. There are many different battery chemistries; the most common of which are sodium-sulfur, lead-acid and lithium-ion chemistries. Batteries are charged and discharged with direct current. The power electronics are used to switch the alternating current coming into the battery to direct current and switch the outgoing direct current into alternating. This is done using a bidirectional power electronic interface (Such and Hill, 2012).

During certain weather events, the wind can increase energy generation significantly. This resulting output fluctuation can result in voltage variation if the correct action is not taken. Storage systems such as BESS can provide power quickly to minimize disturbances. Paired with the proper control scheme, BESS can mitigate such challenges while improving system reliability and the economics of the renewable resource. This provides us with a true smart grid solution to the integration of renewable energy sources. A BESS can also be used for solar panels, as shown in Figure 14.2. When the sun is blocked by clouds, there is a power fluctuation resulting in a decrease in power outage. When such an event occurs, the energy stored within a battery will be used to help account for the decrease in power.

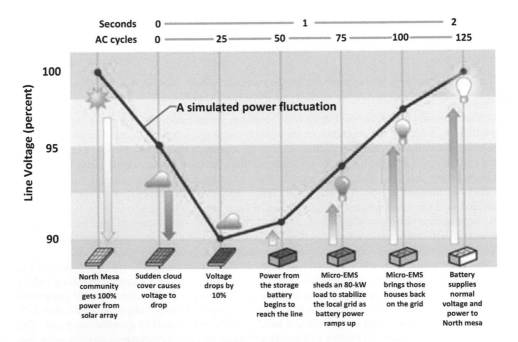

Figure 14.2.   How batteries can help solar power generation.

Many different energy storing technologies are available, such as superconducting magnetic energy storage (SMES), supercapacitor energy storage (SCES) and flywheel energy storage (FES) (Molina, 2012). An SMES is a device that stores magnetic energy. It generates the magnetic field by the flow of a direct current through a superconducting coil. The coil is cryogenically cooled below its critical temperature to exhibit superconductivity. The basic principle is that when the coil is fully charged, the current will not decay, thus keeping the magnetic energy stored.

An SCES consists of a porous structure of activated carbon for one or both electrodes. The electrodes are submersed into an electrolytic solution and a separator. The separator does not allow the two electrodes to make physical contact but permits for ion transfers. This gives rise to two capacitors at the end of each electrode. This allows for a significantly high capacitance per unit of volume.

A flywheel device stores electrical energy as kinetic energy. This is done using a rotor that spins at high speeds. It transmits energy through an electric machine acting as a motor or a generator. When the flywheel is charging, the electric machine is being used as a motor to increase the speed of the rotor. When the flywheel is discharging, the electric machine is considered a generator and decelerates the rotor. The stored energy is directly proportional to the rotor momentum.

## 14.3 CHALLENGES

### 14.3.1 *Storage, communication and integration with the current power grid*

One of the biggest challenges electrical engineers face today is the storage of energy. Some of the main concerns are technological and economic. The technological challenges include increasing the capacity and efficiency of current technologies and developing new technologies. Currently, energy storage is aiding the integration of renewable sources, but in the long term it may not be able to handle decentralized generation.

The second challenge is economic. The economic challenges vary based on the functionality of the energy storage, whether it is to be used in generation, distribution, or at the consumer level. Compensation schemes may be drawn up to give incentives for more energy storage, but at the same time it is unclear how that would affect the field. Also, ownership of the storage devices will be a battle. Should the utilities be allowed ownership or the transmission system operators?

Another challenge with the smart grid is communication. Communication is a very important part of the smart grid design. For the smart grid to be a functional system all the different parts and subsystems should be able to communicate with each other. As specified earlier, the sensors must be able to send information back to the central control system for the grid to be able to make the proper adjustments to the system.

There are many different ways for the smart grid to communicate. There are four main challenges with the communication layer of a smart grid; the first of which is the fragmented architecture (Gungor *et al.*, 2013). There are many ways that the communication of a smart grid can be set up, but there has yet to be a clear leader in the race. The grid in the USA, for example, is connected from the east coast to the Rocky Mountains and then from the other side of the mountains to the west coast. As you can see there is a big gap in the grid, which is the main concern of connecting all of America.

The second challenge is the lack of adequate bandwidth. This lack of bandwidth prevents us from achieving the two-way communication that is needed for the smart grid. This problem can be solved by simply increasing the bandwidth; however, increasing the bandwidth is no simple task.

The lack of interoperability between system components is the third challenge. If the smart grid is to be a complete system, all the components within it must be able to communicate properly and efficiently. This is also something that cannot be solved overnight. This requires all parts of the system to be able to both send out information/commands and receive them.

The final challenge is the inability to handle the increasing amount of data from smart devices. Today's world is filled with many smart devices from phones to generators. Since they are smart, they deal with lots of data in order for all the different applications to function properly. Furthermore, distribution transformers (DTR) are one of the most important elements of the electrical distribution network. In order for the smart grid to be efficient and reliable, the distribution transformers must become smart themselves. In order to do so, the transformer must be given monitoring and communication capabilities. This will help to minimize DTR failures (Kempton and Tomić, 2005b).

Overloading is one of the major concerns of DTRs. To help prevent such incidents, DTRs should not only monitor and communicate but allow two-way power flows, maintain voltage and frequency levels of the connected generation, and provide efficient routing of electricity interconnections (Rao et al., 2013). DTRs are considered as the joints of a grid. How can the system work when the joints connecting generation and distribution are outdated?

### 14.3.2   Security and interoperability

The smart grid is a more efficient electrical grid that will require the transfer of large amounts of data. This data is highly sensitive and should be shared based on the user's intent. For this reason, security systems are a big challenge. According to Speiser et al. (2013), web technologies should be used. The smart grid was likened to the internet based on their similar architecture. The smart grid must be flexible, accessible, reliable and economic; it must allow open data access in the same way as the internet.

With all the new connections that will be made with the smart grid there will be more security measures needed. When complexity and integration increase, so does the communication between different generators and consumers. With the increasing use of wireless communication technology, there will be a further increase in the difficulty of protection. Also, with many more smart meters and appliances, the border of the network extends further into the user's domain. This calls for an increase in user security.

The smart grid protection system is meant to be like the rest of the smart grid. It should have an early warning system, timely and accurate discovery, active response and automatic recovery. One must never forget, however, the importance of securing the generation, transmission, substation, distribution, consumption and scheduling operational systems (Zhang et al., 2010).

Likewise the challenges of interoperability are also tremendous. The main problem with this is that we do not know what will happen until the technologies are implemented. For that reason the best choice in solving problems in interoperability is trial and error. One great example of this is the Intelligent Energy System (IES) in Singapore (Koh et al., 2012). On 19 November 2009, the IES pilot test bedding was announced at the Smart Grids 2009 Summit by Energy Market Authority (EMA) (Kempton and Tomić, 2005b). To test different technologies and analyze their benefits, EMA supplied Nanyang Technological University (NTU) with 4500 smart meters (Koh et al., 2012). NTU was able to use IES to enable the testing of various grid applications and solutions. Many applications have been tested on IES, such as: a home automation system and home electronics scheduling, monitoring and set up from either locally installed video display unit (VDU), or internet portal applications via personal computers or mobile devices (Koh et al., 2012).

Although the smart grid will allow the seamless integration of renewable resources, there are some challenges involved as well. Investigation into how to integrate these sources properly means the investigation of their controls and hardware. Proper investigation can be done through two methods: formula analysis and simulation studies. A field test, however, is time and cost consuming. There are also not enough distributed energy resources penetrating into the grid to have the appropriate conditions for a field test. Furthermore, the investigation on a downscaled energy resource in a lab is a very important way of gathering information (Fuchs et al., 2012).

### 14.3.3   Standardizing the smart grid

In order to standardize and guide the construction of the smart grid system, the International Electrical Commission (IEC), the National Institute of Standards and Technology (NIST) and the

State Grid Corporation of China (SGCC) have developed roadmaps for the smart grid. In April 2009, IEC set up a smart grid strategy working group, which completed and released the "IEC Smart Grid Standardization Roadmap Edition 1.0" (Miao *et al.*, 2012). However, a comparison of IEC, NIST and SGCC standards is needed. All three roadmaps are meant to effectively guide the planning, design, construction and operation of a smart grid. There is also a smart grid technology standards roadmap offered by all three. The technology roadmap is mainly used to realize smart grid interoperability. IEC has a more technical perspective on the matter with a bottom to top view.

There are five aspects to the IEC roadmap: overview, demand, existing standards, gaps and suggestions. The NIST, on the other hand, has a top-down view according to United States market conditions and stakeholder perspectives. The SGCC roadmap is built according to the Chinese power industry. It has eight focused domains including general planning, generation, transmission, substations, distribution, utilization and dispatching. The SGCC is more concerned with smart grid technology. Furthermore, NIST has security standards along with SCADA (supervisory control and data acquisition). The standards are categorized in three different ways. The first is in terms of terminal security, communication security and system security. The second is in terms of power generation, power transmission, substations, power distribution, utilization and power dispatch. The third is in terms of system planning, system analysis, system design, system implementation, system running and maintenance, and system obsolescence. The security standards related to each topic within each categorization are well-studied. The top ten major smart grid standardization roadmaps are:

- United States: NIST IOP Roadmap.
- Microsoft: SERA.
- IEEE: P2030.
- European Union: Mandate CEN/CENELEC M/441.
- Germany: BMWi E-Energy Program, BDI initiative - Internet der Energie.
- China: SGCC Framework.
- Japan: Ministry of Economy, Trade and Industry (METI) Smart Grid Roadmap.
- Korea: Smart Grid Roadmap 2030.
- IEC SMB: SG 3 Roadmap.
- CIGRE: D2.24.

### 14.3.4  *Electric vehicles*

Although EVs are a great way to save the environment, there are concerns about potential stresses on the grid. These stresses may arise from the random uncoordinated charging of electric vehicles. A reliable method for co-ordinated charging is to enable utilities to directly control household electric devices, such as the electric vehicle. This will be implemented through the smart grid communication infrastructure using an algorithm. The algorithm will decide when the vehicle can be charged based on certain priorities. The algorithm will update the status of the electric vehicle and determine when it is time to charge it (Masoum *et al.*, 2012).

Electric vehicles can also have an impact on the grid's power quality. The power quality is affected by non-linear current consumption of loads. A new smart charger was developed to help mitigate the impact on power quality. This smart charger has sinusoidal current consumption and unitary power factor. It allows for control of the voltage and current in the batteries for a maximum battery lifespan (Monteiro *et al.*, 2011). Table 14.2 summarizes some of the important projects, solutions and technologies used to overcome the challenges involved with the smart grid.

### 14.4  CONCLUSIONS

There are many opportunities and challenges in the development and applications of smart grids. Allowing for the seamless integration of renewable energy resources and ushering in a new era

Table 14.2.    Examples of projects, solutions and technologies used to overcome the challenges involved with the smart grid.

| Area | Challenges | Examples of solutions, projects and technologies |
|---|---|---|
| Communication | – Fragmented architecture<br>– Lack of adequate bandwidth<br>– Lack of interoperability between components<br>– Inability to handle large amounts of data | Distribution line carrier: verification, integration and test of PLC technologies and Internet protocol (IP) communication for utilities. (DLC+VIT4IP): This project develops, verifies and tests a high-speed narrow-band power line communications infrastructure using Internet protocol (IP), which can support existing and extending multiple communication applications. |
| Integration | – Outdated infrastructure<br>– Distribution transformer must become "smart" | European Distributed Energy Project (EU-DEEP): This project combines eight European energy utilities and targets the removal of most barriers that prevent deployment of distributed energy resources in Europe.<br>Flexible electricity networks to integrate the expected energy evolution (Fenix): The main objective is to boost distributed energy resources by maximizing their contribution to the electric power system, through virtual power plants and decentralized management. |
| Security | – Securing the data being communicated | Cognitive radio: This technology has proved to be a great way for the smart grid to communicate based on its safe and secure methods. |
| Interoperability | – Do not know what will happen until tested | INOVGRID: This project aims to replace the current low voltage meters with electronic devices, known as "energy boxes", using automated meter management standards.<br>Grid4EU: This project is led by a group of European distribution system operators and aims to test some innovative system concepts and technologies in real size in order to highlight and aid in the removal of some barriers concerned with smart grid deployment.<br>Model City Mannheim: This project concentrates on urban areas in which distributed renewable energy resources are used more frequently. A representative trial is being conducted both in Mannheim and Dresden to demonstrate how the project can be applied to other regions. |
| Standards | – Smart grid is not fully developed<br>– Standards will change as the smart grid becomes closer to implementation | United States: NIST IOP Roadmap, Microsoft: SERA, IEEE: P2030<br>European Union: Mandate CEN/CENELEC M/441<br>Germany: BMWi E-Energy Program, BDI initiative Internet der Energie<br>China: SGCC Framework<br>Japan: METI Smart Grid roadmap<br>Korea: Smart Grid Roadmap 2030, IEC SMB: SG 3 Roadmap, CIGRE: D2.24. |

of consumer choice is a great prospect. The smart grid will also enable us to use plug-in hybrids nationwide without disturbing the power quality. However, although energy storage is a big obstacle, smart grids can make large-scale energy storage a reality. Communication, however, is still a grey area as there are many ways to achieve it but no clear front runner at the moment. So, this is a niche area where further research is needed. One of the biggest challenges is integrating the new smart grid with the old power grid, as it would be too expensive and time-consuming to replace the entire grid. Security is another much needed research area, as the information being sent throughout the system is very personal, along with the standards a smart grid should have. Interoperability is also a challenge, although this may be fixed through experiments and trial and error.

## REFERENCES

Fand, X., Misra, S., Xue, G. & Yang, D. (2012) Smart grid-the new and improved power grid: a survey. *Communications Surveys & Tutorials*, 14(4), 944–980.

Fuchs, F.W., Gebhardt, F., Hoffman, N., Knop, A., Lohde, R., Reese, J. & Wessels, C. (2012) Research laboratory for grid integration of distributed renewable energy resources: design and realization. *IEEE Energy Conversion Congress and Exposition (ECCE), 15–20 September 2012, Raleigh, NC*. pp. 1974–1981.

GAO (2011) Electricity grid modernization progress being made on cybersecurity guidelines, but key challenges remain to be addressed. United States Government Accountability Office, Washington, DC.

Guan, X., Xu, Z. & Jia, Q.S. (2010) Energy-efficient buildings facilitated by microgrid. *IEEE Transactions on Smart Grid*, 1(3), 243–252.

Gungor, V.C., Sahin, D., Kocak, T., Ergut, S., Buccella, C., Cecati, C. & Hancke, G.P. (2013) A survey on smart grid potential applications and communication requirements. *IEEE Transactions on Industrial Informatics*, 9(1), 28–42.

Kempton, W. & Tomić, J. (2005a) Vehicle-to-grid power fundamentals: calculating capacity and net revenue. *Journal of Power Sources*, 144(1), 268–279.

Kempton, W. & Tomić, J. (2005b) Vehicle-to-grid power implementation: from stabilizing the grid to supporting large-scale renewable energy. *Journal of Power Sources*, 144(1), 280–294.

Koh, L.H., Tan, Y.K., Wang, P. & Tseng, K.J. (2012) Renewable energy integration into smart grid: problems and solutions, Singapore experience. *IEEE Power and Energy Society General Meeting, 22–26 July 2012, San Diego, CA*. pp. 1–7.

Masoum, M.A.S., Moses, P.S. & Hajforoosh, S. (2012) Distribution transformer stress in smart grid with coordinated charging of plug-in electric vehicles. *IEEE PES Innovative Smart Grid Technologies (ISGT)*, 1–8.

Miao, X., Chen, X., Ma, X., Liu, G., Feng, H. & Song, X. (2012) Comparing smart grid technology standards roadmap of the IEC. NIST and SGCC. *China International Conference on Electricity Distribution (CICED), 10–14 September 2012, Shanghai, China*.

Molderink, A., Bakker, V., Bosman, M.G.C., Hurink, J.L. & Smit, G.J.M. (2010) Management and control of domestic smart grid technology. *IEEE Transactions on Smart Grid*, 1(2), 109–119.

Molina, M.G. (2012) Distributed energy storage systems for applications in future smart grids. *IEEE/PES Transmission and Distribution: Latin America Conference and Exposition (T&D-LA), 3–5 September 2012, Montevideo, Uruguay*. pp. 1–7.

Monteiro, V., Concalves, H. & Afonso, J.L. (2011) Impact of electric vehicles on power quality in a smart grid context. *International Conference on Electrical Power Quality and Utilisation (EPQU), 17–19 October, 2011, Lisbon, Portugal*. pp. 1–6.

Potter, C.W., Archambault, A. & Westrick, K. (2009) Building a smarter grid through better renewable energy information. *IEEE/PES Power Systems Conference and Exposition, 15–18 March 2009, Seattle, WA*. pp. 1–5.

Rao, N.M., Narayanan, R., Vasudevamurthy, B.R. & Das, S.K. (2013) Performance requirements of present-day distribution transformers for smart grid. *IEEE Innovative Smart Grid Technologies – Asia (ISGT Asia), 10–13 November 2013, Bangalore, India*. pp. 1–6.

Roozbehani, M., Dahleh, M. & Mitter, S. (2010) Dynamic pricing and stabilization of supply and demand in modern electric power grids. *IEEE International Conference on Smart Grid Communications (SmartGridComm), 4–6 October, Gaithersburg, MD*. pp. 543–548.

Speiser, S., Wagner, A., Raabe, O. & Harth, A. (2013) Web technologies and privacy policies for the smart grid. *IEEE Industrial Electronics Society, 10–13 November 2013, Vienna, Austria*. pp. 4809–4814.

Such, M.C. & Hill, C. (2012) Battery energy storage and wind energy integrated into the smart grid. *IEEE PES Innovative Smart Grid Technologies (ISGT), 16–20 January 2012, Washington DC*. pp. 1–4.

Tushar, M.H.K., Assi, C., Maier, M. & Uddin, M.F. (2014) Smart microgrids: optimal joint scheduling for electric vehicles and home appliances. *IEEE Transactions on Smart Grid*, 5(1), 239–250.

Zhang, T., Lin, W., Wang, Y., Deng, S., Shi, C. & Chen, L. (2010) The design of information security protection framework to support smart grid. *International Conference on Power System Technology (POWERCON), 24–28 October 2010, Zhejiang, China*. pp. 1–5.

# CHAPTER 15

## Current trends and future prospects of renewable energy-driven desalination (RE-DES)

Lourdes García-Rodríguez

### 15.1 INTRODUCTION

Everybody recognizes that access to freshwater is one of the biggest challenges for the human being, a challenge that gets ever more acute due to population increase. The World Health Organization estimates that 1.2 billion people in the world have no secure access to safe drinking water. Furthermore, inadequate sewage treatment and impure water can cause severe diarrhea, leading to dehydration and potential death. Indeed, this is a significant cause of death in many of the world's children and these problems also affect many coastal regions and areas with brackish water available, making desalination a key option.

The only bottlenecks that limit the use of desalination processes are high costs and high energy requirements. Due to their high energy consumption, the installation of desalination plants in isolated regions with no electricity grid available and difficulties in fossil fuel supply frequently requires renewable energy sources. Thus, *renewable energy-powered desalination* (RE-DES) is the only option for some rural communities and hence the importance of further developing such technologies.

From the point of view of using water desalination to provide safe and secure access to freshwater wherever there is no alternative drinking-water resource, the development of RE-DES is a responsibility for all of us. The availability of small-capacity, portable systems would permit its use elsewhere, for example, in coastal areas when infrastructure destruction has been caused by a natural disaster or war occurs. A hybrid wind-photovoltaic (PV) reverse osmosis (RO) system was designed by the Instituto Tecnológico de Canarias (ITC), a public research company of the Canary Islands (Spain), as a portable system with easy installation. This is referred to as MORENA. In addition, RE-DES plants sometimes become the only option for health protection. An exemplary case study relates to the small village of Ksar Guilenne, Tunisia, where a PV-RO plant was installed by the ITC (Subiela *et al.*, 2009). This village is 150 km away from the electrical grid and 60 km from a freshwater source. The high investment costs should not be a barrier to providing a healthy freshwater source in such cases by using RE-DES.

With regards to bigger communities, in developing countries there are frequently no adequate infrastructures for water distribution, electrical grid or wastewater treatment. This fact not only limits quality of life, but also opportunities for socioeconomic development. Therefore, a key idea would be to develop an integrated solution that includes not only energy and water production to satisfy basic needs, but also considers the planning of economic activities, wastewater treatment, healthcare and scholarship within the community. Energy generation systems based primarily on renewable energies, including hybrid energy systems and microgrids, offer the best option.

Furthermore, promotion of large-capacity desalination plants driven, either totally or partially, by renewable energies is essential to the sustainability of freshwater production. The retrofit of existing reverse osmosis (RO) plants through incorporation of a renewable energy source could be a key idea in avoiding electrical consumption from the grid during peak hours. In principle, the use of oceanic energy, such as waves, tides and the thermal gradient of the sea, would be especially suitable for seawater desalination plants. However, at the present time, the best option is wind energy, which is more cost-effective and requires a smaller land area than PV systems.

Within the context described above, this chapter deals with an assessment of current trends and future prospects in RE-DES from the author's perspective and builds on previous reviews (Blanco-Gálvez et al., 2009a; Delgado-Torres and García-Rodríguez, 2012; García-Rodríguez, 2002, 2003, 2004, 2007; García-Rodríguez and Gómez-Camacho, 2001a; García-Rodríguez et al., 2002; Peñate and García-Rodríguez, 2012a). The major objectives are to highlight the RE-DES technologies having the best opportunities to create market demand in specific scenarios, as well as to outline the principal research topics required to achieve these ends.

## 15.2   ASSESSMENT OF CURRENT TRENDS IN RE-DES TECHNOLOGIES

The selection of candidate desalination technologies mainly depends on:

- The raw water resource available, either seawater or brackish water.
- The renewable energy sources available.
- The freshwater demand.

The following subsections present an assessment of the latest, state-of-the-art RE-DES technologies, arranged in terms of the volume of freshwater demand to which they cater.

### 15.2.1   Water demands up to 1000 $m^3$ $day^{-1}$

These days, water demands of up to 1000 $m^3$ $day^{-1}$ normally correspond to a remote plant location with no electrical grid available. If the access to the plant location is difficult, the supply and storage of fossil fuels could be expensive or not feasible. This is the best scenario in which to use RE-DES, since it would be the only option by which to provide drinking water from an existing raw water resource. Within this capacity range, key issues would be to improve:

- *Batch operation*. The desalination systems that have been most developed to operate in batches of small capacity are electrodialysis (García-Rodríguez, 2003), for brackish water desalination, and membrane distillation, used for seawater desalination (Blanco-Gálvez et al., 2009a). Electrodialysis systems are suitable even in the smallest capacity systems, which is an advantage in some scenarios. PV energy is the most appropriate renewable energy source for coupling to this process (see examples in Table 15.1).
- *Variable load operation*. Specific designs, referred to as gradual-capacity systems (Miranda and Infield, 2003; Peñate et al., 2011; Pohl et al., 2009; Subiela et al., 2009), can enhance such desalination systems. The main objectives are to increase efficiency outside nominal conditions and to lower the power requirement to permit longer and cheaper operation of the desalination unit.

Systems designed for such operational modes permit the production of some water on days where energy sources are limited, even without any form of energy storage.

As the water demand increases, a key idea to partially overcome the bottleneck of costs is to consider an integrated solution in which not only water and energy production are combined, but there is also generation of economic activities that may emerge from the availability of water and energy. Thus, in addition to research and development activities that focus on the cost and efficiency of desalination components, units and plants, this could be a complementary point of view from which to promote the development of RE-DES technologies.

When it comes to the selection of RE-DES technologies, this depends on the renewable energy source(s) available. The following options should be considered:

- *Geothermal energy* is frequently used to drive an organic Rankine cycle, which can be coupled to a RO system. This option is significantly more efficient than any desalination process based

Table 15.1. Some of the first PV-driven electrodialysis plants (García-Rodríguez, 2007).

| Plant location | Plant capacity |
|---|---|
| Thar desert, India | $0.120 \, m^3 \, h^{-1}$ |
| Ohsima island, Nagasaki, Japan | $10 \, m^3 \, day^{-1}$ |
| Fukue city, Nagasaki, Japan | $8.33 \, m^3 \, h^{-1}$ |
| Spencer Valley, New Mexico, USA | $2.8 \, m^3 \, day^{-1}$ |

Table 15.2. Thermal energy consumption of desalination technologies at different top temperatures (calculated from Sharaf *et al.*, 2010a; Delgado-Torres and García-Rodríguez, 2010a).

| Technology | Thermal energy consumption [kJ kg$^{-1}$] | Process top temperature [°C] |
|---|---|---|
| Multi-effect distillation | | |
| – [1]DEAHP ($PR = 20$) | 117 | 180 |
| – Thermocompressors ($PR = 14$) | 166 | 225 |
| – $PR = 10$ | 233 | 70 |
| Reverse osmosis (main consumption: 2–3 kWh m$^{-3}$) | | |
| – Organic Rankine cycle, 36% (hexamethyldisiloxane MM) | 2–2 | 380 |
| – Organic Rankine cycle, 16% (R245ca) | 3–5 | 145 |
| – Organic Rankine cycle, 10% (R245ca) | 5–8 | 95 |

[1]Double-effect absorption heat pump.

on distillation. Table 15.2 compares the main energy consumption of both distillation and thermally driven RO based on organic Rankine cycles. In addition, note that desalination processes that involve phase changes are only appropriate to treat seawater. This is attributable to the fact that their energy consumption for brackish water desalination would be similar to that for seawater desalination. With regards to auxiliary consumption, distillation processes require a seawater coolant flow in addition to the seawater desalination feed itself. This results in auxiliary electrical consumption within the same range as the main electrical consumption for seawater desalination based on the RO process. This problem can only be overcome by coupling a heat pump with the process.

Geothermal energy-powered systems require neither energy storage nor energy backup to guarantee water production. In addition, as a continuous energy source, the plant design is based on matching: (i) the thermal power available and the desalination power consumption; (ii) the nominal plant capacity and the daily average water demand. This gives geothermal energy significant advantages in comparison to discontinuous energy sources. The latter require oversizing of the energy subsystem with respect to the desalination unit, and installation of energy storage in order to increase the daily operating hours of water production. Moreover, periods with no water production have to be balanced by oversizing the desalination unit to compensate at other times.

- *Wind energy* is the best option if there are exploitable wind resources at the plant location. As an example of system sizing, Table 15.3 and Table 15.4 show simulation results obtained for Pozo Izquierdo (Gran Canaria), Canary Islands (Fig. 15.1). A thorough description of the procedure for calculation of wind resource from experimental data is reported by Peñate *et al.* (2011).
- *Solar energy* is normally abundant in arid regions, thus making it the best option if neither geothermal nor wind energies are available. However, distillation using solar power is expensive

Table 15.3. Results of annual simulation for three wind turbines coupled to a $1000\,\mathrm{m^3}\,\mathrm{day^{-1}}$ seawater RO plant with nominal consumption of $2.78\,\mathrm{kWh\,m^{-3}}$ ($112\,\mathrm{kW}$ nominal; $82.8\,\mathrm{kW}$ minimal operation point) (adapted from Peñate *et al.*, 2011).

| Results | Fuhrlander FL100 (100 kW) | Vestas V27 (250 kW) | Enercon E32 (300 kW) |
|---|---|---|---|
| Total water produced [m³] | 172896 | 230318 | 250865 |
| Annual average product flow [m³ day⁻¹] | 47369 | 631.01 | 687.30 |
| Total plant operation hours [h] | 4487 | 5688 | 6135 |
| Annual operation rate [%] | 51 | 65 | 70 |
| Total plant stop hours [h] | 4273 | 3072 | 2625 |
| Energy produced by wind turbine [kWh] | 468674 | 886511 | 1299430 |
| Energy consumed by the desalination plant [kWh] | 467161 | 620901 | 676008 |
| Excess energy [kWh] | 1512 | 265610 | 623421 |
| Productivity ratio, $q_{vp}/P_{WT}$ | 0.37 | 0.26 | 0.19 |

Table 15.4. Results of annual simulation for three wind turbines coupled to a $1000\,\mathrm{m^3}\,\mathrm{day^{-1}}$ seawater RO plant designed for gradual capacity (adapted from Peñate *et al.*, 2011).

| Results | Fuhrlander FL100 (100 kW) | Vestas V27 (250 kW) | Enercon E32 (300 kW) |
|---|---|---|---|
| Total water produced [m³] | 159166 | 223090 | 245540 |
| Annual average product flow [m³ day⁻¹] | 436.07 | 611.21 | 672.71 |
| Total plant operation hours [h] | 6464 | 6844 | 7062 |
| Annual operation rate [%] | 74 | 78 | 81 |
| Total plant stop hours [h] | 2296 | 1916 | 1698 |
| Energy produced by wind turbine [kWh] | 468674 | 886511 | 1299430 |
| Energy consumed by the desalination plant [kWh] | 468599 | 660850 | 728179 |
| Excess energy [kWh] | 74 | 225661 | 571250 |
| Productivity ratio, $q_{vp}/P_{WT}$ | 0.34 | 0.25 | 0.19 |

and inefficient in comparison to solar-driven PV-RO (Table 15.5). In addition, specific devices to couple solar thermal concentrators to RO plants have been developed (Childs *et al.*, 1999; García-Rodríguez, 2003), as well as to solar organic Rankine cycles (Delgado-Torres and García-Rodríguez, 2007a, 2007b, 2007c, 2007d, 2010b; Delgado-Torres *et al.*, 2007; García-Rodríguez and Blanco-Gálvez, 2007; García-Rodríguez and Delgado-Torres, 2007; Peñate and García-Rodríguez, 2012b; Sharaf *et al.*, 2010a). Finally, although dish-Stirling collectors can also power RO plants, no pilot plant has thus far been established to the knowledge of the author.

In terms of the sizing of a solar-driven RO system, sun-tracking systems generally permit longer operational hours than stationary systems and more stable energy production. In stand-alone plants, to satisfy a given water demand, this results in a requirement for lower energy storage and a smaller nominal capacity. The use in a desalination system of the recommended size of stationary thermal collector results in operating hours equivalent to 30-35% of the year. By comparison, use of parabolic troughs similar to those used in electricity generation would require operation for 50-55% of the year.

- *Wave energy* is, in principle, a good option for coupling to RO (Clément *et al.*, 2002; Cormick *et al.*, 1997; García-Rodríguez, 2003; Hicks *et al.*, 1989). However, it does not compete in seawater desalination with mature technologies such as wind- or PV-driven RO.

Figure 15.1.  Instituto Tecnológico de Canarias (ITC) test facilities at Pozo Izquierdo (courtesy of ITC).

Table 15.5.  Solar energy consumption of seawater desalination (compiled and calculated from Delgado-Torres and García-Rodríguez, 2007c, 2010a, 2012; García-Rodríguez, 2007) (*PR*: performance ratio between 2330 kJ kg$^{-1}$ and the specific thermal consumption; dimensionless).

| Technology | Solar collector | Solar energy consumption [kJ kg$^{-1}$] | Top temperature within the solar collector [°C] |
|---|---|---|---|
| Multi-effect distillation | | | |
| – [1]DEAHP (*PR* = 20) | Parabolic troughs, 73% | 161 | 270 |
| – Thermocompressors (*PR* = 14) | Parabolic troughs, 71% | 234 | 315 |
| – *PR* = 10 | Parabolic troughs, 74% | 317 | 230 |
| – *PR* = 10 | [2]LTC – Evacuated | 333–369 | 80 |
| – *PR* = 10 | [2]LTC – Non-evacuated | 545–1600 | 80 |
| Reverse osmosis, 2–3 kWh m$^{-3}$ (main) | | | |
| – Solar organic Rankine cycle, 20.6% (siloxane MM) | Parabolic troughs | 3–4 | 337 |
| – Solar organic Rankine cycle, 8.4% (R245ca) | [2]LTC – Evacuated | 7–10 | 150 |
| – Solar organic Rankine cycle, 4.4% (R245ca) | [2]LTC – Non-evacuated | 13–19 | 100 |
| – Solar PV conversion, 16% | PV field | 3–5 | – |

[1]Double-effect absorption heat pump; [2]Low-temperature collectors.

The concept of gradual capacity in RO plants was developed to accommodate intermittent wind and PV energy sources, but it is also applicable to any discontinuous energy source. In cases where water demand ranges from 50 to 1000 m$^3$ day$^{-1}$, gradual-capacity systems are recommended. The opposite to this gradual-capacity concept is really the conventional PV-RO design with batteries, in which the RO plant operates at or close to its nominal conditions at all times.

In principle, there are two options available to design a gradual-capacity RO desalination system. In the first option, the plant consists of a number of RO units operated in parallel, which could be connected or disconnected independently. Assimacaopoulos (2001) describes such a RO pilot plant that is able to adapt to variable wind power. The capacity range varies from 60 m$^3$ day$^{-1}$ to 900 m$^3$ day$^{-1}$ with eight modules operating in parallel. Another pilot plant with eight parallel RO modules was installed at the ITC within the framework of the SDAWES project (Carta *et al.*, 2003; Subiela *et al.*, 2004). Each of the RO modules has a nominal capacity of 25 m$^3$ day$^{-1}$.

Further, another pilot RO desalination plant of analogous design consists of three identical sub-units. This was experimentally analyzed by Kosmadakis *et al.* (2015). In addition, the ITC has also developed another gradual-capacity concept based on a $1000 \, m^3 \, day^{-1}$ plant consisting of two RO racks, one with capacity of $400 \, m^3 \, day^{-1}$ and one with $200 \, m^3 \, day^{-1}$. A thorough analysis of wind-driven desalination is reported in Peñate *et al.* (2011).

The second option by which an RO plant can adapt its energy consumption to variable input power is changing its operating parameters, although it should be recognized that product quality is strongly affected by operational conditions. Moreover, the availability of energy recovery devices capable of operating in variable conditions has, historically, been limited. To this end, the Clark pump energy recovery system, specifically developed for RO desalination powered by renewable energies (Thomson *et al.*, 2002) is noteworthy. Miranda and Infield (2002) reported on a RO pilot plant that is able to adjust to variable wind power without the use of batteries.

### 15.2.2   *Water demands from 1000 $m^3$ day$^{-1}$ to 10,000 $m^3$ day$^{-1}$*

For the range of water demands between 1000 and $10,000 \, m^3 \, day^{-1}$, quite different RE-DES technologies are suitable. The best renewable energy conversion for water desalination is mechanical energy to mechanical energy, namely wind, wave and tidal energy. However, other energy conversions have been considered too.

In the 20th century, special attention was paid to solar thermal collectors for driving conventional distillation units, multi-effect distillation (MED) and multi-stage flash (MSF) distillation (Blanco-Gálvez *et al.*, 2009b; García-Rodríguez and Gómez-Camacho, 1999a, 1999b, 1999c, 1999d, 2001b; García-Rodríguez *et al.*, 2001). When driven by conventional energy, distillation only made sense in seawater desalination integrated into power plants. This is mainly attributable to the low maximum temperature needed to avoid scaling and the cooling requirements. The MSF process requires the circulation of a relatively large mass of brine along with a cooling flow of seawater. This results in extremely high auxiliary power consumption. MED requires a smaller flow of seawater coolant because it is more efficient than MSF plants. If thermocompressors are coupled to the MED unit, the cooling seawater flow is even lower or not necessary at all, but thermocompressors require a source of higher pressure and temperature steam. In spite of these drawbacks, MED and MSF processes were considered for solar-powered desalination, not least because, in the past, RO exhibited high energy consumption and today's efficient energy recovery devices were not available. Nevertheless, nowadays, MED and MSF distillation processes have been rendered obsolete in this context, mainly by high primary energy consumption in support of the temperatures required and by high capital costs. The thermal consumption of the current distillation technology is about $230 \, kJ \, kg^{-1}$ (at 70°C) or $180 \, kJ \, kg^{-1}$ (at $\sim$200°C). The smallest main energy consumptions were obtained in the following units:

- *SOL-14 plant coupled to a double-effect absorption heat pump (DEAHP)*. A prototype of a DEAHP was coupled to a MED unit called the SOL14 plant. A thermal energy consumption of about $115 \, kJ \, kg^{-1}$ (saturated steam at 180°C) was experimentally obtained with an advanced process, which was developed at the Plataforma Solar de Almería (PSA) research centre by the Spanish government's Centre for Investigation of Energy, Environment and Technology (CIEMAT) in cooperation with the company Entropie (Alarcón-Padilla and García-Rodríguez, 2007; Alarcón-Padilla *et al.*, 2007, 2008, 2010a, 2010b, 2010c; Palenzuela *et al.*, 2014; García-Rodríguez *et al.*, 2001). This technology did not reach the pre-commercial stage as even such low consumption was not able to compete with RO technology (Table 15.5).
- *Barge MED-thermocompression unit*. This exhibited a thermal energy consumption of about $105 \, kJ \, kg^{-1}$ (Darwish and Abdulrahim, 2008).

In addition to their comparatively low efficiency, distillation units should not be intermittently operated due to fouling and scaling on heat exchanger surfaces and further efficiency losses outside nominal conditions. Thus, the use of solar thermal energy to drive distillation processes should not be considered when in competition with solar thermal RO.

A salinity-gradient solar pond is a solar system that integrates solar thermal energy conversion and thermal storage within a single device. They are natural or artificial lakes in which gradients of temperature and salt concentration are maintained as a result of the corresponding gradient in water density. The bottom of the pond exhibits the highest salt concentration and temperature. Three layers can be distinguished, with the uppermost one being called the upper convective layer and consisting of freshwater. Its main role is to isolate the next layer from atmospheric phenomena. A supply of freshwater is required in order to balance water loss due to evaporation. This freshwater requirement is the main drawback of artificial solar ponds. Nevertheless, treated wastewater could be used for this purpose, thus resulting in a level of water consumption that does not reduce that produced in desalination. The second layer is called the non-convective layer. A gradient of salt concentration prevents convective mass transfer because the saline water is stratified due to the increase of density with depth. This layer acts as thermal insulation for the last and lowest layer, referred to as the lower convective layer. This is on the bottom and consists of water with a salt concentration of about 20% or higher. Temperatures as high as 100°C can be reached in artificial solar ponds, while in natural solar ponds temperatures of about 50–65°C have been described (Hull, 1989). Solar ponds provide continuous and inexpensive thermal energy and exhibit a good economy of scale. Nevertheless, the temperature of the heat storage and also the recommended power of heat extraction vary depending on the season. Therefore, desalination systems driven by solar ponds should be able to operate at variable levels of temperature and thermal power. An MSF unit called the Atlantis "AutoFlash" was designed for this purpose, and a 300 m$^3$ day$^{-1}$ plant was installed on the Cape Verde islands (Szacsvay *et al.*, 1999).

Today, salinity-gradient solar ponds are well-understood, that is, the influence of their main parameters on operation and maintenance, the technology for heat extraction, modeling, etc. Examples of systems that could be driven by salinity-gradient solar ponds include thermal-powered desalination, absorption chillers, and absorption heat pumps or organic Rankine cycles to produce electricity, which may also be used to drive desalination and wastewater treatment processes. Therefore, the adequate use of solar ponds permits the fulfillment of basic needs, namely, water, electricity, heating and cooling. In addition, since electricity and thermal energy is made available, diverse economic activities could be generated as well. Their application could enable a significant improvement in quality of life. In addition to the aforementioned plant installed on the Cape Verde islands, other distillation plants driven by salinity-gradient solar ponds have been implemented in the following locations:

- *Margarita de Savoya, Italy:* Delyannis (1987) reports on this plant, which is based on an MSF unit with a capacity of 50–60 m$^3$ day$^{-1}$.
- *El Paso, Texas:* This installation is described by Lu and Swift (1998). One of the desalination systems implemented consists of a MSF plant of 19 m$^3$ day$^{-1}$ capacity.
- *University of Ancona, Italy:* In this test facility (Caruso and Naviglio, 1999), the distillation process used is MED with thermocompression. The MED unit, with a capacity of 30 m$^3$ day$^{-1}$, is directly driven by the solar pond. Because the solar pond temperature is not high enough to drive the thermocompression process, conventional energy is also used. This pilot plant thereby demonstrated the concept of using conventional energy as backup by means of ther-mocompressors. Note that solar parabolic troughs are also able to drive thermocompressors, which is a feasible concept too: a plant based on a MED unit with thermocompression driven by parabolic troughs was installed at the PSA to demonstrate this concept (Alarcón-Padilla and García-Rodríguez, 2007).
- *Laboratoire de Thermique Industrielle, Tunis:* Safi (1998) describes an MSF unit at laboratory scale.
- *Near Dead Sea:* A pilot plant was installed with a MED unit of 3000 m$^3$ day$^{-1}$ (EC, 1998).

Solar parabolic concentrators can drive a RO plant by means of an organic Rankine cycle. This technology was mainly developed in the first decade of the 20th century (Bao and Zhao, 2012; Chen *et al.*, 2010, 2011; Gang *et al.*, 2011; Manolakos *et al.*, 2009; Quoilin *et al.*, 2011a,

Table 15.6.  Some early RO plants driven by means of an organic Rankine cycle powered by solar thermal energy (García-Rodríguez and Delgado-Torres, 2007).

| Plant location | Solar system |
| --- | --- |
| Los Baños California, USA | Solar pond |
| El Paso, Texas, USA | Solar pond |
| Cadarache, France | Flat-plate collector (3 kW) |
| El Hamrawin, Egypt | Flat-plate collector (10 kW) |

2011b; Rayegan and Tao, 2011; Shengjun *et al.*, 2011; Tchanche *et al.*, 2009, 2010; Twomey *et al.*, 2013; Wang JL *et al.*, 2010, 2012; Wang M *et al.*, 2013; Wang XD *et al.*, 2010), although it was scarcely implemented. Table 15.6 shows some of the first plants implemented. The overall performance of the solar organic Rankine cycle is obtained by considering both the solar thermal energy conversion and the performance of the Rankine cycle. At the beginning of the 21st century, seawater desalination by solar organic Rankine cycle-RO was significantly superior in efficiency to PV-RO desalination. Nevertheless, the continuing evolution in the efficiency of solar PV conversion and costs make PV-RO technology the more suitable solar technology. PV systems require less area than solar thermal-powered organic Rankine cycles, but the latter have been more cost-effective so far. Nevertheless, it is expected that PV-RO systems will continue to reduce in cost. In both technologies, the energy backup should be electricity, since the thermal performance of the organic Rankine cycle is so low as to need an energy backup. Another issue that should be considered is that coastal areas might not be suitable to install large parabolic trough collectors. This is attributable to damage of the reflective surfaces and lower direct normal irradiance, among other issues. In addition, parabolic troughs of smaller size normally have protective covers but exhibit limited maximum temperatures, which in turn limits their power cycle performance. Therefore, PV systems would, in principle, be the more suitable for seawater desalination. Eventually, advanced thermal energy storage and advanced batteries will enable the respective progress of both technologies.

### 15.2.3   *Water demands greater than 10,000 m³ day⁻¹*

Only mature RE technologies are useful for desalination plants in meeting water demands above $10,000\,\text{m}^3\,\text{day}^{-1}$. If exploitable wind resources are available, wind-powered RO is the best option for seawater desalination. In the case of brackish water desalination, RO and electrodialysis compete, depending on the salt concentration. Electrodialysis exhibits the lowest energy consumption at low salt concentrations. In addition, wave energy systems, when fully developed should, in principle, be one of the best options to power large-scale seawater desalination.

Another alternative is to integrate the desalination plant with an existing power plant driven by renewable energy based on either concentrated solar power or the oceanic gradient. The latter is a cost-effective alternative for large-scale seawater desalination. Currently, RO is the only technology recommended for seawater desalination.

Regarding solar power plants, they require two or three different product qualities: water for the power cycle – with a salt concentration within the range of several mg $\text{L}^{-1}$, and freshwater for human consumption and for cleaning the solar mirrors, depending on the cleaning method. Therefore, the plant should be located near fresh or brackish water resources if possible. Solar power plants are not normally located in coastal areas, although it would be feasible. This is attributable to corrosion problems on mirrors and lower values of direct normal irradiance, as previously described. The integration of desalination with solar power plants has been analyzed in the literature (Palenzuela *et al.*, 2011a, 2011b, 2013; Sharaf *et al.*, 2010b, 2011a, 2011b). Specific research on this particular desalination integration is also being conducted within the framework of the European Energy Research Alliance's Joint Program on Concentrating Solar Power, Sub-program 4: Concentrated Solar Power and Desalination (EERA, 2015). Presently, desalination driven by solar power plants may consist of a brackish water plant based on either the RO or the electrodialysis process. If RO is selected, the end-product requires further treatment before use in the power cycle: electrodeionization and ionic exchange resins could be used. Normally, a brackish resource is not sufficient for large-capacity desalination. Hence, the integration of a desalination process into solar power plants is not usually a solution for meeting a large freshwater demand. In such cases, an additional desalination plant installed near the sea would be required. Finally, another critical issue is that even if heat storage based on molten salts is used (Torresol Energy, 2015), solar power plants operate intermittently throughout the year unless they use fossil fuel as an energy backup. Thus, distillation would only be reliable in plants with energy backup. These drawbacks, together with their relative inefficiency, make distillation processes inadequate for these higher levels of demand.

With regards to solar electricity production, driving RO plants by means of solar PV or dish-Stirling concentration systems is a feasible option. It is applicable to both seawater and brackish water desalination. The efficiency and cost evolution of PV systems make this technology the best. Unsurprisingly, the interest in developing gradual-capacity RO systems in order to minimize the requirement for battery capacity is even higher here than in smaller-scale desalination plants. However, the problems of concentrated solar collectors in coastal areas previously described also be a drawback for dish-Stirling collectors. Desalination based on solar organic Rankine cycles is also less suitable than PV-driven desalination for these same reasons. Finally, there are no solar thermal technology options for driving distillation processes due to its low efficiency, requirement for both continuous operation and seawater cooling, and high capital costs.

## 15.3   FUTURE PROSPECTS FOR RE-DES TECHNOLOGY DEVELOPMENT

The main recommendations for RE-DES research from the point of view of the author are presented in the following subsections, ordered by the capacity range of their application, from small-to large-scale desalination.

### 15.3.1   *Desalination plants based on batch operation*

In principle, the use of desalination processes that are able to operate in batch mode permits plant designs without energy storage along with the full use of the energy produced. Therefore, this concept is a key issue in small-capacity systems driven by renewable energies.

A few desalination systems are able to operate in batches, such as electrodialysis and membrane distillation (MD). However, electrodialysis is only suitable for brackish water desalination. In addition, MD is not efficient, which is inherent to the thermodynamics of the process (Vega-Beltrán *et al.*, 2010). Therefore, the availability of efficient seawater desalination systems would permit significant capital cost savings in cases of low water demand. RO desalination might be reliable, efficient and cost-effective here, and forward osmosis might also be a candidate process.

### 15.3.2   *Development of brine concentration and crystallization based on membrane distillation (MD)*

Crystallization based on small-scale MD is feasible and reliable, according to the literature (Creusen *et al.*, 2012, 2013; Edwie and Chung, 2012; Ge *et al.*, 2014; Ji *et al.*, 2010; Macedonio *et al.*, 2007; Mariah *et al.*, 2006; Meng *et al.*, 2010, 2014, 2015; Prince *et al.*, 2012). This enables the extraction of seawater components with high economic value. Brine concentration and crystallization are applications that make MD attractive, since it does not compete with any other technology in these areas. The input to the MD system is the brine output of another desalination process, thus resulting in a hybrid desalination system. Although crystallization of seawater components is not new, the sparse literature mainly deals only with laboratory-scale tests. Therefore, a thorough model of the process, and the development of a prototype at pre-commercial scale, would be innovative.

### 15.3.3   Emergent processes and disruptive concepts

Hybrid desalination plants and hybrid energy plants, in particular, make quite different features of desalination processes useful. Therefore, in addition to improving mature desalination systems, all emergent technologies and new disruptive concepts should be analyzed case-by-case. However, the main requirement of renewable energy-driven processes is high efficiency because the energy subsystems exhibit high capital costs.

Special attention should be paid to assessing theoretical and actual driving forces and efficiency limits in order to avoid overestimation of efficiency. Moreover, thorough analysis and comparison of technologies within the framework of specific target markets is important to establish the actual benefit of developing the new technology or concept being assessed.

In terms of non-mature technologies, as well as the existing forward osmosis systems, from the point of view of the author there are still possibilities for development of innovative system design concepts based on membranes semipermeable to water passage. The main objective is to operate at as low a pressure as possible while achieving reasonable product flux and quality in the system as a whole. In terms of RO, highly innovative membranes based on recognition of water molecules have been developed, such as Aquaporin (Mangrove Membranes, 2013), as well as other advanced membranes. Nevertheless, designs based on innovative membranes might not be fully optimized as yet.

### 15.3.4   Gradual-capacity RO desalination plants

In order to avoid or, at least, minimize capital costs associated with energy storage, an RO system should be able to operate at partial load with high efficiency. As freshwater demand increases, the practice of batch operation in the desalination process in order to minimize energy storage needs should be replaced by gradual-capacity operation.

Some innovative gradual-capacity configurations have been identified that seem to be feasible and reliable. They are applicable to diverse renewable energies: wind, wave, tidal, solar PV and systems based on solar thermally driven power cycles. Moreover, in the case of microgrids, the gradual-capacity feature of the desalination process permits its use as a managed load.

With regards to distillation, the configuration of MED units normally permits a modified control strategy in order to operate them at "gradual capacity". This is an innovative concept in MED technology operation that would result in the ability to operate continuously throughout the year if the plant was powered by salinity-gradient solar ponds, in which not only the temperature of the heat source, but also the power extraction are variable depending on the season. A preliminary test campaign has been conducted at PSA-CIEMAT (Fernández-Izquierdo et al., 2012).

### 15.3.5   Innovative configurations of RO desalination plants

RO is the dominant technology in the desalination market. Currently, the main electricity consumption for seawater RO desalination in the Canary Islands is in the range 1.85–2.2 kWh m$^{-3}$. However, target consumption is set as 1 kWh m$^{-3}$ by the European Commission's NMP-24-2015 research program, 'Low-energy solutions for drinking water production' (EC, 2015). Only highly innovative concepts are likely to achieve such an ambitious target.

There are relevant research activities on membranes at laboratory scale (Buonomenna, 2013; Fathizadeh et al., 2011; Kang and Cao, 2012; Kazemimoghadam, 2010; Kim et al., 2010; Kurt et al., 2009; Lee et al., 2011; LG Chem, 2012a, 2012b; Lin et al., 2010; Macedonio et al., 2012; Pendergast et al., 2010, 2013; Xu et al., 2012). In order to promote the manufacture of innovative membranes, not only is systematic membrane testing important, but also assessment of their prospective impact on plant efficiency and costs.

Conceptually, inefficiencies within a plant are attributable not only to individual components, but also to the overall plant configuration. Thus, if advanced commercial products simply adopt existing plant configurations, this might result in an incomplete efficiency improvement. This is the key idea behind identifying opportunities for innovative design in seawater RO plants. The designs of such plants have only changed slightly over the last 15 years (Peñate and García-Rodríguez, 2011a, 2011b, 2011c), despite the great advances exhibited by membrane elements and changes in standards of product quality, such as boron content (Rodrigo and Peñate, 2006). Therefore, there are still opportunities for plant improvement (Pérez-González *et al.*, 2012; Sassi and Mujtaba, 2010, 2011) and enhanced designs.

### 15.3.6   *Hybrid desalination plants*

As desalination plant capacity increases, the possibility of taking advantage of several processes appears through the design of hybrid desalination plants. Innovative plant designs could be created, especially when the ability of several processes to use the concentrate of the existing desalination process is considered. This increases the overall recovery rate of the desalination plant, thus resulting in capital and operational cost savings due to reductions in seawater intake and the chemicals required. Moreover, a related concept is the zero liquid discharge (ZLD), which could be required in specific plants, for example, some brackish water desalination plants, in order to avoid salinization of the brackish source well. ZLD plants are now feasible, based on either conventional or renewable energy sources, although high investment costs are currently required. A cost-efficient plant should make use of the most efficient process within the specific salt concentration range that pertains.

### 15.3.7   *Design for process integration aimed at promoting socioeconomic development*

To improve the opportunities for socioeconomic development, freshwater production and the parallel energy generation, along with wastewater treatment of effluents from human and economic activities, should be optimized as a whole. It is innovative to analyze the global concept of integrated design of desalination, wastewater treatment and energy production, in order to create sustainable solutions for socioeconomic development in quite different scenarios. Key issues could be:

- *Desalination and wastewater processes:* Desalination processes could take advantage of wastewater treatment effluents in different ways. This subject should be further investigated.
- *Economic activities:* A thorough assessment of the energy and water consumption of different activities that could provide socioeconomic development should be carried out in order to elaborate recommendations in this regard. In addition, economic activities based on using the concentrate generated by the desalination process should be similarly assessed.
- *Transport:* Since many economic activities, including supply of food and other basic needs, require adequate transport media, sustainable vehicles should also be analyzed with a view to integrating them into the proposed solutions.

## 15.4   MARKET OPPORTUNITIES

At the current time, RE-DES technologies mainly focus on small-size plants to supply a basic need where no conventional energies are available. However, research and development in this area may also be applicable to conventional plants and may have a significant influence on cost reduction, thereby promoting RE-DES.

Moreover, since renewable energy systems involve high capital costs, the efficiency of the desalination processes used is even more important than in conventional energy-driven desalination plants. Therefore, advances achieved in desalination efficiency within the framework of RE-DES could be applicable to the conventional desalination market as well.

Finally, a small-capacity RE-DES plant is an ideal scenario in which to demonstrate technologies and acquire operational experience with relatively low investment. This fact could be a key aspect in obtaining private funding to develop emergent technologies and demonstrate disruptive technology concepts.

The following subsections discuss different scenarios of interest in RE-DES development.

### 15.4.1  Sustainable solutions based on renewable energies for rural communities in developing countries

An integrated solution to supply basic needs and to generate economic activities based on renewable energies and microgrids may promote the demand for RE-DES in rural communities in developing countries.

A key issue in small water demands is the development of discontinuous and efficient desalination processes in order to minimize energy storage, since storing freshwater is much more cost-effective than storing generated energy. Moreover, in the case of microgrids, desalination and wastewater treatment systems could take on the role of managing load. Wastewater treatment processes with low energy requirements, in combination with natural depuration, could be applied in small villages. Gradual-capacity RO plants are also useful in these scenarios.

Finally, focusing on the added value of installing the RE-DES plant could be an essential point in promoting these technologies: not only to produce excess energy for other purposes by oversizing the energy system, but also to develop economic activities. It would be important to assess the suitability of economic activities that could be generated from the desalination process itself, such as:

- *Crystallization* to obtain seawater components with economic value by using membrane distillation;
- *Microalgae growth* from effluents of the desalination process.

### 15.4.2  Sustainable solutions based on renewable energies for developed countries, islands, isolated communities and resorts

In developed countries, it is not uncommon to find isolated communities, natural parks and, of course, resorts by the coast, or sometimes on islands. RE-DES could provide sustainable solutions in such situations. Some key developments in RE-DES that may be needed in order to encourage take-up and achieve significant advances from the current state-of-the-art may include:

- *Sustainable integrated solutions* of water and electricity production and wastewater treatment driven by renewable energy, from which RE-DES may be able to take advantage of process synergies.
- *Batch operation of desalination processes* to minimize energy storage in small-capacity systems. This is applicable to the use of systems based on a single renewable or hybrid energy systems, including microgrids.
- *Gradual-capacity RO plants* that can minimize the use of batteries in wind- and PV-powered systems. They could also be usefully driven by wave and tidal energy, by hybrid energy systems or by microgrids.
- *Gradual-capacity distillation systems* coupled to solar ponds.
- *Integration of desalination processes with conventional power production systems* to use a few megawatts of generated power, if required, as energy backup.
- *Integration of solar thermal-driven desalination systems with industrial processes where waste heat is available* in order to use it as energy backup. Depending on the temperature and power

available, membrane distillation could be a candidate technology for small-capacity plants. However, solar organic Rankine cycles are the better option, with selection of an appropriate solar collector and working fluid.

### 15.4.3   *Sustainable solutions based on existing desalination plants*

In terms of the target of sustainable desalination development, the key issues are the avoidance of fossil fuel depletion and the elimination of hazardous effluents. Advanced gradual-capacity concepts that are being developed to work with wind and PV installations could also be useful in existing large-capacity RO plants. This would permit the disconnection from the grid of the desalination plant at peak hours, and the use of wind and PV energy without having to connect the solar field or the wind farm to the grid. On islands, this would enable a larger contribution by renewable energies to the island's overall consumption with no effect on the grid.

Furthermore, advances achieved within the framework of RE-DES are also applicable to conventional desalination processes. Retrofits of existing RO plants represent an excellent scenario by which to introduce highly innovative components to the marketplace and improve desalination efficiency, as the corresponding capital costs are easily amortized by the savings in energy consumption. This is exactly analogous to the historical case in which obsolete Pelton turbines in the RO market were replaced with isobaric chambers. Moreover, immature desalination processes developed under RE-DES could then have application in RO plants, giving rise to a hybrid desalination process.

## REFERENCES

Alarcón-Padilla, D.C. & García-Rodríguez, L. (2007) Application of absorption heat pumps to multi-effect distillation: a case study of solar desalination. *Desalination*, 212, 294–302.
Alarcón-Padilla, D.C., García-Rodríguez, L. & Blanco-Gálvez, J. (2007) Assessment of an absorption heat pump coupled to a multi-effect distillation unit within AQUASOL project. *Desalination*, 212, 303–310.
Alarcón-Padilla, D.C., Blanco-Gálvez, J., García-Rodríguez, L., Gernjak, W. & Malato-Rodríguez, S. (2008) First experimental results of a new hybrid solar/gas multi-effect distillation system: the AQUASOL project. *Desalination*, 220(1), 619–625.
Alarcón-Padilla, D.C., García-Rodríguez, L. & Blanco-Gálvez, J. (2010a) Design recomendations for a multi-effect distillation plant connected to a double-effect absorption heat pump: a solar desalination case-study. *Desalination*, 262(1–3), 11–14.
Alarcón-Padilla, D.C., García-Rodríguez, L. & Blanco-Gálvez, J. (2010b) Connection of absorption heat pumps to multi-effect distillation systems: pilot test facility at the Plataforma Solar de Almería (Spain). *Desalination and Water Treatment*, 18(1–3), 126–132.
Alarcón-Padilla, D.C., García-Rodríguez, L. & Blanco-Gálvez, J. (2010c) Experimental assessment of connection of an absorption heat pump to a multi-effect distillation unit. *Desalination*, 250, 500–505.
Assimacaopoulos, D. (2001) Desalination powered by renewable energy sources. *REFOCUS The International Renewable Energy Magazine*, July/August 2001, 38–43.
Bao, J. & Zhao, L. (2012) Exergy analysis and parameter study on a novel auto-cascade Rankine cycle. *Energy*, 48, 539–547.
Blanco-Gálvez, J., Alarcón-Padilla, D.C., Malato-Rodríguez, S., Fernandez-Ibañez, P., Gernjak, W. & Maldonado-Rubio, M.I. (2009a) Review of feasible solar energy applications to water processes. *Renewable and Sustainable Energy Reviews*, 13(6–7), 1437–1445.
Blanco-Gálvez, J., García-Rodríguez, L. & Martín-Mateos, I. (2009b) Stand-alone seawater desalination by innovative solar-powered membrane-thermal distillation system: MEDESOL project. *Desalination*, 246, 567–576.
Buonomenna, M.G. (2013) Nano-enhanced reverse osmosis membranes. *Desalination*, 314(2), 73–88.
Carta, J.A., González, J. & Subiela, V. (2003) Operational analysis of an innovative wind powered reverse osmosis system installed in the Canary Islands. *Solar Energy*, 75, 153–168.
Caruso, G. & Naviglio, A. (1999) A desalination plant using solar heat as heat supply, not affecting the environment with chemicals. *Desalination*, 122(2–3), 225–234.

Chen, H., Goswami, D.Y. & Stefanakos, E.K. (2010) A review of thermodynamic cycles and working fluids for the conversion of low-grade heat. *Renewable and Sustainable Energy Reviews*, 14, 3059–3067.

Chen, H., Goswami, D.Y., Rahman, M.M. & Stefanakos, E.K. (2011) A supercritical Rankine cycle using zeotropic mixture working fluids for the conversion of low-grade heat into power. *Energy*, 36, 549–555.

Childs, W.D., Dabiri, A.E., Al-Hinai, H.A. & Abdullah, H.A. (1999) VARI-RO solar-powered desalting technology. *Desalination*, 125, 155–166.

Clément, A., McCullen, P., Falccao, A., Fiorentino, A., Gardner, F., Hammarlund, K., Lemonis, G., Lewis, T., Nielsen, K., Petroncini, S., Pontes, M.T., Schild, P., Sjöström, B.O., Sørensen, H.C. & Thorpe, T. (2002) Wave energy in Europe: current status and perspectives. *Renewable and Sustainable Energy Reviews*, 6, 405–431.

Creusen, R.J.M., van Medevoort, J., Roelands, C.P.M. & van Renesse van Duivenbode, J.A.D. (2012) Brine treatment by a membrane distillation-crystallization (MDC) process. *Procedia Engineering*, 44, 1756–1759.

Creusen, R., van Medevoort, J., Roelands, M., van Renesse van Duivenbode, A., Hanemaaijer, J.H. & van Leerdam, R. (2013) Integrated membrane distillation-crystallization: process design and cost estimations for seawater treatment and fluxes of single salt solutions. *Desalination*, 323, 8–16.

Darwish, M.A. & Abdulrahim, H.K. (2008) Feed water arrangements in a multi-effect desalting system. *Desalination*, 228, 30–54.

Delgado-Torres, A. & García-Rodríguez, L. (2007a) Status of solar thermal-driven reverse osmosis desalination. *Desalination*, 216, 242–251.

Delgado-Torres, A. & García-Rodríguez, L. (2007b) Preliminary assessment of solar organic Rankine cycles for driving a desalination system. *Desalination*, 216, 252–275.

Delgado-Torres, A. & García-Rodríguez, L. (2007c) Comparison of solar technologies for driving a desalination system by means of an organic Rankine cycle. *Desalination*, 216, 276–291.

Delgado-Torres, A. & García-Rodríguez, L. (2007d) Double cascade organic Rankine cycles for solar-driven reverse osmosis desalination. *Desalination*, 216, 306–313.

Delgado-Torres, A.M. & García-Rodríguez, L. (2010a) Analysis and optimization of the low-temperature solar organic Rankine cycle (ORC). *Energy Conversion and Management*, 51(12), 2846–2856.

Delgado-Torres, A. & García-Rodríguez, L. (2010b) Preliminary design of seawater and brackish water desalination system driven by low-temperature solar organic Rankine cycles. *Energy Conversion and Management*, 51(12), 2913–2920.

Delgado-Torres, A. & García-Rodríguez, L. (2012) Design recommendations for solar ORC-powered reverse osmosis desalination. *Renewable and Sustainable Energy Reviews*, 16(1), 44–53.

Delgado-Torres, A., García-Rodríguez, L. & Romero Ternero, V. (2007) Preliminary design of a solar thermal-powered seawater reverse osmosis system. *Desalination*. 216, 292–305.

Delyannis, E.E. (1987) Status of solar assisted desalination. *Desalination*, 67, 3–19.

EC (1998) A guide to desalination using renewable energies. Thermie Programme, Directorate General for Energy (DG XVII), European Commission, Brussels.

EC (2015) NMP-24-2015 – Low-energy solutions for drinking water production. European Commission, Brussels. Available from: http://cordis.europa.eu/programme/rcn/665024_en.html [accessed November 2016].

Edwie, F. & Chung, T.S. (2012) Development of hollow fiber membranes for water and salt recovery from highly concentrated brine via direct contact membrane distillation and crystallization. *Journal of Membrane Science*, 421–422, 111–123.

EERA (2015) Concentrated solar power (CSP). European Energy Research Alliance, Brussels. Available from: http://www.eera-set.eu/eera-joint-programmes-jps/concentrated-solar-power-csp/ [accessed November 2016].

Fathizadeh, M., Aroujalian, A. & Raisi, A. (2011) Effect of added NaX nano-zeolite into polyamide as a top thin layer of membrane on water flux and salt rejection in a reverse osmosis process. *Journal of Membrane Science*, 375(1–2), 88–95.

Fernández-Izquierdo, P., García-Rodríguez, L., Alarcón-Padilla, D.C., Palenzuela, P. & Martín-Mateos, I. (2012) Experimental analysis of multi-effect distillation unit operated out of nominal conditions. *Desalination*, 284, 233–237.

Gang, P., Jing, L. & Jie, J. (2011) Design and analysis of a novel low-temperature solar thermal electric system with two-stage collectors and heat storage units. *Renewable Energy*, 36, 2324–2333.

García-Rodríguez, L. (2002) Seawater desalination driven by renewable energies, a review. *Desalination*, 143(2), 103–113.

García-Rodríguez, L. (2003) Renewable energy applications in desalination. State of the art. *Solar Energy*, 75, 381–393.

García Rodríguez, L. (2004) Desalination by wind power. *Wind Engineering*, 28, 453–466.

García-Rodríguez, L. (2007) Assessment of most promising development in solar desalination. In: Rizzuti, L., Ettouney, H. & Cipollina, A. (eds) *Solar Desalination for the 21st Century*. Springer, Dordrecht, pp. 355–369.

García-Rodríguez, L. & Blanco-Gálvez, J. (2007) Solar-heated Rankine cycles for water and electricity production: POWERSOL project. *Desalination*, 212, 311–318.

García-Rodríguez, L. & Delgado-Torres, A. (2007) Solar-powered Rankine cycles for freshwater production. *Desalination*, 212, 319–327.

García-Rodríguez, L. & Gómez-Camacho, C. (1999a) Design parameters selection of a distillation system coupled to a solar parabolic trough collector field. *Desalination*, 122, 195–204.

García-Rodríguez, L. & Gómez-Camacho, C. (1999b) Thermoeconomic analysis of a solar multieffect distillation plant installed at the Plataforma Solar de Almería (Spain). *Desalination*, 122, 205–214.

García-Rodríguez, L. & Gómez-Camacho, C. (1999c) Thermoeconomic analysis of a solar parabolic trough distillation plant. *Desalination*, 122, 215–224.

García-Rodríguez, L. & Gómez-Camacho, C. (1999d) Preliminary design and cost analysis of a solar distillation system. *Desalination*, 126(1–3), 109–114.

García-Rodríguez, L. & Gómez-Camacho, C. (2001a) Perspectives of solar-assisted desalination. *Desalination*, 136(1–3), 213–218.

García-Rodríguez, L. & Gómez-Camacho, C. (2001b) Exergy analysis of the SOL-14 plant (Plataforma Solar de Almería, Spain). *Desalination*, 137(1–3), 251–258.

García-Rodríguez, L., Palmero-Marrero, A.I. & Gómez-Camacho, C. (2001) Thermoeconomic optimisation of the Sol-14 plant (Plataforma Solar de Almería, Spain). *Desalination*, 136(1–3), 219–223.

García-Rodríguez, L., Palmero-Marrero, A.I. & Gómez-Camacho, C. (2002) Comparison of solar thermal technologies for applications to seawater desalination. *Desalination*, 142(2), 135–142.

Ge, J., Peng, Y., Li, Z., Chen, P. & Wang, S. (2014) Membrane fouling and wetting in a DCMD process for RO brine concentration. *Desalination*, 344, 97–107.

Hicks, D.C., Mitcheson, G.R., Pleass, C.M. & Salevan, J.F. (1989) Delbouy: ocean wave-powered seawater reverse osmosis desalination system. *Desalination*, 73, 81–94.

Hull, J.R., Nielsen, C.E. & Golding, P. (1989) *Salinity-Gradient Solar Ponds*. CRC Press, Boca Raton, FL.

Ji, X., Curcio, E., Al Obaidani, S., Di Profio, G., Fontananova, E. & Drioli, E. (2010) Membrane distillation-crystallization of seawater reverse osmosis brines. *Separation and Purification Technology*, 71(1), 76–82.

Kang, G.D. & Cao, Y.M. (2012) Development of antifouling reverse osmosis membranes for water treatment: a review. *Water Research*, 46(3), 584–600.

Kazemimoghadam, M. (2010) New nanopore zeolite membranes for water treatment. *Desalination*, 251(1–3), 176–180.

Kim, M., Lin, N.H., Lewis, G.T. & Cohen, Y. (2010) Surface nano-structuring of reverse osmosis membranes via atmospheric pressure plasma-induced graft polymerization for reduction of mineral scaling propensity. *Journal of Membrane Science*, 354(1–2), 142–149.

Kosmadakis, G., Manolakos, D., Ntavou, E. & Papadakis, G. (2015) Multiple reverse osmosis sub-units supplied by unsteady power sources for seawater desalination. *Desalination and Water Treatment*, 55(11), 3111–3119.

Kurth, C.J., Burk, R. & Green, J. (2009) Leveraging nanotechnology for seawater reverse osmosis. *International Desalination Association (IDA) World Congress, 7–12 November 2009, Dubai, UAE*.

Lee, K.P., Arnot, T.C. & Mattia, D. (2011) A review of reverse osmosis membrane materials for desalination – development to date and future potential. *Journal of Membrane Science*, 370(1–2), 1–22.

LG Chem (2012a) Lanzarote, Spain: LG Water Solutions hybrid design lowers specific energy. Available from: http://www.lgwatersolutions.com/?page_id=493 [accessed November 2016].

LG Chem (2012b) Las Salinas, Spain: LG Water Solutions minimizes boron for strict irrigation needs. Available from: http://www.lgwatersolutions.com/?page_id=502 [accessed November 2016].

Lin, N.H., Kim, M., Lewis, G.T. & Cohen, Y. (2010) Polymer surface nano-structuring of reverse osmosis membranes for fouling resistance and improved flux performance. *Journal of Materials Chemistry*, 20(22), 4642–4652.

Lu, H. & Swift, A.H.P. (1998) An update of the El Paso Solar Pond Project. *Proceedings of the International Solar Energy Conference, Solar Engineering 1998, 14–17 June, Albuquerque, NM*. ASME, New York. pp. 333–338.

McCormick, M.E. & Kim, Y.C. (1997) Ocean wave-powered desalination. *Proceedings of the 27th Congress of the International Association of Hydraulic Research (IAHR), 10–15 August 1997, San Francisco, CA.* ASCE, New York. pp. 577–582.

Macedonio, F. & Drioli, E. (2010) Membrane systems for seawater and brackish water desalination. In: Drioli E. & Giorno, L. (eds) *Comprehensive Membrane Science and Engineering.* Elsevier, Oxford. pp. 241–257.

Macedonio, F., Curcio, E. & Drioli, E. (2007) Integrated membrane systems for seawater desalination: energetic and exergetic analysis, economic evaluation, experimental study. *Desalination*, 203(1–3), 260–276.

Macedonio, F., Drioli, E., Gusev, A.A., Bardow, A., Semiat, R. & Kurihara, M. (2012) Efficient technologies for worldwide clean water supply. *Chemical Engineering and Processing: Process Intensification*, 51, 2–17.

Mangrove Membranes (2013) Aquaporin production: in-house production. Mangrove Membranes, Inc., Nordborg, Denmark. Available from: http://www.mangrovemembranes.com/?page_id=741 [accessed November 2016].

Manolakos, D., Kosmadakis, G., Kyritsis, S. & Papadakis, G. (2009) On site experimental evaluation of a low-temperature solar organic Rankine cycle system for RO desalination. *Solar Energy*, 83(5), 646–656.

Mariah, L., Buckley, C.A., Brouckaert, C.J., Curcio, E., Drioli, E., Jaganyi, D. & Ramjugernath, D. (2006) Membrane distillation of concentrated brines – role of water activities in the evaluation of driving force. *Journal of Membrane Science*, 280(1–2), 937–947.

Meng, S., Ye, Y., Mansouri, J. & Chen, V. (2014) Fouling and crystallization behaviour of superhydrophobic nano-composite PVDF membranes in direct contact membrane distillation. *Journal of Membrane Science*, 463, 102–112.

Meng, S., Ye, Y., Mansouri, J. & Chen, V. (2015) Crystallization behavior of salts during membrane distillation with hydrophobic and superhydrophobic capillary membranes. *Journal of Membrane Science*, 473, 165–176.

Miranda, M. & Infield, D. (2003) A wind powered seawater RO system without batteries. *Desalination*, 153, 9–16.

Palenzuela, P., Zaragoza, G., Alarcón-Padilla, D.C. & Blanco, J. (2011a) Simulation and evaluation of the coupling of desalination units to parabolic-trough solar power plants in the Mediterranean region. *Desalination*, 281, 379–387.

Palenzuela, P., Zaragoza, G., Alarcón-Padilla, D.C., Guillén, E., Ibarra, M. & Blanco, J. (2011b) Assessment of different configurations for combined parabolic-trough (PT) solar power and desalination plants in arid regions. *Energy*, 36, 4950–4958.

Palenzuela, P., Zaragoza, G., Alarcón-Padilla, D.C. & Blanco, J. (2013) Evaluation of cooling technologies of concentrated solar power plants and their combination with desalination in the Mediterranean area. *Applied Thermal Engineering*, 50, 1514–1521.

Palenzuela, P., Roca, L., Zaragoza, G., Alarcón-Padilla, D.C., García-Rodríguez, L. & de la Calle, A. (2014) Operational improvements to increase the efficiency of an absorption heat pump connected to a multi-effect distillation unit. *Applied Thermal Engineering*, 63, 84–96.

Peñate, B. & García-Rodríguez, L. (2011a) Reverse osmosis hybrid membrane inter-stage design: a comparative performance assessment. *Desalination*, 281(1), 354–363.

Peñate, B. & García-Rodríguez, L. (2011b) Energy optimisation of existing SWRO (seawater reverse osmosis plants) with ERT (energy recovery turbines): technical and thermoeconomic assessment. *Energy Journal*, 36, 613–626.

Peñate, B. & García-Rodríguez, L. (2011c) Retrofitting assessment of Lanzarote IV seawater reverse osmosis desalination plant. *Desalination*, 266(1–3), 244–255.

Peñate, B. & García-Rodríguez, L. (2012a) Current trends and future prospects of seawater reverse osmosis desalination. *Desalination*, 284, 1–8.

Peñate, B. & García-Rodríguez, L. (2012b) Seawater reverse osmosis desalination driven by a solar organic Rankine cycle: design and technology assessment for medium capacity range. *Desalination*, 284, 86–91.

Peñate, B., Castellano, F., Bello, A. & García-Rodríguez, L. (2011) Assessment of a stand-alone gradual capacity reverse osmosis desalination plant to adapt to wind power availability: a case study. *Energy*, 36(7), 4372–4384.

Pérez-González, A., Urtiaga, A.M., Ibáñez, R. & Ortiz, I. (2012) State of the art and review on the treatment technologies of water reverse osmosis concentrates. *Water Research*, 46(2), 267–283.

Pendergast, M.T.M., Nygaard, J.M., Ghosh, A.K. & Hoek, E.M.V. (2010) Using nanocomposite materials technology to understand and control reverse osmosis membrane compaction. *Desalination*, 261(3), 255–263.

Pendergast, M.T.M., Ghosh, A.K. & Hoek, E.M.V. (2013) Separation performance and interfacial properties of nanocomposite reverse osmosis membranes. *Desalination*, 308, 180–185.

Pohl, R., Kaltschmitt, M. & Holländer, R. (2009) Investigation of different operational strategies for the variable operation of a simple reverse osmosis unit. *Desalination*, 249(3), 1280–1287.

Prince, J.A., Singh, G., Rana, D., Matsuura, T., Anbharasi, V. & Shanmugasundaram, T.S. (2012) Preparation and characterization of highly hydrophobic poly(vinylidene fluoride) – clay nanocomposite nanofiber membranes (PVDF-clay NNMs) for desalination using direct contact membrane distillation. *Journal of Membrane Science*, 397–398, 80–86.

Quoilin, S., Declaye, S., Tchanche, B.F. & Lemort, V. (2011a) Thermo-economic optimization of waste recovery organic Rankine cycles. *Applied Thermal Engineering*, 31, 2885–2893.

Quoilin, S., Orosz, M., Hemond, H. & Lemort, V. (2011b) Performance and design optimization of a low-cost solar organic Rankine cycle for remote power generation. *Energy*, 85, 955–966.

Rayegan, R. & Tao, Y.X. (2011) A procedure to select working fluids for solar organic Rankine cycles (ORCs). *Renewable Energy*, 36, 659–670.

Rodrigo, M. & Peñate, B. (2006) An introduction to the boron problem and its relevant in desalination technologies in the Canary Islands. *6th ANQUE International Congress of Chemistry, 5–7 December 2006, Puerto de la Cruz, Spain.*

Safi, M.J. (1998) Performance of a flash desalination unit intended to be coupled to a solar pond. *Renewable Energy*, 14(1–4), 339–343.

Sassi, K.M. & Mujtaba, I.M. (2010) Simulation and optimization of full scale reverse osmosis desalination plant. *Computer Aided Chemical Engineering*, 28, 895–900.

Sassi, K.M. & Mujtaba, I.M. (2011) Optimal design and operation of reverse osmosis desalination process with membrane fouling. *Chemical Engineering Journal*, 171(2), 582–593.

Sharaf, M.A., Nafey, A.S. & García-Rodríguez, L. (2010a) Thermo-economic analysis of a combined solar organic Rankine cycle-reverse osmosis desalination process with different energy recovery configurations. *Desalination*, 261, 138–147.

Sharaf, M.A., Nafey, A.S. & García-Rodríguez, L. (2010b) A new visual library for design and simulation of solar desalination systems (SDS). *Desalination*, 259, 197–207.

Sharaf, M.A., Nafey, A.S. & García-Rodríguez, L. (2011a) Thermo-economic analysis of solar thermal power cycles assisted multi effect distillation-vapor compression (MED-VC) desalination processes. *Energy*, 36(5), 2753–2764.

Sharaf, M.A., Nafey, A.S. & García-Rodríguez, L. (2011b) Exergy and thermo-economic analysis of a combined solar organic cycle with multi effect distillation (MED) desalination process. *Desalination*, 272, 135–147.

Shengjun, Z., Huaixin, W. & Tao, G. (2011) Performance comparison and parametric optimization of subcritical organic Rankine cycle (ORC) and transcritical power cycle system for low-temperature geothermal power generation. *Applied Energy*, 88, 2740–2754.

Subiela, V.J., Carta, J.A. & González, J. (2004) Operational analysis of an innovative wind powered reverse osmosis system installed in the Canary Islands. *Desalination*, 168, 39–47.

Subiela, V., de la Fuente J.A., Piernavieja, G. & Peñate, B. (2009) Canary Islands Institute of Technology (ITC), experiences in desalination with renewable energies (1996–2008). *Desalination and Water Treatment*, 7(1–3), 220–235.

Szacsvay, T., Hofer-Noser, P. & Posnansky, M. (1999) Technical and economic aspects of small-scale solar-pond-powered seawater desalination systems. *Desalination*, 122, 185–193.

Tchanche, B.F., Papadakis, G., Labrinos, G. & Frangoudakis, A. (2009) Fluid selection for a low-temperature solar organic Rankine cycle. *Applied Thermal Engineering*, 29, 2468–2476.

Tchanche, B.F., Labrinos, G., Frangoudakis, A. & Papadakis, G. (2010) Exergy analysis of micro-organic Rankine power cycles for a small scale solar driven reverse osmosis desalination system. *Applied Energy*, 87, 1295–1306.

Thomson, M., Miranda, M.S. & Infield, D. (2002) A photovoltaic-powered seawater reverse-osmosis system without batteries. *Desalination*, 153, 229–236.

Torresol Energy (2015) Central-tower technology. Available from: http://www.torresolenergy.com/TORRESOL/central-tower-technology/en [accessed November 2016].

Twomey, B., Jacobs, P.A. & Gurgenci, H. (2013) Dynamic performance estimation of small-scale solar cogeneration with an organic Rankine cycle using scroll expander. *Applied Thermal Engineering*, 51(1–2), 1307–1316.

Vega-Beltrán, J.C., García-Rodríguez, L., Martín-Mateos, I. & Blanco-Gálvez, J. (2010) Solar membrane distillation: theoretical assessment of multi-stage concept. *Desalination and Water Treatment*, 18, 133–138.

Wang, J.L., Zhao, L. & Wang, X.D. (2010) A comparative study of pure and zeotropic mixtures in low-temperature solar Rankine cycle. *Applied Energy*, 87, 3366–3373.

Wang, J.L., Zhao, L. & Wang, X.D. (2012) An experimental study on the recuperative low-temperature solar Rankine cycle using R245fa. *Applied Energy*, 94, 34–40.

Wang, M., Wang, J., Zhao, Y., Zhao, P. & Dai, Y. (2013) A Thermodynamic analysis and optimization of a solar-driven regenerative organic Rankine cycle (ORC) based on flat-plate. *Applied Thermal Engineering*, 50, 816–825.

Wang, X.D., Zhao, L., Wang, J.L., Zhang, W.Z., Zhao, X.Z. & Wu, W. (2010) Performance evaluation of a low-temperature solar Rankine cycle system utilizing R245fa. *Solar Energy*, 84, 353–364.

Xu, J., Feng, X. & Gao, C. (2011) Surface modification of thin-film-composite polyamide membranes for improved reverse osmosis performance. *Journal of Membrane Science*, 370(1–2), 116–123.

# Subject index

α-Fe, 58, 61–68
AD (*see* desalination technology, adsorption distillation)
adsorption-desalination (AD) (*see* desalination technology, adsorption-desalination)
aerodynamics (*see also below main entries* wind *and* wind-desalination), 149, 154
aerofoil, 154–156, 164
AFC (*see* fuel cell, alkaline)
AGMD (*see* air gap membrane distillation)
agriculture, 131, 211, 215, 216, 218, 220, 222, 223
air, 6, 18, 45, 47, 75, 112, 132, 133, 135, 136, 154, 183, 202
    compressor, 201, 203, 205
air gap membrane distillation (AGMD), 111, 137
airflow, 152, 154
    windmill, 73, 74, 91, 149, 188, 189
Algeria, 237, 241, 242
alkaline fuel cell (*see* fuel cell, alkaline)
anatase, 45, 48
anode, 132, 133, 141, 170
aquifer, 109, 114, 124
aquifer thermal energy storage (ATES), 114
asynchronous machine, 83, 93
ATES (*see* aquifer thermal energy storage), 114
auxiliary power unit, 9, 169, 174, 180

balance of energy production, 36, 37
base station, 184, 185, 195
battery, 1, 27, 28, 39, 40, 74, 84, 89, 91–93, 97–100, 135, 183, 185, 254, 257, 265, 266, 272
    bank, 27, 33, 39, 92
    NaS, 98
    storage, 6, 58, 94, 98, 99, 254, 255, 258, 262, 263, 269, 270, 272
        systems, 84, 89, 97, 98
        technologies, 97, 98
battery energy storage system (BESS), 254
BEM (*see* blade element momentum)
BESS (*see* battery energy storage system)
biodiversity, 219, 221–223, 227
    approaches, for food security (*see* food security, solutions, biodiversity approaches)
biofouling, 44, 45, 54
biological oxygen demand (BOD), 50
blade (*see also entries below* wind), 75, 79–81, 148, 150–157, 159, 160, 163, 164, 188
    deformation, 161, 163
    design, 147, 155

element, 148, 159, 160
    momentum (BEM), 148
    rotor, 82
    tip, 151, 157, 159, 160
    wind turbine, 80, 147
blade element momentum (BEM), 148
BMWi E-Energy Program, 257, 258
BOD (*see* biological oxygen demand)
brackish water desalination (*see* desalination, brackish water)
brine, 16, 44, 266
    heater, 117, 137
buyer country, 225

capacity
    cumulative global wind energy, 148
    installed wind energy, 148, 149
capital, 195, 214, 223, 235, 238, 241–243, 246, 271
    cost, high, 4, 266, 269, 270, 272
carbon, total organic, 44, 49, 50, 52–54
catalyst, 43, 46–49, 51, 177
cathode, 132, 133, 170
CFD (*see* computational fluid dynamics)
chemical oxygen demand (COD), 50
chilled water stream, 121
China, 73, 87, 88, 148, 188, 189, 220, 227, 257
chloride, 44, 49, 50, 54
climate change, 213, 217–219, 221, 223, 227, 228, 235, 243
coal, 9, 125, 172, 187, 237, 240
coastal area, 261, 268, 269
COD (*see* chemical oxygen demand)
cogeneration system, 133–135, 142, 202–204
combustor, 200, 202, 203, 205
compound, organic, 43, 45–48
compressor, 77, 140, 141, 200, 202
computational fluid dynamics (CFD), 164
concentrated solar power (CSP), 4, 237, 240, 241, 243, 246, 268, 269
    plant, 4
    project, 241, 242
condensate, 16, 18, 109, 111
condenser, 7, 18, 20–22, 24, 112, 118, 201, 202
    temperature, 18, 21
conservation of latent heat, 121, 122
contaminant, 57, 59
continuous reactor, 52, 53
control
    strategy, 151, 153

# Sustainable Water Developments

*Book Series Editor: Jochen Bundschuh*

ISSN: 2373-7506

Publisher: CRC Press/Balkema, Taylor & Francis Group

1. Membrane Technologies for Water Treatment: Removal of Toxic Trace
   Elements with Emphasis on Arsenic, Fluoride and Uranium
   Editors: Alberto Figoli, Jan Hoinkis & Jochen Bundschuh
   2016
   ISBN: 978-1-138-02720-6 (Hbk)

2. Innovative Materials and Methods for Water Treatment:
   Solutions for Arsenic and Chromium Removal
   Editors: Marek Bryjak, Nalan Kabay, Bernabé L. Rivas & Jochen Bundschuh
   2016
   ISBN: 978-1-138-02749-7 (Hbk)

3. Membrane Technology for Water and Wastewater Treatment, Energy and Environment
   Editors: Ahmad Fauzi Ismail & Takeshi Matsuura
   2016
   ISBN: 978-1-138-02901-9 (Hbk)

4. Renewable Energy Technologies for Water Desalination
   Editors: Hacene Mahmoudi, Noreddine Ghaffour, Mattheus A. Goosen & Jochen Bundschuh
   2017
   ISBN: 978-1-138-02917-0 (Hbk)

5. Application of Nanotechnology in Membranes for Water Treatment
   Editors: Alberto Figoli, Jan Hoinkis, Sacide Alsoy Altinkaya & Jochen Bundschuh
   2017
   ISBN: 978-1-138-89658-1 (Hbk)

Printed and bound by CPI Group (UK) Ltd, Croydon, CR0 4YY

24/10/2024

01778290-0002